Maple V:

Mathematics and its Application

OTHER MAPLE WORKSHOP TITLES:

MATHEMATICAL COMPUTATION WITH MAPLE V:
IDEAS AND APPLICATIONS

Proceedings of the Maple Summer
Workshop and Symposium,
University of Michigan,
Ann Arbor, June 28-30, 1993

Thomas Lee
Editor

Maple V:
Mathematics and its Application

*Proceedings of the Maple Summer
Workshop and Symposium,
Rensselaer Polytechnic Institute,
Troy, New York, August 9-13, 1994*

Robert J. Lopez
Editor

Springer Science+Business Media, LLC

Robert J. López
Rose-Hulman Institute of Technology
5500 Wabash Avenue
Terre Haute, IN 47803

Library of Congress Cataloging-in-Publication Data

Maple Summer Workshop and Symposium (1994 : Rensselaer Polytechnic
    Institute, Troy)
        Maple V: Mathematics and its application : proceedings of the Maple
    Summer Workshop and Symposium, Rensselaer Polytechnic Institute, Troy,
    New York, August 9-13, 1994 / Robert J. Lopez, editor.
                p.   cm.
        Includes bibliographical references.
        ISBN 978-0-8176-3791-0   ISBN 978-1-4612-0263-9 (eBook)
        DOI 10.1007/978-1-4612-0263-9
        1. Maple (Computer file) -- Congresses. 2. Mathematics--Data
    processing--Congresses.   I. Lopez, Robert J., 1941-   II. Title.
    QA76.95M35   1994   94-21604
        510'285'53--dc20   CIP

Printed on acid-free paper
© Springer Science+Business Media New York 1994 ®
Originally published by Birkhäuser Boston in 1994

ISBN 978-0-8176-3791-0

Camera-ready text prepared by the authors.

9 8 7 6 5 4 3 2 1

Contents

Preface

The Maple Summer Workshop and Symposium, MSWS '94, reflects the growing community of Maple users around the world. This volume contains the contributed papers. A careful inspection of author affiliations will reveal that they come from North America, Europe, and Australia. In fact, fifteen come from the United States, two from Canada, one from Australia, and nine come from Europe. Of European papers, two are from Germany, two are from the Netherlands, two are from Spain, and one each is from Switzerland, Denmark, and the United Kingdom.

More important than the geographical diversity is the intellectual range of the contributions. We begin to see in this collection of works papers in which Maple is used in an increasingly flexible way. For example, there is an application in computer science that uses Maple as a tool to create a new utility. There is an application in abstract algebra where Maple has been used to create new functionalities for computing in a rational function field. There are applications to geometrical optics, digital signal processing, and experimental design.

We begin, I believe, to see that Maple is understood as more than just a new tool for accomplishing specific tasks within a world of traditional practice. There are hints that users have begun to see Maple as a new working environment wherein whole segments of professional activity are carried out. The paper by Taylor and Atherley suggests that the chemical engineering student should be functioning in a Maple envelope, and the paper by Monagan points to a similar direction in mathematics.

The philosopher and theologian Thomas of Aquinas points out that all things are known according to the mode of the knower. It is entirely possible that as Editor, I am simply finding evidence of my own perspective on Maple. That is why forums like the Maple Summer Workshop and Symposium are so essential to the development of change in teaching, learning, and practice. It is by the sharing and exchanges which characterize a conference that insights will be appraised, and their implications nurtured. The final judgments are up to the conference participants, the readers of this volume, and history.

The organization of this volume is programmatic. The sections in the text correspond to the sessions scheduled on the program. This layout was adopted for the convenience of the conference attendees. We apologize to readers who would have preferred a stricter thematic organization.

Finally, we have included brief abstracts of the talks by the invited speakers. For readers who were not able to attend MSWS '94 these abstracts might help delineate the experience of those who did attend. And for those who did attend, they might serve as a reminder of all the good ideas the conference generated.

The only fitting conclusion to this preface is an acknowledgment of the efforts of all those who helped make the MSWS '94, and this volume, a success. Lest I omit anyone's name, I will issue a blanket sigh of gratitude for the help from the folks at Waterloo Maple Software, and at Birkhäuser, the publishers, and from the referees whom I cajoled into reviewing the contributed papers. Finally, I thank the authors themselves for their efforts at writing and formatting these papers, and for their promptness and patience in meeting all the deadlines.

Robert J. Lopez
Rose-Hulman Institute of Technology
Terre Haute, Indiana

Abstracts of Invited Presentations

Experimental Mathematics: Prospects and Pitfalls

Jonathan Borwein
Shrum Professor of Science
Centre for Experimental and Constructive Mathematics
Simon Fraser University

The use of sophisticated symbolic, graphic and other packages allows mathematicians (applied or pure) to do mathematics from an experimental perspective until recently far from the experience of most of us. I will present case studies (from linear algebra, analysis and combinatorics) that attempt to illuminate the enormous possibilities and the potential pitfalls of doing empirical mathematics.

I will also discuss some of the related projects being undertaken at the Simon Fraser Centre for Experimental and Constructive Mathematics.

Dr. Charles E. Campbell Jr.
NASA Goddard Space Flight Center
cec@gryphon.gsfc.nasa.gov

Goddard Space Flight Center has been pursuing robotics to help the next service flight for the Hubble Space Telescope. To do this requires dexterity of the robot, kinematically and dynamically. A seven degree of freedom RRC arm provides a model of kinematic dexterity, and the new GSFC-invented capaciflector sensor provides control systems with the data to gain dynamic dexterity. The talk will cover the computer architecture at the Intelligent Robotics Laboratory used to support robotics, some kinematics, and control problems involving use of the capaciflector: virtual force, imaging, and 6-DOF pre-contact alignment.

Theorist in the Classroom

Dr. Donald Hartig
Mathematics Department
California Polytechnic State University - San Luis Obispo
San Luis Obispo, CA

Theorist is a WYSIWYG computer algebra system featuring a user-friendly interface, incredibly good graphics, and the seamless integration of equations, tables, matrices, and number crunching. This presentation will focus on the use of Theorist in the classroom as a demonstration platform and in the computer lab as a resource for the student. Recent student Notebooks will be examined as well as Notebooks used in calculus, differential equations, and linear algebra. Special attention will be given to the use of Theorist in conjunction with the development of calculus from the numerical, graphical, and analytical point of view as advocated in the Harvard Core Curriculum.

Biology, Mathematics, and Maple

James V. Herod
School of Mathematics
Georgia Institute of Technology
Atlanta, GA 30332-0160

Mathematical biology, resource modeling, and population dynamics are rich sources of problems in mathematics. Undergraduates in biology often finish their studies with only limited exposure to the richness of these three areas. This happens in part because of a lack of training that biology students have in mathematics. Their introduction to a tool such as Maple allows them to understand more deeply some of the issues in these research areas and to examine current work, even with a limited mathematical background. On the other hand, mathematically sophisticated students may find a fascination in their introduction to the problems of biology. Mathematical biology may use mathematics that the students already know, but in ways that they had not imagined. We present some mathematical models that are accessible to undergraduate biology students.

Advanced Control and Engineering Computations Using Maple V

Dr. Ayowale B. Ogunye
Computing and Information Technology-CR & D
The Dow Chemical Company
Midland, Michigan

In recent years, considerable interest has been generated in the areas of symbolic computation for advanced control and engineering computations. The driving force for this interest is the potential for solving complex problems which were hitherto impossible or difficult to undertake. The analysis and design of control systems for multivariable processes described by polynomial matrices were practically non-existent before the advent of symbolic computation. Manipulation of polynomial matrices by hand-calculations is very complex and not amenable to computer implementation. Consequently, the numerical methods developed to perform the necessary manipulations are computationally intensive and often suffer from numerical instability. The numerical solution of differential equations modeling physical phenomena such as diffusion, reaction, heat transfer and fluid flow processes are of paramount importance. One of the established methodologies for this class of problems involves the use of polynomial approximations using orthogonal collocation, finite elements, etc. In this talk, we describe the use of Maple V for solving the above mentioned problems. The advantages of error-free symbolic manipulations and numerically stable results, obtained quickly and efficiently, with a tremendous gain in time and minimal effort, are demonstrated.

Using Maple To Teach Physics

Glenn Sowell
Department of Physics
University of Nebraska at Omaha
Omaha, NE 68182-0266
USA
(402) 554-3724
sowell@unomaha.edu

Computer algebra software represents a mature technology that has already made significant contributions to physics research. It is only just beginning to impact physics education. And we have a long way to go before that impact will be significant. I will discuss some of the problems we face in incorporating this software into the curriculum and suggest some of the solutions.

I A. MAPLE IN COMPUTER SCIENCE

"TURTLE GRAPHICS" IN MAPLE V.2

Eugenio Roanes Lozano, Eugenio Roanes Macías
Faculty of Education, Universidad Complutense de Madrid

Abstract

We think the "Turtle Geometry" [1] is a useful and convenient way of doing graphics programming. Recently, we have developed brand new implementations of the "Turtle Geometry" in Turbo Pascal [12] and in Maple V [14]. Our implementations include the usual primitives with Turbo-Pascal-like syntax: Forwd(dist), TurnLeft(ang), SetPenColor(color), SetPosition(x1,x2), etc. We shall begin with a brief resume of the implementation developed in Maple V.

Further on we shall show how using an implementation that is based upon floating point arithmetic, can be unsuitable even for dealing with problems in elementary Geometry. Details concerning how exact arithmetic is treated in our implementation are given. Clarifying examples are also included.

1 An Introduction to the "Turtle Geometry"

The "turtle" is a mobile prompt on the graphics screen. As an animal, it can move forwards or backwards and it can turn. By default it leaves a trail behind. See [1] for details.

The "turtle" is connected to Cartesian coordinate axes x, y. The x coordinate axis is horizontal, the y coordinate axis is vertical, and the point $(0,0)$ is the centre of the screen. Headings are measured clockwise from the $y+$ axis. All the angles are measured in degrees.

Below there is a list of procedures and accessible variables (global variables). We shall classify the procedures according to their purpose.

Primitives that change the position of the turtle:

Forwd(n:algebraic): Moves the turtle forward the number of steps given.
Back(n:algebraic): Moves the turtle forward $(-1)\cdot n$.
SetPosition(x1:algebraic,x2:algebraic): Takes the turtle to the specified position.
SetX(x1:algebraic): As SetPosition, but only changing the x coordinate.
SetY(x2:algebraic): As SetPosition, but only changing the y coordinate.
Home(): We implement it as a SetHeading(0) followed by a SetPosition(0,0).
ClearScreen(): As Home, plus clearing the previous drawings.

Primitives that change the orientation of the turtle:

TurnRight(ang:algebraic): It turns the turtle clockwise, by adding the given angle to its orientation.
TurnLeft(ang:algebraic): It turns the turtle counterclockwise, by subtracting the given angle to its orientation.
SetHeading(ang:algebraic): The turtle is turned until the new Heading is "ang".
SetHeadingTowards(x1:algebraic, x2:algebraic): The turtle is oriented towards the given point.

Other primitives:

PenUp(): Upon command, the turtle doesn't leave a trail behind.

PenDown(): Upon command, the turtle leaves a trail behind.

SetPenColor(color): Lets the user choose the pen color. The usual Maple inputs COLOUR-(HUE(a)) and COLOUR(RGB,a,b,c) are admitted.

Dot(x1:algebraic, x2:algebraic): The computer draws the given point.

FullScreen(): Changes to the Graphics Screen. The computer makes a "PLOT" of the "CURVES" stored in the list "dib".

Global variables:

i) Current x coordinate: **XCor** (type: algebraic)
ii) Current y coordinate: **YCor** (type: algebraic)
iii) Current orientation: **Heading** (type: algebraic)
iv) Current color for the "pen" the turtle carries: **PenColor**.

2 Floating Point Arithmetic vs. Exact Arithmetic

Observation.- In this section the examples are implemented in IBM LOGO, because of the similarity of syntax. Nevertheless, the kind of problems that we present have nothing to do with the kind of language or the version. They are a consequence of the assumptions made when representing numbers in floating point arithmetic.

Note.- As there is no terminator in LOGO, to distinguish inputs from outputs, we shall represent by "?" the LOGO prompt in interactive mode and by ">" the LOGO prompt inside the body of the procedures.

2.1 Does the Pythagorean theorem hold in floating point arithmetic?

The following program clears the screen and orders the turtle to draw a right angled isosceles triangle.

```
?TO TRI
>CLEARSCREEN
>FORWARD 100
>RIGHT 90
>FORWD 100
```

```
>RIGHT 135
>FORWD 100*SQRT 2
>RIGHT 135
>END
```

The turtle should have returned to the center of the screen. But, surprisingly, if we ask for the position now, it is not $(0,0)$

```
?PRINT POS
0.002 0.002
```

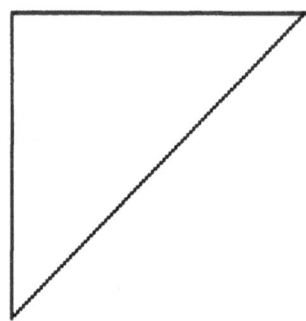

It is clear that the turtle is very close to point (0,0). But what would have happened if we had checked whether the turtle was in the center of the screen or not?

```
?IF AND XCOR = 0 AND YCOR = 0
  [PR "YES][PR "NO]
NO
```

The answer would have been completely wrong, not "approximately" right.

2.2 Rediscovering Incommensurability?

We could think that the reason for what has happened is that $\sqrt{2}$ is an irrational, and as a consequence its decimal representation has an infinite number of non periodic ciphers, so it has been truncated. This is the correct explanation.

However, let's consider the diagonals of a regular pentagon. According to [3], the observation of this drawing could be what led to the discovery of incommensurability in ancient Greece: the ratio of a diagonal to a side in a regular pentagon is an irrational number (they are incommensurable).

The following procedure

```
?TO PENTAG
>CS
```

```
>RIGHT 90
>REPEAT 5 [FORWARD 80 LEFT
   72]
>LEFT 36
>REPEAT 5 [FORWARD 80*
   (1+2*SIN 18) LEFT 144]
>PR POS
>END
```

draws a regular pentagon with sides 80 steps long and its diagonals. So we could expect that the same error as in 2.1 would take place. Surprisingly, in this case, the turtle returns safely to point (0,0). As all depends on the "luck" of the roundings, the behaviour of the floating point arithmetic is unpredictable.

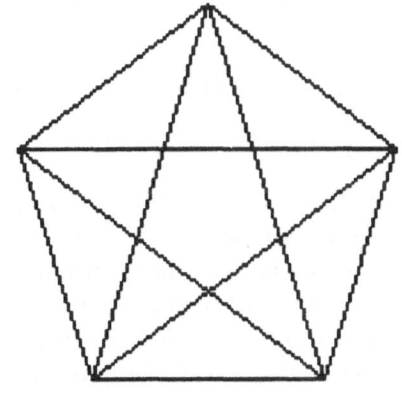

2.3 Relative position of points and circumferences

Let $\overline{PQ}$ be the diameter of a circumference, c, and R a point. The following is a well known result in Elementary Geometry:

i) $R \in c$ iff $\widehat{PRQ} = 90^0$

ii) R is in the interior of c iff $\widehat{PRQ} > 90^0$

iii) R is in the exterior of c iff $\widehat{PRQ} < 90^0$

Let's consider the following LOGO-program that draws the triangle PQR and calculates $\widehat{PRQ}$

```
?TO ANG :R1 :R2 :P1 :P2 :Q1 :Q2
>CLEARSCREEN
>PENUP
>SETPOS SE :R1 :R2
>PENDOWN
>SETPOS SE :P1 :P2
>SETPOS SE :Q1 :Q2
>SETPOS SE :R1 :R2
>SETH TOWARDS SE :P1 :P2
```

```
>MAKE "A1 HEADING
>SETH TOWARDS SE :Q1 :Q2
>MAKE "A2 HEADING
>MAKE "A COMPL(ABS(:A1-:A2))
>PRINT :A
>END
```

```
?TO ABS :AN
>IF :AN > 0 [OP :AN] [OP -:AN]
>END
```

```
?TO COMPL :AN
>IF :AN > 180 [OP 360 - :AN]
   [OP :AN]
>END
```

ABS calculates the absolute value and COMPL calculates, if necessary, $(360 - angle)$.

If the points are $P = (0, 10)$, $Q = (10*\sqrt{3}, 10)$ and $R = (0,0)$, the angle obtained is 60.001^0. It is a close approximation.

But what would happened if we wanted to study whether or not the point R belongs to the circumference c (whose diameter is $\overline{PQ}$)? If, for instance, $P = (-100, 0)$, $Q = (100, 0)$ and $R = (0, 100)$, the result is 90^0 and as a consequence, $R \in c$. But if $R = (0, 100.001)$ or $R = (0, 99.999)$ the result is also 90^0. We would find that there are three points aligned in c !!!.

2.4 The dangers of iteration

Let's return to the procedure described in 2.1. If we type

?REPEAT 1000 [TRI]

instead of drawing the same triangle 1000 times, the turtle draws several very close triangles. The final image is a "thick sided" triangle. And at the end of the process, the distance to the center of the screen (the point where the turtle should be) is greater than 2.

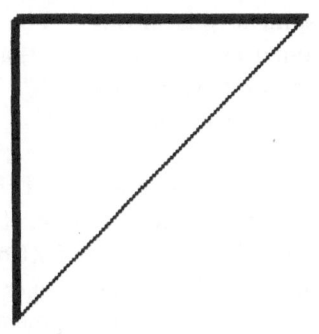

2.5 Small numbers and periodic numbers

If the turtle moves $\frac{10}{3}$ of a step forwards 30 times (in floating point arithmetic) it is not the same as moving forward 100 steps.

?REPEAT 30 [FORWARD (10/3)]
?PRINT POS
0 99.999

In this case it is clear that the reason of the inaccuracy is the truncation of 3.3333.

Very small numbers are not handled correctly either.

?FORWARD 0.0001
?PRINT POS
0 0

The reason in this case is that very close numbers are represented the same way. This also occurs, for instance, if we type

?PRINT 30000000001 - 30000000000
0

Both numbers have been represented as 3×10^{10}. And to increase the number of significant digits is not the right solution. It is clear how, given any number of significant digits, a similar example can be found such that the result is not correct.

3 The design of our implementation

The idea behind our implementation is that the turtle draws a sequence of segments. A segment can be described to Maple with the command "CURVES". This way, all of the orders that require the computer to draw, are stored in a list of "CURVES", each member of which includes an option where the desired color of the correspondent segment is stored (the active color when the segment was drawn). The final drawing is obtained as a "PLOT" of this list of "CURVES", that we have named "dib". This variable is reset by ClearScreen().

The current position of the turtle is stored in the variables XCor,YCor. If the "turtle's pen" is "up", the old values of XCor and YCor are just substituted by the new ones. And if the "turtle's pen" is "down", the segment

$$\overline{(old\_XCor, old\_YCor)(new\_XCor, new\_YCor)}$$

is also added to the list "dib".

For instance, if a Forwd(distance) is typed, the new XCor is the previous one plus the distance we want the turtle to move multiplied by the cosine of the angle $(90^0 - Heading)$. The new YCor is calculated similarly.

All the procedures involving changes of the turtle's position are implemented in a similar way. Those which involve changes of the turtle's orientation produce only changes in the value stored in the variable Heading. We have tried to avoid similar code been written in different places. The whole file occupies about 5k in its readable form.

4 The arithmetic in our implementation

In the first version "dib" was a list of numbers in exact arithmetic. The disadvantage is that, in exact arithmetic, the length of the expressions grows very quickly (the length of the representations of the numbers is not fixed and not even related to the size of the numbers). The reason for this growth is that turns generally involve irrational numbers. Compare, for instance

$$165 - 3 \wedge (1/2) * Pi \wedge 4 + 23 \wedge (1/2)$$

with its representation in floating point arithmetic

$$1.078336623$$

This way "dib" could reach huge sizes. Moreover, the change from the text screen to the graphics screen was extremely slow.

Therefore we decided to approach the problem in a different way. The coordinates of the endpoints of the segments to be drawn can be stored in floating point. The reason is that the screen is a discrete world (the pixels are not points from a mathematical point of view). What is really important is that the current position of the turtle is maintained in exact arithmetic. This way both advantages are obtained:

- At any moment, one can confidently check whether the turtle is at a certain point or not.

- The speed of the plotting is increased.

4.1 The Pythagorean theorem (in Maple)

We shall represent by ">" the Maple prompt. Let's consider again the example of the right angled isosceles triangle of section 2.1. In Maple we could type

```
>Forwd(100);
>TurnRight(90);
>Forwd(100);
>TurnRight(135);
>Forwd(100*sqrt(2));
>TurnRight(135);
>FullScreen();
```

for the same triangle to be drawn. Now, the question about the final position of the turtle is answered correctly:

```
>XCor;
  0
>XCor;
  0
```

The list "dib" is:

```
{CURVES([[0,0],[0,100.]],
    COLOUR(RGB,0,0,0)),
    ([[0,100.],[100.,100.]],
    COLOUR(RGB,0,0,0)),
    ([[100.,100.],[0,0]],
    COLOUR(RGB,0,0,0))}
```

Let's look at another example. We want to draw the segment $\overline{PQ}$, $P = (100 \times \sqrt{2}/2, 100 \times \sqrt{2}/2)$, $Q = (100 \times \sqrt{2}, 100 \times \sqrt{2})$.

```
>TurnRight(45);
>PenUp();
>Forwd(100);
>PenDown();
>Forwd(100);
>FullScreen();
```

The list "dib" is:

```
{CURVES([[70.71067810,70.71067810],
    [141.4213562,141.4213562]],
    COLOUR(RGB,0,0,0))}
```

If we ask for the current position, we obtain the correct answer

```
>XCor;
  100 2^{1/2}
>XCor;
  100 2^{1/2}
```

Numbers in floating point arithmetic usually include a decimal point. Those in exact arithmetic never do. It is clear how $100 \times 2^{1/2}$ has been approximated by 141.4213562.

4.2 Relative position of points and circumferences (in Maple)

Let's compare 2.3 with the corresponding translation to Maple.

```
>Ang:=proc(r1,r2,p1,p2,q1,q2)
>    ClearScreen();
>    PenUp();
>    SetPosition(r1,r2);
>    PenDown();
>    SetPosition(p1,p2);
>    SetPosition(q1,q2);
>    SetPosition(r1,r2);
>    SetHeadingTowards(p1,p2);
>    a1 := Heading;
>    SetHeadingTowards(q1,q2);
>    a2 := Heading;
>    a := Compl(abs(a1-a2));
>  end; #Ang

>Compl := proc(an:algebraic)
>    if evalf(an) > 180 then (360 - an)
>      else an;
>    fi;
>  end; #Compl
```

If $P = (0, 10)$, $Q = (10 * \sqrt{3}, 10)$ and $R = (0, 0)$, the angle obtained is 60^0. That is the correct value.

If, for instance: $P = (-100, 0)$, $Q = (100, 0)$ and $R = (0, 100)$, the result is 90^0 and as a consequence, $R \in c$. If $R = (0, 100.001)$ the result is $135 - \frac{141.3707694}{Pi}$. Using **evalf** we can approximate it: $90.0028649 > 90$ (observe that this approximation takes place on the final value, not throughout the whole process). If $R = (0, 99.999)$ the result is $135 - \frac{141.3725694}{Pi}$, that is, approximately: $89.9971354 < 90$. Therefore, using exact arithmetic, the position of the point with respect to the circumference can be successfully determined.

See [9] for a clear and sensible introduction to the importance of exact arithmetic.

5 Some recursive advanced examples

The possibilities of the turtle are definitely boosted by the use of recursion. As we have seen, the fact of using an implementation based upon exact arithmetic is key to working safely with very small numbers, very big numbers or very close numbers. This is specially important when dealing with recursive procedures. Below we shall develop some advanced examples for our implementation of "Turtle Graphics".

Example 1: Given n and l, draw n squares $S_1, S_2, ..., S_n$ such that:

- The sides of the first one are l steps long

- The vertices of S_n are the midpoints of the sides of S_{n-1}

```
>polycuad:=proc(l:algebraic,niv:integer);
>    PenUp();
>    Forwd(l/2);
>    TurnRight(90);
>    Forwd(l/2);
>    PenDown();
>    for i to 4 do
>       TurnRight(90);
>       Forwd(l);
>       od;
>    PenUp();
>    Back(l/2);
>    TurnLeft(90);
>    Back(l/2);
>    TurnRight(45);
>    PenDown();
>    if niv > 1 then
>       polycuad(l/sqrt(2),niv-1);
>       fi;
>    end; #polycuad
```

A nice example can be obtained, for instance, typing

```
>ClearScreen();
>polycuad(300,15);
>FullScreen();
```

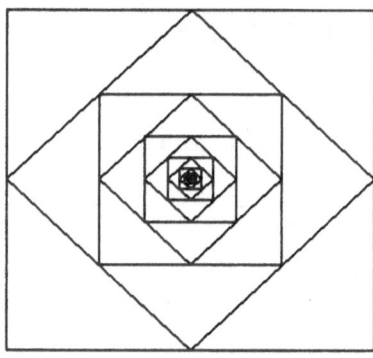

Example 2: The shell of a nautiloid is a really curious example of growth in animals. A nautiloid lives just outside the shell and uses the different chambers, as the depots of a submarine, to obtain the desired density. From time to time a new chamber grows. The longitudinal section shows a spiral. We can plot an approximate drawing very simply:

- Draw an interpolation of a spiral. The first side is l steps long, the ratio of one side to the next one is inc and the number of sides is $numl$. We have chosen to turn an angle of 20 degrees from one side to the next one. Sides 1 to 18 belong to the border of the embryonic chamber.

- From vertex 18 onwards we draw the segment that connects the nth vertex to the $(n-18)^{th}$.

To draw a new chamber, all we have to do is draw another side of the spiral and the segment whose endpoints are the $(n+1)^{th}$ vertex and the $(n+1-18)^{th}$ vertex.

```
>nautiloid:=proc(l:algebraic,
            inc:algebraic,numl:integer)
>    TurnRight(180);
>    lpun:=[[evalf(XCor),evalf(YCor)]];
>    nautiloid_recurs(l,inc,numl);
>    end; #nautiloid

>nautiloid_recurs:=proc(l:algebraic,
            inc:algebraic,numl:integer)
>    local x,y;
>    Forwd(l);
>    TurnRight(20);
>    lpun:=[[evalf(XCor),evalf(YCor)],
            op(lpun)];
>    x:=XCor;
>    y:=YCor;
>    if nops(lpun) > 17 then
>       SetPosition(op(1,op(18,lpun)),
```

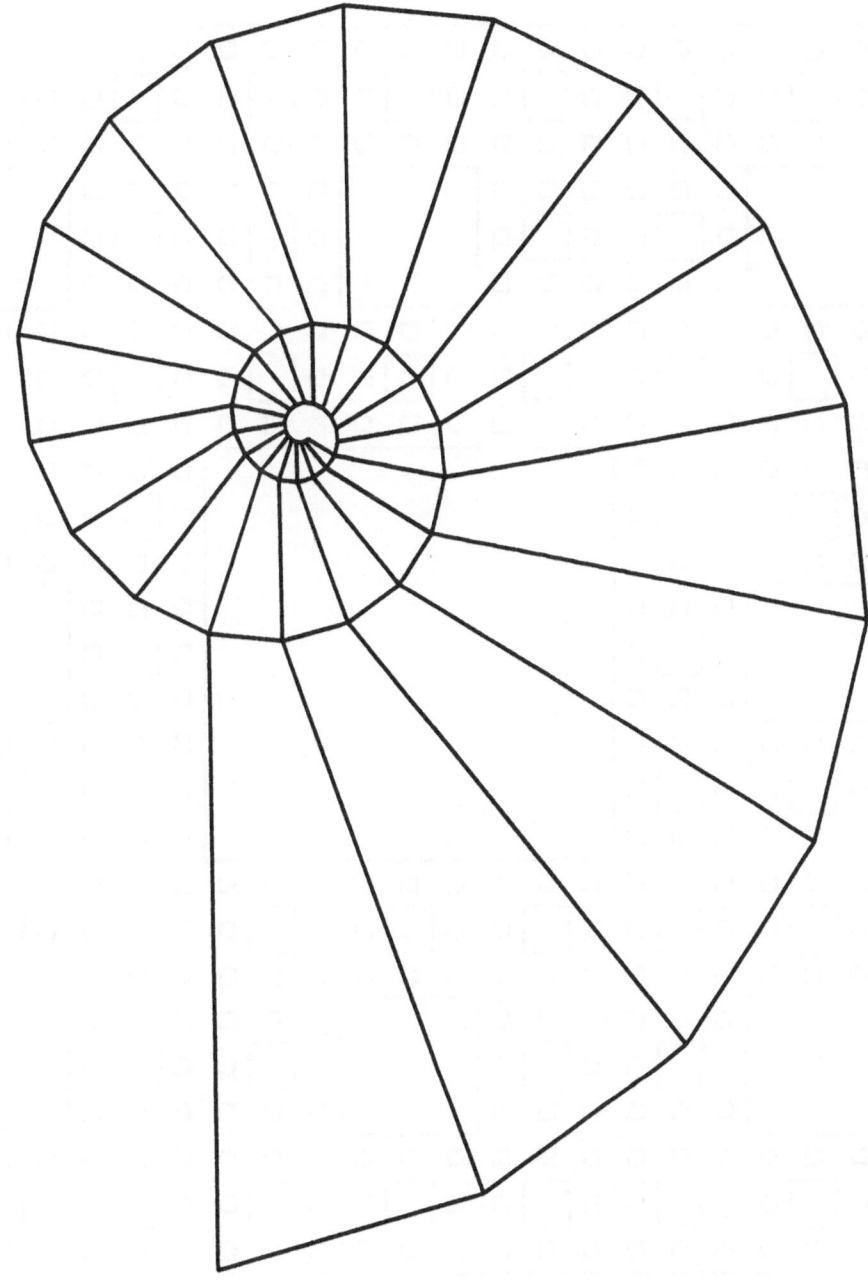

Example 2: Longitudinal section of a nautiloid

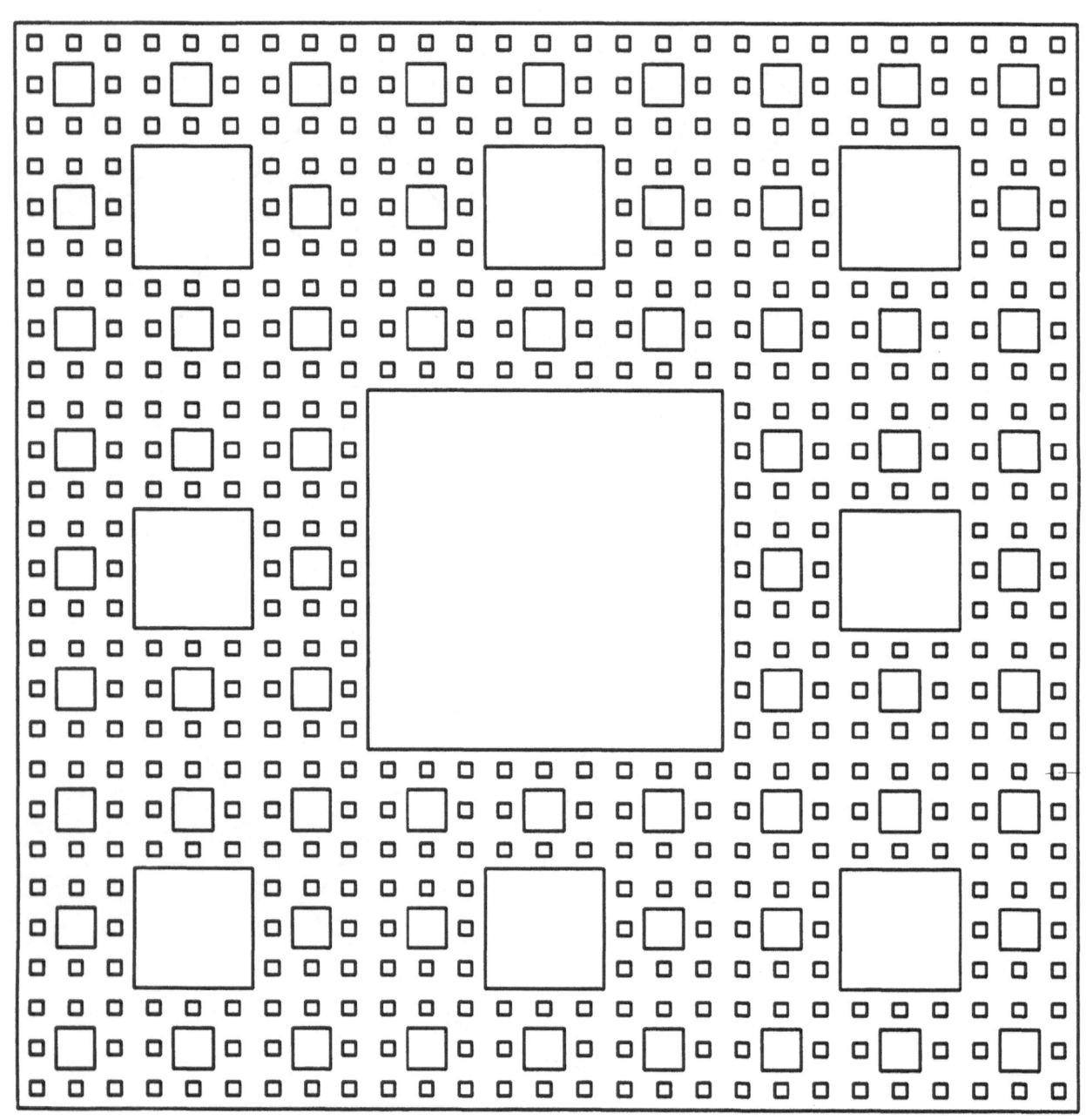

Example 3: Third step in the construction of a 2-D Sierpinski sponge

```
        op(2,op(18,lpun)));
>     PenUp();
>     SetPosition(x,y)
>     PenDown();
>    fi;
>   if numl > 0 then
>     nautiloid_recurs(l*inc,inc,numl-1);
>    fi;
>   end; #nautiloid_recurs
```

An example of a realistic nautiloid can be obtained for $l = 1$ and ratio $inc = \frac{108}{100}$.

```
>ClearScreen();
>nautiloid(1,108/100,55);
>FullScreen();
```

3) Another typical example of recursion are sponges. In this case the program is not as short as the previous ones, so we shall not include the code. The plotting of the third step in the construction of a 2-D Sierpinski sponge is included above ($side = 300$).

6 Conclusions

Firstly, our implementation allows us to plot turtle drawings in Maple. Additionally, it permits us to do the plottings in exact arithmetic, which can be extremely important when dealing with many problems.

7 Obtaining the code of our implementation

The code is going to be incorporated to the Maple Share Library. Alternatively, you can contact the authors.

References

[1] **H. Abelson, A. diSessa**: *Turtle Geometry. The Computer as a Medium for exploring Mathematics*; M.I.T., 1981.

[2] **H. Abelson**: *Apple Logo*. Byte Books / McGraw-Hill, 1982.

[3] **C.B. Boyer**: *A History of Mathematics*. John Wiley, 1968.

[4] **A. Heck**: *Introduction to Maple*. Springer-Verlag, 1993.

[5] **B.W. Char, K.O. Geddes, G.H. Gonnet, B.L. Leong, M.B. Monagan, S.M. Watt**: *Maple V Language Reference Manual*. Springer-Verlag, 1991.

[6] **B.W. Char, K.O. Geddes, G.H. Gonnet, B.L. Leong, M.B. Monagan, S.M. Watt**: *Maple V Library Reference Manual*. Springer-Verlag, 1991.

[7] **B.W. Char, K.O. Geddes, G.H. Gonnet, B.L. Leong, M.B. Monagan, S.M. Watt**: *First Leaves: A Tutorial Introduction to Maple*. Springer-Verlag, 1991.

[8] **A. Montes, M. Noguera, A. Grau**: *Apuntes de Maple V*. Universitat Politècnica de Catalunya, 1994.

[9] **R. Pavelle, M. Rothstein, J. Fitch**: Algebra por ordenador. Investigación y Ciencia, Feb. 1982 (spanish version of the article in Scientific American num. 245, Dic. 1981).

[10] **E. Roanes Lozano**: *Automatización e implementación de algunos problemas algebraicos y geométricos*. (Ph.D. Thesis). Univ. Politécnica de Madrid, 1993.

[11] **E. Roanes Lozano**: *"Precisión Indefinida" y Matemática Elemental*. Boletín de la Sociedad "Puig Adam" de Profesores de Matemáticas, núm. 31, Mayo 1992.

[12] **E. Roanes Macías, E. Roanes Lozano**: *Una implementación mejorada y flexible de la "Turtle-Geometry"*. II Encuentro de Geometría Computacional. Zaragoza, 1992.

[13] **E. Roanes Macías, E. Roanes Lozano**: *Matemáticas con ordenador*. Ed. Síntesis, 1988.

[14] **E. Roanes Macías, E. Roanes Lozano**: *An Implementation of "Turtle Graphics" in Maple V.2*. Computer Algebra Nederland (CAN), Nieuwsbrief 12, March 1994.

[15] -: *LOGO Reference Manual*. IBM/LCSI, 1984.

[16] -: *Manual de Referencia y del Usuario ACTI-LOGO*. Idealogic, 1984.

[17] -: *Guia de Referencia y Manual del usuario WIN-LOGO*. Idea, Investigación y Desarrollo S.A., 1991.

[18] -: *Turbo Pascal v.6.0 User's Guide, Programmer's Guide & Library Reference*. Borland International, 1990.

[19] -: *Turbo Pascal v.3.0 Reference Manual (Cap. 19 PC-Goodies)*. Borland International.

Eugenio Roanes Lozano (eroanes@fi.upm.es) studied Mathematics at the Universidad Complutense de Madrid (UCM), received his Ph.D. in Mathematics from the Universidad de Sevilla in 1991 and received a second Ph.D. (in Computer Science) from the Universidad Politécnica de Madrid in 1993. His research interests include Computer Algebra, Computational Geometry and Algebraic aspects of A.I. He is presently a Lecturer at the Faculty of Education (Sec. Dep. Algebra) of the UCM.

Eugenio Roanes Macías (roanes@mat.ucm.es), father of the former author, received his Ph.D. in Mathematics from the UCM in 1973. His research interests include Algebraic Geometry, Computer Algebra and Computation in Mathematics Teaching. He is presently a Professor at the Faculty of Education (Sec. Dep. Algebra) of the UCM.

We can be reached by the following address:
Eugenio Roanes
Sec. Dep. Algebra (Fac. of Education)
Edificio Pablo Montesino
Universidad Complutense
c/ Santísima Trinidad 37
E-28010-Madrid (SPAIN)

A DISTRIBUTED APPROACH TO PROBLEM SOLVING IN MAPLE

K. C. Chan, A. Díaz, and E. Kaltofen
Department of Computer Science, Rensselaer Polytechnic Institute, Troy, NY

Abstract

A system is described whereby a Maple computation can be distributed across a network of computers running Unix. The distribution is based on the DSC system, which can ship source code and input data to carefully selected computers for execution and which can retrieve the produced output data. Our code is fully portable and requires no changes of the underlying Maple or Unix systems. Speedup over Maple's built-in sequential procedure is demonstrated when computing determinants of integer matrices.

1. Introduction

The paradigm of distributing a compute-intensive program over a network of computers is becoming commonplace in today's problem solving by computer. Several systems are available for support of distributed computation, among them Gelernter's Piranha Linda, PVM (Parallel Virtual Machine) by Geist et al., and the Unix system's remote shell command. In 1991, we announced a platform for distributing symbolic code, called the Distributed Symbolic Computation tool—DSC (Díaz et al. 1991). The DSC system has been significantly improved over the past three years (Díaz et al. 1993). This paper describes our new interface to Maple, which permits the automatic distribution of a Maple computation over a network of computers.

The following important DSC features are available to Maple users through the new interface.

— The distribution of so-called parallel subtasks is performed from the Maple environment by a system call to a DSC program which communicates with the concurrent server daemon process. That process, which has established IP/TCP/UDP connections to equivalent daemon processes on the participating compute nodes, handles the call and sends the subtask to one of them. Similarly, the control flow of the application program is synchronized by library calls that wait for the completion of one or all subtasks. Participating compute nodes can be anywhere on the Internet.

— DSC can distribute a Maple source file and the corresponding input data file. The remote computer starts a Maple shell and executes the source file, which is assumed to read from the input data file and produce an output data file, which is returned to the parent process. Note that the distribution of source code, which is not restricted to Maple but can be in C or Lisp instead, allows the parent process to dynamically construct programs for distribution thus permitting the use of so-called "black box" data types. Furthermore, on a shared file system file transfer can be performed by path name and no physical file movement needs to take place.

— The master-slave paradigm for distribution can be relaxed by making use of a co-routine-like distribution mechanism. In that case, the parallel subtask can exchange information with the parent task in the middle of a computation.

— The interface from Maple to DSC consists of 10 library functions. Processor allocation and interprocess communication is completely hidden from the Maple programmer. Indeed, DSC has a fairly sophisticated scheduler that tries to match the subtask's resource demands, which are given in rough estimates as (optional) arguments to the DSC call, with a suitable computer on the network. CPU and memory usage of participating compute nodes is estimated by having a resident daemon process probe them in 10 minute intervals and communicate the CPU and memory load to the DSC server daemons. If no computer meets certain threshold requirements, the parallel subtask gets queued for later distribution under presumably better load conditions on the network. The scheduler makes DSC a truly heterogeneous parallel system.

— The progress of a distributed computation can be monitored by an independently run controller program. This controller also initializes the DSC environment by establishing server daemons on the participating computers.

— It is possible to run several distributing Maple programs simultaneously by specifying distinct UDP port numbers on start-up. DSC and the Maple interface do not require any changes to existing Unix or Maple setups. Security is guaranteed because DSC tags the messages which it sends through unprotected ports with a secret key.

Note that the design of the features discussed above has been extensively tested in our symbolic applications, which are: the parallel Cantor/Zassenhaus polynomial factorization algorithm (Díaz et al. 1991), the Goldwasser-Kilian/Atkin primality prover for titanic primes, i.e., prime numbers with more than 1000 decimal digits (Valente 1992), and the block Wiedemann algorithm for the solution of sparse linear systems over finite fields (Díaz et al. 1993).

The main design goal for the DSC/Maple interface was that it had to be completely portable. Therefore, we chose a mechanism by which the DSC library functions are called through a system call. That call executes an interface program that sends a signal to a concurrent daemon process. It is that single daemon process which calls the C language DSC user library function which in turn communicates with the server process via a user datagram. Thus we avoid any dependence on calls to functions written in C from within Maple, a feature that is lacking in older Maple installations.

We demonstrate the usage of our interface by a standard example, namely the computation of the determinant of an integer matrix by homomorphic imaging and Chinese remaindering (McClellan 1973). We can report that determinants of 100×100 integer matrices with single digit en-

tries are computed faster by distribution. We note that all of the needed computer algebra technology, such as finding prime numbers, computing matrix determinants modulo prime numbers, and the Chinese remainder algorithm, is provided by Maple. There are more sophisticated coarse grain parallel algorithms, such as root finding or sparse interpolation (Char 1990), which would require more custom-made Maple programming. In our opinion, an important application of the DSC interface is the parallelization of our Maple code for the solution of sparse linear systems. This code was developed as a prototype for our superfast C++ implementation (Kaltofen and Lobo 1994). Nonetheless, the distributed version of our prototype code would make parallel sparse linear system solving available to the entire Maple world.

We hope to add several new features to DSC and the Maple interface in the future. First, we are planning to build a graphical user interface to the control program which monitors all DSC processes and computers. Second, we are planning to implement some form of process migration. Although our scheduler attempts to find the best compute node for a given parallel subtask, often the choice is not optimal because unexpected load levels appear later at the selected computer. In that case, it would be very helpful if a partially completed task could be moved to another node. And finally, the Maple interface could be enhanced by high level data types that hide the actual explicit DSC calls, such as the "bags of incomplete futures" that we have implemented in our Lisp interface (Díaz et al. 1991).

2. System Layers

In Díaz et al. 1991 we described how conceptually the DSC system itself is organized in a multi-layered fashion, each layer drawing from its predecessor. Referring to Figure 1, the bottom layer consists of the interprocess communication using DARPA Internet standard protocols IP/TCP/UDP. Built on this layer lies the first DSC level which includes the internal DSC routines, the daemons, and the C library functions. The second layer consists of the Lisp/C interface, the controller program, the new Maple/C interface and C user programs that use only the seven basic C library functions. The third layer draws on the Lisp/C interface and the Maple/C interface. This layer contains the Lisp library, basic Lisp functions and basic Maple functions. At the topmost layer lies the implementation of high level

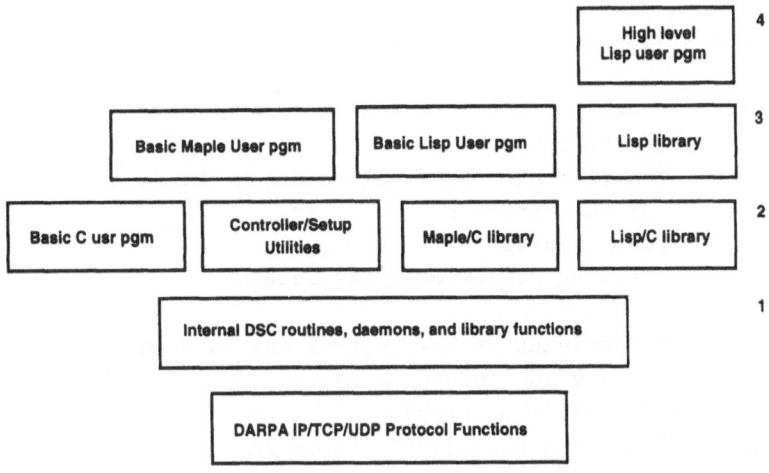

Figure 1: System Layers

Lisp functions which utilize the support routines contained in the Lisp library. Such a Library is being developed for Maple.

3. Maple Interface

The Maple User Interface Library contains ten functions that can be invoked from a user's Maple program or session. The standard example of computing the determinant of an integer matrix in parallel (by homomorphic imaging and Chinese remaindering) illustrates the interaction between DSC and Maple.

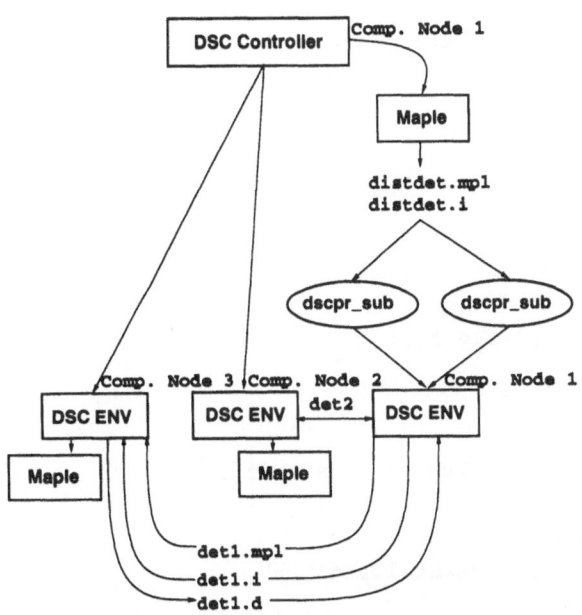

Figure 2: Simple Distribution Scenario

Before a Maple application can distribute parallel subtasks over the network, the database of available compute nodes needs to be initialized

and DSC server daemons have to be started. The DSC Controller program supports the user in configuring the network and in the monitoring of computational tasks (Díaz et al. 1991).

Figure 2 portrays the Scenario of a distributed Maple determinant computation. After starting the controller program on compute node 1, and DSC servers on compute nodes 1, 2, and 3, the main Maple task `distdet.mpl` is initiated. It first reads in an integer matrix from `distdet.i` and selects appropriate moduli. For each modular image, `distdet.mpl` calls the DSC Maple library function `dscpr_sub`, passing Maple code in a file `det1.mpl` and the residue matrix in `det1.i` to DSC for distribution. The function `dscpr_sub` communicates the distribution request to the local server running on compute node 1. The local server locates a suitable processor, in this case compute node 3, and sends both files to the corresponding remote server. The server on compute node 3 starts a Maple session for `det1.mpl`. That program then reads its data from the remote copy of `det1.i` and writes its output to the remote file `det1.d`. When the parallel subtask completes the file, `det1.d` is copied back to compute node 1. In the meantime, the application program has initiated distribution of another parallel subtask, `det2`, for which the local server selected compute node 2. A call to the DSC Maple library function `dscpr_wait` or `dscpr_next` can be used to block the main task `distdet.mpl` on the completion of its children.

The source code of `distdet.mpl` found in Figures 3 through 9 illustrates the usage of the DSC Maple library functions.

```
with(linalg);
#################################################
## distdet(subprobsf,solfile,A,iprod)
## input -
##          subprobsf - source file (without path information)
##                      of the subproblems
##          solfile   - the file name that will contain the determinant
##                      of the matrix A
##          A         - square matrix
##          iprob     - starting point for prime moduli
## description -
##          compute the determinant of the integer matrix A by
##          homomorphic imaging and Chinese remaindering
distdet := proc(subprobsf,solfile,A,iprod)
 local LOW_CPU, MEM, PWD, SOURCE_FILE, INDEX,fid,build,exec,y,
       NUMPROBS,moduli,residue,tmod,ubound,prod,inc,u;
  ## LOW_CPU      - estimated CPU usage
  ## MEM          - estimated memory usage in megabytes
  ## PWD          - real path
  ## SOURCE_FILE  - source file with real path information
  ## INDEX        - subproblem index file
  ## NUMPROBS     - the number of primes needed given the initial value
  ## fid          - temporary file id
  ## build        - Maple build command
  ## exec         - Maple execute command
  ## y            - loop counter
  ## moduli       - the array of moduli
  ## residue      - the array of determinant residues
  ## tmod         - moduli for each subroblem (save can only save names)
  ## ubound       - the upper bound for the product of the moduli
  ## prod         - the product of the moduli
  ## inc          - the next starting point for the moduli search
  ## u            - determinant
  # Read in DSC Maple User Functions
  read('dsc_maple.mpl');
  # Set problem parameters
  subprobsf:= subprobsf;
  LOW_CPU:= 240;
  MEM:= 5;
  # Start dsc_maple daemon
  system('dsc_maple commandfile messagepid dscmaplepid &');
  # Wait a while for startup
  system('sleep 2');
  # Get real path information
  PWD := dscpr_pwd('commandfile','messagepid','dscmaplepid','answerfile');
  SOURCE_FILE:= cat(PWD,'/',subprobsf);
  # Log problem in local server problem database
  INDEX := dscdbg_start('commandfile','messagepid','dscmaplepid',PWD,
                         'answerfile');
```

Figure 3: Interface Start-Up

Figure 3 displays the initialization steps that are needed when distributing Maple code. Once the **dsc_maple.mpl** interface library has been loaded, a dsc_maple daemon can be started via a system call to the host operating system from the Maple environment. After startup of the dsc_maple daemon, calls to the functions in the Maple User Interface Library can be made. In order to assure distinct file names when running multiple distributed Maple applications, the dsc_maple daemon program must receive three command line arguments, which are file names representing a dsc_maple command file, a dsc_maple_message temporary daemon process id file, and a dsc_maple daemon process id file. These files, along with a fourth string denoting the dsc_maple_message answer file, are passed as parameters to all of the subsequent Maple User Interface Library functions. These four files are only used by the different layers of the Maple Interface and are never accessed directly from the user's Maple application. The function dscpr_pwd is used to determine the current path of the local Maple session. This information is used later as a parameter to all remaining Maple User Interface Library functions. The function dscdbg_start in Figure 3 is the first maple interface call that communicates a request to the local DSC server. This function can be used to track a problem and is useful when one wishes to debug distributed tasks using Maple interactively. Arguments to this function include the four interface specific file names and the current path of the local Maple session.

```
ubound := evalf(2*hadamard(A));
```

```
# Compute the Moduli
print('Compute_the_Moduli');
prod := 1; NUMPROBS:= 0;
inc :=iprod;
```

```
while prod < ubound do
 NUMPROBS := NUMPROBS+1;
 moduli[NUMPROBS] := nextprime(inc);
 inc := moduli[NUMPROBS]+1;
 prod := prod * moduli[NUMPROBS];
od;
```

Figure 4: Compute Necessary Moduli

One way of computing the determinant of a matrix in a distributed fashion is to first compute the determinant of the input matrix modulo n relatively prime numbers in parallel and then to construct the corresponding true integer value by Chinese remaindering. Figure 4 shows the computation of the moduli by using the Hadamard inequality to get an upper bound for the determinant.

```
# Prepare corresponding input files
for y from 1 to NUMPROBS do
fid:=cat(SOURCE_FILE,y,'.','i');
tmod := moduli[y];
save A , tmod, fid;
od;
```

```
# Make copies of the subproblem
copyfiles(NUMPROBS,SOURCE_FILE,'mpl');
```

Figure 5: Prepare Parallel Subtasks

After the moduli have been computed, the parallel subtasks can be prepared for distribution. Figure 5 details how the parallel subtask input files and copies of the parallel subtasks are created. The copyfiles function is provided in the Maple User Functions Library. Eventually, we hope to provide additional high level library functions.

Figure 6 demonstrates the use of the function dscpr_sub to start a parallel subtask within the DSC environment. Additional arguments to this function are four strings denoting, respectively, the path to the local files, the types of the local files to be sent, the build (compile, for instance) command, and the exec (load and run) command required for the parallel subtasks; two additional integer arguments represent the CPU and memory requirements of the parallel subtasks. It is possible to use default build/exec commands by supplying null strings for the corresponding arguments. It is also possible to bypass C and Lisp compilation by supplying dsc_c.build or dsc_lisp.build as the build command. The shell script dsc_maple.int applies the source and input files to a Maple session.

After the determinant parallel subtasks have been spawned Figure 7, shows the Maple code necessary for blocking the execution of the main task and after synchonization the code required for performing the shutdown of the DSC Maple interface. The functions dscpr_wait and dscpr_kill are used, respectively, to wait on and to kill specific subtasks. In both cases, the additional argument is the index of the subtask in question. The dscpr_wait function is also used to wait on the completion of *all* outstanding subtasks when a value of −1 is supplied for the index. The function dscpr_next is used to wait for the completion of the next parallel subtask. Its additional argument is filled with the name of the solution file corresponding to the completed subtask. The function dscpr_maple_kill, will terminate the dsc_maple daemon in an orderly fashion.

```
for y from 1 to NUMPROBS do
    # Prepare arguments
    fid:= cat(SOURCE_FILE,y);
    build:= cat('dsc_maple.build');
    exec:= cat('dsc_maple.int ',subprobsf,y,'.mpl ',subprobsf,y,'.d',
            ' ',subprobsf,y,'.i');
    print('one_level_driver: fid = '.fid);
    print('one_level_driver: build = '.build);
    print('one_level_driver: exec = '.exec);
    print('one_level_driver: cpu requirement = '.LOW_CPU);
    print('one_level_driver: mem requirement = '.MEM);

    # Spawn subproblems
    rval:= dscpr_sub('commandfile','messagepid','dscmaplepid',
                    PWD,fid,'pi',build,exec,LOW_CPU,MEM,
                    'answerfile');
od;
```

Figure 6: Distribute Parallel Subtasks

```
    # Wait on all subproblems
    rval := dscpr_wait('commandfile','messagepid','dscmaplepid',
                    PWD,-1,'answerfile');
dscpr_maple_kill('commandfile','messagepid','dscmaplepid',PWD);
```

Figure 7: Wait On All Parallel Subtasks

```
# get residues from .d files
for y from 1 to NUMPROBS do
 fid:= cat(SOURCE_FILE,y,'.d');
 read(fid);
 residue[y] := ddet;
od;
u := chrem(convert(residue,list),
        convert(moduli,list));
save u , prod ,solfile;
RETURN(u);
end;
```

Figure 8: Compute Determinant

The code in Figure 8 computes the integer determinant of A from the n residues of the determinants by Chinese remaindering.

```
read('IM_10_10_1');
answer := distdet('det','distdet.d',
                A,10000000);
quit;
```

Figure 9: Initial Startup Mechanism

Since the file is executed in a interpreted manner the call to distdet is made after the function definition and the reading of the matrix. Finally the session is terminated via a quit command as seen in Figure 9.

Figure 10 is the subproblem file copied and distributed by distdet.mpl which computes the determinant of the integer matrix by homomorphic imaging.

The input file and output file specified by the call to dscpr_sub in the main task are put in the variables DSC_IFILE and DSC_OFILE by the Maple interface.

The control program allows the user to inspect the progress of an application program and its parallel subtasks. This will include the main application program and the parallel subtasks that have been spawned in the current DSC environment.

Figure 11 is an example of the parallel subtask database. The main Maple program prefixed as distdet (running on troi.cs.rpi.edu) registered with the DSC server via dscdbg_start, was given the task name DSC40e0_1. This task distributed Maple code (and an input file) to pleiades.cs.rpi.edu. This is indicated by both maple: 1 and input: 1 being set. The shell script dsc_maple.int is in the exec argument of the call to dscpr_sub. The parallel subtask has not spawned children. All processes run on Sun compute nodes, as CPU 3 indicates.

For 100×100 integer matrices with single digit entries the computation of the the distributed version of the determinant computation program on 22 processors was 1.5 times faster than the standard Maple determinant function (det). For 200×200 matrices on 18 processors the speed up was a factor of 1.9. The speed up becomes more

```
with(linalg);
#######################################################
## ddet(one_input,one_output)
## input -
##          one_input  - input file for subproblem
##          one_output - output file for subproblem
ddet := proc(one_input,one_output)
 local commandfile,messagepid,dscmaplepid,answerfile,msgstr,PWD;
 ## commandfile  - dsc_maple DSC command file
 ## messagepid   - dsc_maple_message process id file
 ## dscmaplepid  - dsc_maple process id file
 ## answerfile   - subtask solution file
 ## msgstr       - temporary string
 ## PWD          - real path
 # Read in DSC Maple User Functions
 read('dsc_maple.mpl');
 # Start dsc_maple daemon
 commandfile := cat(one_input,'.cmd');
 messagepid  := cat(one_input,'.mpid');
 dscmaplepid := cat(one_input,'.ppid');
 answerfile  := cat(one_input,'.ans');
 msgstr := cat('dsc_maple ',commandfile,' ',messagepid,' ',dscmaplepid,' &');
 print(msgstr);
 system(msgstr);
 # Wait a while for startup
 system('sleep 2');
 # Get real path information
 PWD := dscpr_pwd(commandfile,messagepid,dscmaplepid,answerfile);
 # Kill dsc_maple_daemon
 dscpr_maple_kill(commandfile,messagepid,dscmaplepid,PWD);
 # Wait a while for shut down
 system('sleep 2');
 # Remove files
 msgstr := 'rm -f '; msgstr := cat(msgstr,commandfile); system(msgstr);
 msgstr := 'rm -f '; msgstr := cat(msgstr,messagepid);  system(msgstr);
 msgstr := 'rm -f '; msgstr := cat(msgstr,dscmaplepid); system(msgstr);
 msgstr := 'rm -f '; msgstr := cat(msgstr,answerfile);  system(msgstr);
 # Read in matrix and modulo
 msgstr := PWD;
 msgstr := cat(msgstr,'/',one_input);
 read(msgstr);
 # compute the determinant of A modulo tmod
 ddet := Det(A) mod tmod;
 save ddet , one_output;
end;
ddet(DSC_IFILE,DSC_OFILE);
quit;
```

Figure 10: Parallel Subtask

significant with larger problems.

The Maple interface library also allows the user to implement co-routines. To distribute Maple co-routine applications, the function dscpr_cosetup must be called at the beginning of any parallel subtask that is to be treated as a co-routine. Its only arguments are the four interface specific file names and the current path of the local Maple ses-

```
 problem name: distdet
 problem  PID:  16596   node: troi.cs.rpi.edu   index   0        CPU  3
 parent   PID:  16477   naddr: 128.213.2.43      index  -1
 source  name: /fs5/misc1/ugrads/chank/dsc/source/distdet
            c: 0  lisp: 0  math: 0  maple: 1 obj: 0  input:0
build command: dsc_maple.build
 exec command: dsc_maple.int /fs5/misc1/ugrads/chank/dsc/source/distdet.mpl
                             /fs5/misc1/ugrads/chank/dsc/source/distdet.d
 blocked sock:  -1      block condition -1
 completion :   0       status: 00000000 build:   1
 peer link  :  -1       child list:  -1         # subprob:   0
 state      :   0
 estimated cpu use : LOW_CPU     estimated mem need (megs) : 5
MORE
 problem name: DSC40e0_1
 problem  PID:  16608   node: troi.cs.rpi.edu   index   1        CPU  3
 parent   PID:     -1   naddr: 128.213.2.43      index  -1
 source  name: dsc_prob_debug
            c: 0  lisp: 0  math: 0  maple: 0 obj: 0  input:0
build command:
 exec command:
 blocked sock:   6      block condition -1
 completion :   0       status: 00000000 build:   1
 peer link  :  -1       child list:   2         # subprob:   2
 state      :   0
 estimated cpu use : -1 (UNKOWN VALUE)
 estimated mem need (megs) : -1
MORE
 problem name: DSC40e02
 problem  PID:     -1   node: pleiades.cs.rpi.edu      index   2       CPU  3
 parent   PID:  16608   naddr: 128.213.2.43      index   1
 source  name: /fs5/misc1/ugrads/chank/dsc/source/det1
            c: 0  lisp: 0  math: 0  maple: 1 obj: 0  input:1
build command: dsc_maple.build
 exec command: dsc_maple.int det1.mpl det1.d det1.i
 blocked sock:  -1      block condition -1
 completion :   1       status: 00000000 build:   0
 peer link  :   3       child list:  -1         # subprob:   0
 state      :   0
 estimated cpu use : HIGH_CPU     estimated mem need (megs) : 5
MORE
```

Figure 11: Problem Data Base

sion. When the Maple parallel subtask calls the dscpr_cowait function with an additional integer argument, it enters a sleep state and optionally transmits a data file back to its parent. A parent task can send a wake up call to a sleeping parallel subtask via the dscpr_coresume function. Additional arguments to this function are an integer which uniquely identifies a parallel subtask (re-turned from the spawning call to dscpr_sub) and a string identifying which input file, if any, should be sent to the co-routine before the parallel subtask is to be resumed.

This material is based on work supported in part by the National Science Foundation under Grant No. CCR-90-06077 and Grant No. CCR-93-

19776, Research Experiences for an Undergraduate supplement (first author), and by GTE under a Graduate Computer Science Fellowship (second author).

The software described in this paper is freely available by anonymous ftp from ftp.cs.rpi.edu in directory dsc.

Literature Cited

Char, B. W., "Progress report on a system for general-purpose parallel symbolic algebraic computation," in *Proc. 1990 Internat. Symp. Symbolic Algebraic Comput.*, edited by S. Watanabe and M. Nagata; ACM Press, pp. 96–103, 1990.

Díaz, A., "DSC Users Manual (2nd ed.)," *Tech. Rep.* **93-11**, Dept. Comput. Sci., Rensselaer Polytech. Inst., Troy, New York, May 1993. 197 pp.

Díaz, A., Hitz, M., Kaltofen, E., Lobo, A., and Valente, T., "Process scheduling in DSC and the large sparse linear systems challenge," in *Proc. DISCO '93*, Springer Lect. Notes Comput. Sci. **722**, edited by A. Miola; pp. 66–80, 1993. Available from anonymous@ftp.cs.rpi.edu in directory kaltofen.

Díaz, A., Kaltofen, E., Schmitz, K., and Valente, T., "DSC A System for Distributed Symbolic Computation," in *Proc. 1991 Internat. Symp. Symbolic Algebraic Comput.*, edited by S. M. Watt; ACM Press, pp. 323-332, 1991. Available from anonymous@ftp.cs.rpi.edu in directory kaltofen.

Kaltofen, E. and Lobo, A., "Factoring high-degree polynomials by the black box Berlekamp algorithm," *Manuscript*, January 1994. Available from anonymous@ftp.cs.rpi.edu in directory kaltofen.

McClellan, M. T., "The exact solution of systems of linear equations with polynomial coefficients," *J. ACM* **20**, pp. 563–588 (1973).

Valente, T., "A distributed approach to proving large numbers prime," *Ph.D. Thesis*, Dept. Comput. Sci., Rensselaer Polytech. Instit., Troy, New York, December 1992. Available from anonymous@ftp.cs.rpi.edu in directory valente.

King Choi Chan currently is an undergraduate student in Computer Systems Engineering at Rensselaer Polytechnic Institute.

Angel Díaz received both his B.S. degree in Computer in 1991 and his M.S. degree in Computer Science in 1993 from Rensselaer Polytechnic Institute. He is currently pursuing the Ph.D. degree at Rensselaer under the direction of Professor Erich Kaltofen. His special fields of interest include parallel methods in symbolic computation. He was a Rensselaer Polytechnic Institute Graduate Fellow from 1991–1992, received support from the National Science Foundation Fellowship Program in 1992, and is currently the recipient of the GTE Fellowship Program. He is also a member of Upsilon Pi Epsilon national computer science honor society.

Erich Kaltofen received both his M.S. degree in Computer Science in 1979 and his Ph.D. degree in Computer Science in 1982 from Rensselaer Polytechnic Institute. He was an Assistant Professor of Computer Science at the University of Toronto and an Assistant and Associate Professor at Rensselaer Polytechnic Institute, where he is now a Professor. His current interests are in computational algebra and number theory, design and analysis of sequential and parallel algorithms, and symbolic manipulation systems and languages. Professor Kaltofen currently is the Chair of ACM's Special Interest Group on Symbolic & Algebraic Manipulation and serves as associate editor on several journals on symbolic computation. From 1985–87 he held an IBM Faculty Development Award. From 1990–91 he was an ACM National Lecturer.

Department of Computer Science
Rensselaer Polytechnic Institute
Troy, New York 12180-3590

Internet: {diaza,kaltofen}@cs.rpi.edu

DENOTATIONAL SEMANTICS APPLIED TO THE TYPESETTING OF MAPLE EXPRESSIONS

Reid M. Pinchback
Academic Computing Services, MIT, Cambridge, MA

Introduction

Using your favourite document preparation package for typesetting mathematical expressions – to include in papers and theses, for example – can be tedious at the best of times. It is particularly frustrating when the desired expressions already exist somewhere else, such as in an active Maple session, but are not in any kind of format that is both transferable and typographically useful. Here, the concepts of denotational semantics are applied to the problem of exporting Maple expressions to FrameMaker via MIF (Maker Interchange Format) files.

The approach shown here involves the following steps:

1. Specify an abstract syntax capable of representing a useful subset of Maple expressions.

2. Specify the meaning of each grammatical production in terms of Maple-coded 'denotations'.

3. Use the grammatical and denotational design information to implement a meta-evaluator in Maple. The resulting program will behave as a transducer for converting Maple expressions into a MIF document.

Denotational Semantics

This is a method useful for specifying the meaning of formal languages in an abstract way, using mathematical constructs called 'denotations'. The theoretical underpinnings of the method are in domain theory, which we will not need to explore here. This is in part because of relative simplicity of our application, and in part because we are not attempting to create a fully abstract implementation-independent specification of meaning. Our objective is to 'operationalize' the semantics in terms of a Maple implementation.

Perhaps the most useful property of denotational semantics is its compositional nature[1], by which we mean:

The meaning of a phrase is determined by its denotation and the meaning of its direct subphrases.

For example, given a grammatical product A ::= B C, the meaning of A is a function of the meanings of B and C. The kind of notation typically used for denotational specifications looks like:

A[A] ::= ... **B[B]** ... **C[C]** ...

where **A**, **B**, and **C** are semantic functions defined over the appropriate syntactic constructs.

While generally denotations are specified in terms of a lambda calculus, such is by no means a requirement of the method. We will specify our denotations in terms of Maple constructs.

Abstract versus Concrete Syntax

Since likely more people are familiar with grammars for concrete syntax than ones for abstract syntax, and since at a glance they are similar in appearance, to avoid later confusion we will briefly contrast the two kinds[2]:

1. Concrete syntaxes are those typically encountered in introductory compiler courses. Their grammars contain many productions useful for resolving precedence relationships and for specifying important syntactic keywords like 'if' and 'then'. All of this is necessary for parsing languages with sentences that are written as character strings.

2. Abstract syntaxes have grammars that are compact and represent the inherent structure of sentences in the language, as represented by tree structures (such as parse trees) instead of character strings. With neither the need to resolve precedence nor to

indicate the syntactic keywords that parsers use for pattern recognition in matching productions, abstract syntax grammars have much simpler and many fewer productions than is the case for concrete syntax grammars.

A particularly useful property of abstract syntax[1] is that it allows us to collapse operator productions. For example, contrast the following concrete and abstract syntaxes for specifying simple arithmetic expressions:

Concrete Syntax	Abstract Syntax
$E ::= E + T \mid T$	$E ::= E \, O \, E$
$T ::= T * P \mid P$	$O ::= + \mid *$
$P ::= (E)$	

Table 1: Example of Concrete vs. Abstract Syntax

The reduced complexity of abstract syntax makes it even easier to utilize compositional approaches for dealing with the meaning of a language. Here is an example of a semantic specification for the above syntax, where the first production indicates that we apply the meaning of the binary operator to the meanings of its two operands:

$E[E \, O \, E] ::= O[O](E[E], E[E])$

$O[+] ::=$ integer addition

$O[*] ::=$ integer multiplication

To use this approach all you need to have (or be able to simulate) is a tree-oriented representation of the language structure for a sentence. When the nodes of the tree contain tag information to indicate the type of each component phrase, it becomes a simple process when traversing the tree to find matching productions of the abstract syntax. Since Maple expressions have this kind of structure[3] and the structure is relatively accessible to us, this is an approach that we can use.

An Abstract Syntax for Maple Expressions

There are very few productions in an abstract syntax describing Maple expressions. We will focus on creating a syntax that does reasonable justice to how Maple types are represented internally. This will allow us to view surface types as different as "+" and 'list' as type construction operators over their component

operands, which are themselves Maple expressions. In the final implementation we will be using the Maple commands 'whattype' and 'op' to parse expressions, instead of using something like the 'hackware' package. This relatively indirect access to the structure of Maple expressions will make the implementation more readable at the cost of having to deal with some special-case situations where 'op' and 'whattype' don't provide us exactly the information we need.

We will restrict consideration to Maple expressions and ignore command constructs like 'if...then', and thus will also ignore the Maple 'procedure' and 'uneval' types. For simplicity of the discussion we will also not include the more complicated aggregate expression types 'series', 'array', and 'table', but by the end of this paper it should become apparent how they can be incorporated.[2]

We first present an over-simplified abstract syntax just to get us started, allowing us discuss some important features before we present a more complete version. The start symbol for this syntax is 'P', and a description of the intended interpretation is included beneath each production:

$P ::= Q$
P (a Maple program) is an expression sequence.
$Q ::= E^*$
Q (an expression sequence) is a sequence of expressions.
$E ::= I \mid N \mid T(X^*)$
E (an expression) is either an integer, a name, or a type construction over a sequence of operands.
$X ::= Q \mid E \mid N$
X (an operand) is either an expression sequence or an expression or a name.
$N ::= S \mid$ function $N(Q) \mid$ indexed $N[Q]$
N (a name) is either a string, a function, or an indexed reference.
$T ::= + \mid * \mid \wedge \mid$ list $\mid$ set $\mid ...$
T is a Maple aggregate type specifier.

One important distinction to make in how Maple represents expressions internally is the difference between an expression sequence (an expression of surface type 'exprseq') versus the sequence of operands for a type. An 'exprseq' internally is a single object, tagged with type EXPSEQ, having a sequence of component objects, each of which in turn is some kind of an expression. Some types (like 'function' and 'array')

---

[1] From this point on we will simply use the term 'syntax', and cease to make the distinction between syntax and grammar.

[2] The results of adding support for these types, excluding indexfcn semantics, can be viewed in the 'maple2mif' code submitted to the Maple share library.

have a sequence of operands, some of which may be expression sequences, but the sequence of operands for the particular type is not itself an 'exprseq'. Such distinctions are important both syntactically and semantically – for parsing expressions properly, for implementing the design cleanly, and for generating the MIF output code correctly.

Another feature of the syntax to notice is the provision for names. These are not quite the same types for which 'type(expr,name)' is true. Maple internal data structures have a different internal concept for a name that has more in common with the concept of Lvalues found in other languages like 'C'. Since 'name' is not actually a surface type in Maple, this is not really a point of confusion for creating an expression parser, but can be confusing to the reader.

As we have stated, the above syntax is over-simplified. Some of the types encompased by the T productions (for example "=" and "^") are binary operators over expressions. Others (such as "+" and "*") are n-ary operators. In other words, we need to deal with the fact that different types have different signatures describing their operands. The solution used here is to add productions to the syntax to deal with grouping operators according to their signature. From the viewpoint of later specifying the semantics, we would be partitioning the types into equivalence classes of signatures, determine the meaning of the operands based on the signature for the class, and then passing the result to the function that denotes the meaning of the particular type. In fact, for consistency, we will do this with the N productions for names as well, since these are also expressions constructed from Maple types and thus generalize to the same treatment. The reason for the N productions is that we use names in circumstances where we don't use non-name expressions, but the productions still describe Maple internal expression types. Integers and strings will also be treated in this type-operator expressive way:

```
P  ::= Q
Q  ::= E*
E  ::= T | N
T  ::= O1(E,E) | O2(E) | O3(E*) | O4(Q)
       | O5(N,E) | O6(I,I) | O7(I)
N  ::= O8(S) | O9(N,Q)
O1 ::= ^ | = | .. | <> | < | <= | and | or
O2 ::= not
O3 ::= + | *
O4 ::= set | list
O5 ::= .
```

```
O6 ::= fraction | float
O7 ::= integer
O8 ::= string
O9 ::= function | indexed
```

Semantic Functions for Maple Expressions

Now that we have a relatively complete and useable syntax for Maple expressions, we need some semantic functions for mapping the syntax into denotations that will represent the MIF code to be generated[4]. These semantic functions, when implemented in Maple, will have the following general structure:

P : maple programs → MIF programs
   P[Q] ::= ... **Q**[Q] ...

Q : expression sequences → a comma-separated sequence of MIF MathFullForm components
   Q[E*] ::= ... map(E,E*) ...

E : expressions → meaning of expressions
   E[N] ::= **N**[N]
   E[T] ::= **T**[T]

T : non-name expressions → meaning of non-name expressions

   T[O1(E,E)] ::= **O**[O1] (**E**[E], **E**[E])
   T[O2(E)]   ::= **O**[O2] (**E**[E])
   T[O3(E*)]   ::= **O**[O3] (map(E,E*))
   T[O4(Q)]   ::= **O**[O4] (**Q**[Q])
   T[O5(N,E)] ::= **O**[O5] (**N**[N], **E**[E])
   T[O6(I,I)] ::= **O**[O6] (**I**[I], **I**[I])
   T[O7(I)]   ::= **O**[O7] (**I**[I])

N : name expressions → meaning of name expressions
   N[O8(S)]   ::= **O**[O8] (**S**[S])
   N[O9(N,Q)] ::= **O**[O9] (**N**[N], **Q**[Q])

O : type operators → meaning of type operators as a MIF MathFullForm component

   O[+] ::= function to map the denotation of the operand sequence to a MIF 'plus[...]' code fragment (etc.)

I : integers → meaning of integers (implemented as the identity operator)

S : strings → meaning of strings (implemented as the identity operator)

In the above specifications, '...' indicates surrounding denotational information. Most of the added material in the denotations comes in dealing with programs, expression sequences, and type operators. For this application the other productions add negligible denotational information beyond retrieval of the denotations of their subphrases.

Implementing the Meta-Evaluator

Four objectives were set for the Maple implementation of the semantics of the meta-evaluator[3]:

1. It must work (obviously!).

2. It must closely follow the notation of the design.

3. It must be easy to modify when implementing simple and obvious extensions.

4. It must be possible to have different instantiations of the meta-evaluator without global naming conflicts.

Objective 1 needs no discussion. Objectives 2 and 3 go hand-in-hand. The desired way to modify the program will be by first determining changes to the abstract syntax, following with changes to the denotational specifications, and then finally by making (presumably manageable) changes to the implementation.

Objective 4 is important if we want to have different MIF converters available to us at the same time. For example we might want different converters for different font sizes, or different converters that typeset rational numbers (type 'fraction') either vertically or horizontally, ie:

$$\left(\frac{1}{2}\right) \text{ versus } (1/2)$$

To allow for these situations the meta-evaluators cannot share, by name, a single global object (like a subroutine) that might need to be modified for differing semantics. The solution is implemented by giving each meta-evaluator its own protected environment of local variable names for internal subroutines. The entire collection of local names is in turn given to the subroutines by such tricky methods as:[4]

```
>M := proc()
>    local f,g;
>    env := {'G'=g,'F'=f};
>    g := subs(env, x -> F(x+1));
>    f := subs(env, y -> sin(y)+G(2));
>    ...
> end;
```

This way the functions 'f' and 'g' are only accessible to other subroutines contained within the procedure 'M'.

---

[3] Terms like 'meta-evaluator' and 'meta-interpreter' may sound esoteric, but simply signify language evaluators implemented in terms of an existing interpreter, often used to modify the default interpretive behaviour.

[4] This technique was found in – and gratefully borrowed from – Michael Monagan's implementation of the Gauss package.

Figure 1 shows a portion of the actual implementation code for 'maple2mif', to give readers a sense of the final result. The routine is easily used by invoking it to construct a procedure to do the conversion, converting an expression to a string, and then writing the results to a file:

```
> read 'maple2mif.txt':
> M := maple2mif():
> X := sin(a+b)^f(pi):
> interface(echo=0);writeto(foo);
> print(printf('%s',M(X)));
> writeto(terminal);interface(echo=1);
```

Limitations

The 'maple2mif' converter works quite well when considered purely as a typesetting tool, with only a few minor caveats:

- Since Maple is an 'eager evaluator' of arguments, the expression you pass to the evaluator may not be the expression that gets typeset. For example, 'maple2mif()(1+2)' would generate MIF code for typesetting '3' in FrameMaker, while for commutative situations like 'x+y', 'maple2mif()(x+y)' may typeset either 'x+y' or 'y+x', varying between Maple sessions. No good solution to this is known at this time, other than perhaps to allow support for some kind of neutralized expression form like that returned by 'procbody'.

- While some support for mapping special symbol names (like 'aleph' and 'pi') to their typeset symbolic has been provided, there is no support for mapping special function names (like 'int') to their typeset equivalent. This should be addressed in the future by modifying the syntax and semantics of the 'function' type, handling it in a method similar to how type operator syntax is treated now.

- There are some Maple expressions that do not have close FrameMaker equation equivalents. The solution used here is to provide MIF code for something that is similar in appearance, for example by using the 'times' specification and character strings. This is not a limitation in 'maple2mif', but in the FrameMaker equation format. It does not have a convenient method for formatting an expression with an arbitrary binary operator.

When we view 'maple2mif' as an exposition of denotational methods, there are real limitations that need addressing:

- While the use of 'whattype' and 'op' allows us reasonable access to the internal structure of Maple expressions, they are not the perfect tools for the job. The result is a slight proliferation of special-case code, for example to distinguish between 'exprseq' and sequences of operands, or to retrieve the occasional zero-th operand that 'op' doesn't return unless specifically asked for. The situation is even less satisfactory for eventually supporting the 'procedure' type, where we would need to use 'procbody' and manipulate neutralized procedure statements. The 'hackware' package provides better access to the internal structures, but at the cost of increased visual complexity, combined with the need to distinguish continually between objects and pointers to objects. What is really needed is a front-end to hackware that would make it more convenient for implementing traversals of arbitrary Maple expressions.

- The methods of denotational semantics have a great deal more muscle than has been explored here[5]. For constructing more complicated meta-evaluators, such methods are necessary for generalizing the implementation to encompass a broader range of possible meanings. The three most important steps to take in the future are:

1. Support domain-oriented structuring of denotations. This is important as a structural design tool, as it serves in the role of a data-type for denotational methods.

2. Support specification of states and environments. This is important for modelling memory and I/O. For example, this could be used to handle treatment of last-name evaluation and 'assume' properties in Maple.

3. Implementation of continuations. This is important for dealing with flow-of-control and error-handling issues, both of which become important when moving beyond evaluation of Maple expressions to include Maple command structures and the 'procedure' type.

Conclusion

The methodology of denotational semantics can be useful for designing programs. For structured programming approaches to problems that can be characterized in terms of formal languages, the compositionality property of denotational semantics allows us to create concise top-down solution methods. We hope

```
>maple2mif := proc()
>    local env,sfP,sfQ,sfE,sfN,sfT,sfO,sfS,sfI,MIFargs,MIFsymbol;
>
>    env := {'P'=sfP,'Q'=sfQ,'E'=sfE,'N'=sfN,'T'=sfT,'O'=sfO,'S'=sfS,'I'=sfI,
>           'MIFARGS'=MIFargs,'MIFSYMBOL'=MIFsymbol};
>
>    # semantic function P
>    # defined over the syntactic domain of programs
>    sfP := subs(env, proc()
>      cat(`<MIFFile 3.00> # generated by maple2mif\n`,
>          `<Math\n`,
>            `<MathFullForm ```,Q(args),`\`>\n`,
>            `<MathOrigin 2.0" 2.0">\n`,
>            `<MathSize MathMedium>\n`,
>          `>\n`);
>    end); # proc sfP
>
>    # semantic function Q
>    # defined over the syntactic domain of expression sequences
>    sfQ := subs(env, proc()
>      if nargs=0 then
>        # NULL exprseq
>        E(``);
>      elif nargs=1 then
>        # single expression
>        E(args[1]);
>      else # nargs > 1
>        # multiple expressions
>        cat(`comma[`,MIFARGS(op(map(E,[args]))),`]`);
>      fi;
>    end); # proc sfQ
```

Figure 1

```
>    # semantic function N
>    # defined over the syntactic domain of typed name expressions
>    # where allowable name operators={string,function,indexed}
>    sfN := subs(env, proc()
>      local o;
>      o := whattype(args[1]);
>      if member(o,{`string`}) then
>        O(o)(S(args[1]));
>      elif member(o,{`function`,`indexed`}) then
>        O(o)(N(op(0,args[1])),Q(op(args[1])));
>      else
>        ERROR(cat(`unsupported expression type, `,o,`, detected`));
>      fi
>    end); # proc sfN
...
>    # semantic function O
>    # defined over the syntactic domain of type operators
>    # where type operators=list+set+`+`+`*`+ (etc).
>    sfO(`string`) := proc()
>      local found;
>      if length(args[1])=1 then
>        cat(`char[`,args[1],`]`);
>      elif MIFSYMBOL(args[1],'found') then    # to deal with symbol names like
>        cat(`char[`,found,`]`);               # aleph, pi, Pi, PI, ...
>      else
>        cat(`string["`,args[1],`"]`);
>      fi;
>    end;
...
>    # return meta-evaluator
>    eval(sfP);
>
> end: # proc maple2mif
```

Figure 1, cont.

that here we have provided a realistic example of such an approach that will motivate others to consider similar methods for solving problems that require the traversal and interpretation of Maple expressions.

References

[1] Tennent, R.D., *Semantics of Programming Languages*, Prentice Hall International, UK (1991).

[2] Allison, L., *A Practical Introduction to Denotational Semantics*, Cambridge University Press, UK (1986).

[3] Char, B.W., et al., *Maple V Language Reference Manual*, Springer-Verlag, New York (1991).

[4] *MIF Reference*, Frame Technology Corporation (1991).

[5] Gordon, J.C.G, *The Denotational Description of Programming Languages*, Springer-Verlag, New York (1991).

Biography

Reid M. Pinchback (reidmp@mit.edu) received his B.Math degree from the University of Waterloo in 1991. In that year he joined the Academic Computing Services department of MIT as a Faculty Liaison, where he currently maintains and supports Unix and Macintosh installations of Maple and other software packages intended for educational use. He has also been serving as Assistant Editor for the Maple Technical Newsletter since July of 1993.

I B. MAPLE IN APPLIED MATHEMATICS

A SYMBOLIC ENGINEERING MECHANICS SYSTEM BUILT ON OEM MAPLE

Philip Todd, Robin McLeod and Marcia Harris
Saltire Software, Beaverton, Oregon

Abstract

In this paper, we present a system which converts graphical descriptions of engineering mechanics problems into closed form equations. We describe the functions we provide for simplifying and manipulating the resulting mathematical equations.

1. Introduction

A major limitation of the use of symbolic mathematics packages, such as Maple, by mechanical engineers is the difficulty of accurately converting a geometry based mechanical engineering problem into a set of algebraic equations which can then be manipulated, solved or visualized by the symbolic mathematics package.

In the last eight years, Variational Geometry has emerged as a natural and powerful language for the expression of mechanical engineering problems [1,2,6], particularly problems in the realm of engineering mechanics. Problems involving the kinematic behavior of mechanisms may be succinctly expressed in the form of a constraint-based variational model of the geometry along with first and second derivatives of the constraints. With an additional capability for handling forces and masses, we add the ability to express problems in the statics and dynamics of mechanisms. The commercial package *Analytix* uses a variational geometry user interface as the means of specifying problems in the dynamic analysis of mechanisms [3-6]. These problems are then solved numerically in the package. Figure 1 shows an Analytix model of a four-bar linkage.

A Constructive Variational Geometry System represents engineering geometry as a sketch with constraints (also called "dimensions") given as labels on the sketch. "Constructive" refers to the classical constructions of Euclid's geometry, for example, the construction of a point of intersection, given two known lines. "Variational" means that numerical information resides in the dimensions, such as distance between points or measure of an angle, which are labeled on the diagram, rather than residing in the coordinates of points; therefore change or motion in the figure can be simply represented as change or variation in a constraint value.

In this paper, we present a system which converts an engineering mechanics problem, expressed as an Analytix model, into a symbolic mathematical representation within Maple. We discuss the functions which were developed to allow the user to manipulate and display this mathematical model. We present several examples to show the process of using the hybrid system.

Our approach allows the engineering mechanics problem to be expressed in the natural graphical terms of a dimensioned sketch, while automatically performing the error prone process of converting the geometrically expressed problem into mathematics.

2. System Architecture

There are three components to the symbolic mechanics package: (i) the front end is a modified version of the Analytix variational geometry package; (ii) a layer of code converts from the Analytix representation of the mechanics problem to a symbolic representation; (iii) a set of Maple functions facilitates user interaction with the symbolic representation.

Component (i) is written in C, component (ii) contains C code and Maple code, component (iii) is entirely Maple code.

(i) Analytix

To create a model of a mechanism in the Analytix package, the user first sketches the mechanism. Typically the sketch contains a simplified representation of the mechanism's geometry. Links are usually represented by straight lines between joints or points of force application. Centers of gravity of any links with non-negligible mass are explicitly sketched.

The user then specifies the exact geometry by adding constraints to the drawing. The constraints specify angles or distances on the drawing. For a kinematic analysis, the constraints should represent quantities whose motion is known. In practice, for a single degree of freedom mechanism, all the constraints but one represent quantities which stay fixed throughout the motion of the mechanism. The final constraint represents the driver, whose motion is known. The Analytix system aids the constraint specification process by giving continuous feedback on whether the model is underconstrained, overconstrained, or consistently constrained. The constraint values may be numbers or expressions. Figure 1 shows an Analytix model for a four bar linkage.

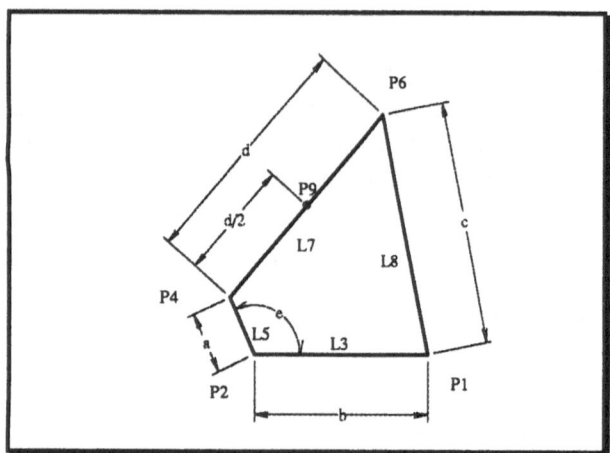

Figure 1: A four bar linkage specified by the lengths of three sides and one angle.

To perform a kinematic analysis, the user specifies the velocities and accelerations of any constraints which are in motion. (In the example of the four bar linkage (figure 1), all constraints would have 0 velocity and acceleration except the driving angle. If the crank is being driven at constant angular velocity v the angle would be given a velocity v and an acceleration 0.) The user also must specify a fixed point and a fixed line.

With the constraint velocities and accelerations given, the model is able to output on demand the position, velocity, and acceleration of any point of the mechanism and the angular velocity and angular acceleration of any line.

Static analysis of a mechanism may be performed in a very natural way in a Variational Geometry system by assuming that constraints carry reactions. Hence not only the motion, but also the statics of the model are defined by the constraints. If loads are applied to a mechanism in Analytix, reaction forces are generated in the constraints. The system will sum the constraint forces which impinge on a point to give the reaction force at that point. The system will integrate the forces on a single line segment to give shear force and bending moment.

Analytix allows external forces to be applied to any point on a mechanism and external torques to be applied to any line. These forces may be constant or they may vary with geometry, time or some other parameter. Analytix also provides conformal force elements: translational and rotational spring-damper-actuators.

To perform inverse dynamics in Analytix, the user specifies the instantaneous motion of the mechanism in the same way as for kinematics, but also applies forces and specifies the mass of points representing mass centers. The user also specifies the moment of inertia of lines which represent rigid links. The user can now generate the same reaction force outputs as for the static analysis, except the inertial forces generated by the accelerating masses will be taken into consideration.

In a dynamic analysis, some constraints represent quantities whose behavior is known in advance - either quantities which remain fixed or drivers, whose motion is given. Other constraints are free to accelerate in response to unbalanced forces present in the model. To set up a dynamic model, the user specifies the fixed constraints and the input drivers in the same way as for kinematic analysis. Free constraints are given initial values and initial velocities and marked as being free to accelerate.

(ii) Creating Symbolic Analytix Models

In a previous paper [4] a method was described for performing a symbolic analysis of dimensioned engineering drawings. Analysis of the constraint graph associated with the drawing led to its description as a sequence of elementary Euclidean constructions, each of which was responsible for constructing a single point or line from its specified relation to known (already constructed) points and lines.

For example, a construction sequence for the four bar linkage of figure 1 is given below.

> Construct a point (point P2) at the origin.
> Construct a line (line L3) through point P2 in the direction of the x-axis.
> Construct a point (point P4) on line L5 distance a from point P2.
> Construct a line (line L5) through point P2 angle e degrees to line L3.
> Construct a point (point P1) on line L3 distance b from point P2.
> Construct a point (point P6) distance c from point P1 and distance d from point P4.
> Construct a line (line L8) through points P1 and P6.
> Construct a line (line L7) through points P4 and P6.
> Construct a point (point P9) on line L7 distance d/2 from point P4.

Each individual construction which is part of the sequence described above has a corresponding set of equations which may be solved to determine the unknown geometrical entity. These are described in [4].

In [4], expressions for the coordinates of the drawing's points and the coefficients of the drawing's lines were established in terms of the independent constraint values alone. With this approach, the system suffered from substantial intermediate expression expansion, and was able to deal only with rather simple drawings. In order to deal with drawings and engineering mechanics problems of arbitrary complexity, it was necessary to maintain a symbolic representation as a nested sequence of expressions with intermediate variables introduced by the system. This eliminated the exponential growth in individual expression complexity. However, as Maple is designed to manipulate single expressions rather than sequences of expressions with collective meaning, this necessitated a layer of code to allow the manipulation of our symbolic representations.

Table 1 shows the system output for the location of point P9. For comparison, table 2 shows the same expressions without intermediate variables.

$$P6d^2 = (a \cos(e) - b)^2 + a^2 \sin(e)^2$$

$$P6e^2 = c^2 - \frac{1}{4} \frac{\left(P6d^2 + c^2 - d^2\right)^2}{P6d^2}$$

$$P6c = \frac{1}{2} \frac{P6d^2 + c^2 - d^2}{P6d}$$

$$P6y = \frac{P6c \, a \sin(e)}{P6d} - \frac{P6e \, (a \cos(e) - b)}{P6d}$$

$$P9y = \frac{1}{2} a \sin(e) + \frac{1}{2} P6y$$

$$P6x = b + \frac{P6c \, (a \cos(e) - b)}{P6d} + \frac{P6e \, a \sin(e)}{P6d}$$

$$P9x = \frac{1}{2} a \cos(e) + \frac{1}{2} P6x$$

Table 1: Equations for the curve traced by point P9 on the four bar linkage shown in Figure 1.

33

```
FullySubstitute(P9x);
```

$$\frac{1}{2}a\cos(e)+\frac{1}{2}b+\frac{1}{4}\frac{\left((a\cos(e)-b)^2+a^2\sin(e)^2+c^2-d^2\right)(a\cos(e)-b)}{a^2\cos(e)^2-2a\cos(e)b+b^2+a^2\sin(e)^2}+\frac{1}{4}\bigg($$

$$2c^2a^2\cos(e)^2-4c^2a\cos(e)b+2c^2b^2+2c^2a^2\sin(e)^2-a^4\cos(e)^4$$

$$+4a^3\cos(e)^3b-6a^2\cos(e)^2b^2-2a^4\cos(e)^2\sin(e)^2+4a\cos(e)b^3$$

$$+4a^3\cos(e)b\sin(e)^2-b^4-2b^2a^2\sin(e)^2-a^4\sin(e)^4+2d^2a^2\cos(e)^2$$

$$-4d^2a\cos(e)b+2d^2b^2+2d^2a^2\sin(e)^2-c^4+2c^2d^2-d^4\bigg)^{1/2}a\sin(e)\bigg/\bigg($$

$$a^2\cos(e)^2-2a\cos(e)b+b^2+a^2\sin(e)^2\bigg)$$

Table 2: Equation for P9x with all the intermediate variables eliminated.

Two different engineering mechanics functions are available to the user. Reaction(x) gives the pseudoforce in parameter x. InstallDiffEq([x_0,...,x_n]) creates new variables A_x_0,...,A_x_n representing the accelerations of x_0,...,x_n. In the process of assigning expressions to these accelerations, a number of new intermediate variables will be introduced (many representing derivatives of existing variables). Velocity variables V_x_0,...,V_x_n will also be introduced and treated as independent input variables. The methods used in these functions are described in [6].

(iii) Interacting With Symbolic Analytix Models

The Intermediate Variable Table (IVT) and the routines to support and manipulate this structure were written in the Maple symbolic programming language and coexist as a package running within Maple. We describe in the following the functions which constitute this library, grouped according to their use.

(a) Adding variables to the IVT

Install(varname,eqn) - Installs a new variable with defining (implicit or explicit) equation eqn.

InstallAreaMoments(name,pointlist) - installs a number of new variables corresponding to area moments of the polygon defined by the points in pointlist.

InstallDerivative(var,dvar) - installs an expression for the derivative of var with respect to dvar.

(b) Extracting Expressions from the IVT

Evaluate(expr) - returns expr with the equation from the IVT substituted for each variable in expr which is in the IVT. This is not done recursively

FullyEvaluate(expr) - A recursive version of the above.

EvaluateF(expr,assgn) - performs a floating point evaluation with numerical assignments given in assgn.

Equations(expr,style) - for each variable v in expr which is in the Intermediate Variable Table, returns the equation defining v. Style specifies whether implicit or explicit equations are desired.

FullEquations(expr,style) - A recursive version of the above

CFunction(fname,varlist) - creates a C function with name fname to evaluate the variables contained in the varlist. Inputs to the function are all the constants in the recursive definition of the variables in varlist.

(c) Displaying the IVT

Show(expr,style) - displays expr and the definitions (recursively) of all the table symbols involved in expr.

(d) Geometric and mechanics functions

Angle(line1,line2) - returns the angle between two lines

Area(pointlist) - returns the signed area of the closed polygon formed by the pointlist.

Distance(entity1,entity2) - returns the distance between the two entities.

InstallDiffEq(parlist) - for each parameter p in parlist, installs equations for the acceleration of p, A_p.

Reaction(var) - returns the reaction pseudoforce in variable var.

(e) Applying Maple functions to individual expressions

Apply(expr,fcn,arg2, ... , argn) - for each variable v in expr which is in the Intermediate Variable Table, applies the function fcn to the expression defining v.

(f) Eliminating Intermediate Variables

Substitute(expr,varlist) - substitutes the definition of each of the variables in varlist into the definition of each of the variables in expr. If no second parameter is given, substitutes all intermediate variables explicitly found in the definition of each variable in expr.

FullySubstitute(expr) - a recursive version of the above which eliminates all intermediate variables from expr.

3. Example Problems

(i) Coupler Curve for a Geared 5-bar linkage

In this example, we derive equations for the curve traced by one point on a geared five bar linkage. A gear pair is modeled in Analytix by creating a formula linking the values of two angles. In Figure 2, one angle is given a value t, while a second angle is given a value r*t+s, where r is the gear ratio and s represents the offset between the gear angles in an initial configuration.

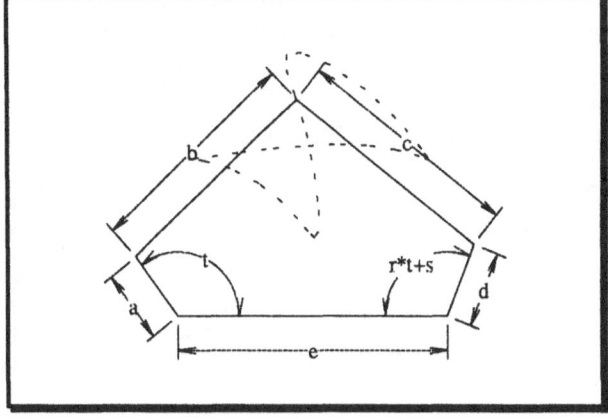

Figure 2: An Analytix model of a geared 5-bar linkage.

Table 3 gives the expressions derived by the system for the coupler curve drawn in figure 2.

Show(P6);

$P8y = d \sin(r\,t + s)$

$L10B = \cos(r\,t + s)$

$P8x = e - d\,L10B$

$P6d^2 = (P8x - a\cos(t))^2 + (P8y - a\sin(t))^2$

$P6c = \dfrac{1}{2}\dfrac{P6d^2 + b^2 - c^2}{P6d}$

$P6e^2 = b^2 - \dfrac{1}{4}\dfrac{\left(P6d^2 + b^2 - c^2\right)^2}{P6d^2}$

$P6y = a\sin(t) + \dfrac{P6c\,(P8y - a\sin(t))}{P6d} + \dfrac{P6e\,(P8x - a\cos(t))}{P6d}$

$P6x = a\cos(t) + \dfrac{P6c\,(P8x - a\cos(t))}{P6d} - \dfrac{P6e\,(P8y - a\sin(t))}{P6d}$

Table 3: Coordinates of the coupler curve for the geared five-bar linkage

(ii) Torque in a Crank / Piston

In our second example, we derive an equation for the torque generated by piston force F in a piston/crank mechanism. The Analytix model is parametrized by the crank length c, the length of the connecting rod L, and the crank angle a. Assuming a force of F (depicted as 1 in figure 3) applied at the piston, we require the torque in the crank.

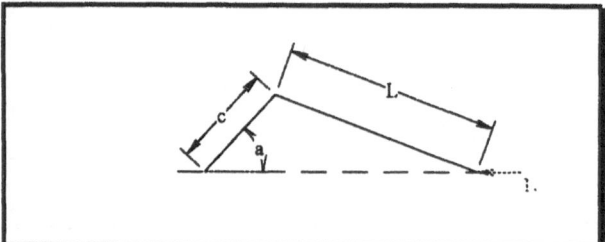

Figure 3: An Analytix model of a Piston Crank Mechanism

The Variational Geometry statics model assumes reaction forces in constraints balance any externally applied forces. Thus a statics question should be translated into a question regarding reaction forces in constraints. In this case, we require the reaction pseudoforce (actually a torque) in parameter a.

Install(torque,torque=Reaction(a));

Show(torque);

$$P8a^2 = L^2 - c^2 \sin(a)^2$$

$$2 P8a \left(\frac{\partial}{\partial a} P8a \right) = -2 c^2 \sin(a) \cos(a)$$

$$torque = \left(-c \sin(a) + \left(\frac{\partial}{\partial a} P8a \right) \right) F$$

FullySubstitute(torque);

$$\left(-c \sin(a) - \frac{c^2 \sin(a) \cos(a)}{\sqrt{L^2 - c^2 \sin(a)^2}} \right) F$$

Table 4: Equations for the torque in the crank generated by an input force F at crank angle a.

In Table 4, a new variable, torque, is introduced into the Intermediate Variable Table with a value equal to the reaction in parameter a. The system returns an expression for the torque which involves the intermediate variable P8a and its derivative with respect to a. To express torque as a single equation in the input parameters, we can apply the function FullySubstitute().

(iii) Dynamics of a Double Pendulum

In our third example, we use the system to derive the equations of motion for a double pendulum with unit masses at the end of each link. Figure 4 is the Analytix model of a double pendulum. The Variational Geometry model of dynamics assumes that one or more of the model's parameters is free to accelerate in response to unbalanced forces. In this case, we wish the angle parameters a and b to be free to accelerate.

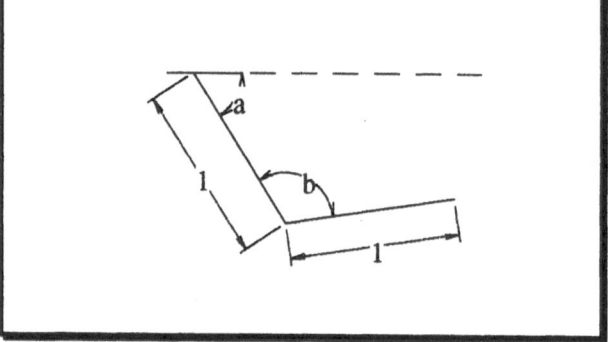

Figure 4: An Analytix model of a double pendulum.

The command to create the differential equations is:

InstallDiffEq([a,b]);

The system introduces new variables A_a and A_b for the accelerations of a and b, and new parameters V_a and V_b for the velocities of a and b. It then derives the equations of motion as expressions for A_a and A_b in terms of the input parameters along with V_a and V_b.

The initial expressions for the accelerations in this example involved a large number of intermediate variables. Application of the function SimplifyTable() resulted in the equations shown in Figure 5. The parameter g seen in Table 5 is the acceleration due to gravity.

$$A\_a = -\left(\frac{1}{2}V\_a^2\sin(2\,b) - 4\cos(b)\cos(a+b)\,g - V\_a^2\sin(b) + 8\cos(a)\,g\right.$$
$$\left. -\,2\,V\_a\,V\_b\sin(b) - V\_b^2\sin(b)\right)\Big/\left(-\frac{3}{2}+\frac{1}{2}\cos(2\,b)\right)$$

$$A\_b = -\left(3\,V\_a^2\sin(b) - 8\cos(a+b)\,g - V\_a^2\sin(2\,b) + 4\cos(b)\,g\cos(a+b)\right.$$
$$+\,8\cos(b)\cos(a)\,g - V\_a\,V\_b\sin(2\,b) - \frac{1}{2}V\_b^2\sin(2\,b) - 8\cos(a)\,g$$
$$\left. +\,2\,V\_a\,V\_b\sin(b) + V\_b^2\sin(b)\right)\Big/\left(-\frac{3}{2}+\frac{1}{2}\cos(2\,b)\right)$$

Table 5: Equations for the accelerations A_a and A_b of the angles a and b. Parameters V_a and V_b are introduced by the system and represent the velocities of a and b, g is the acceleration due to gravity

Table 6 shows the use of CFunction() to create code for evaluating these accelerations. CFunction() automatically derives which variables are inputs and which (any intermediate variables) are locals. The function may be called with the `optimized` option, in which case, further local variables are introduced for any common sub-expressions in the evaluation.

Code generation in CFunction() uses the Maple function C(). In more complex dynamics problems, with dozens of intermediate variables introduced by the mechanics system and dozens more introduced by the use of the `optimized` option, the capability of automatically generating the wrapper code for a C function was essential.

CFunction() Implementation Note

In the case shown above, we desire code for a sequence of equations, namely for an equation representing A_a and an equation representing A_b. The Maple function optimize() will accept a sequence of equations as input, however it is assumed that the last in the sequence of equations is the only one we actually care about. Hence any of the prior equations may be discarded if it is not deemed necessary in computing the final answer. For example, let's assume we want to generate code for x and y from the sequence of equations: [r=2*a, x=r, y=r*sin(t)].

optimize([r=2*a,x=r,y=r*sin(t)]);

$$y = 2\,a\,\sin(t)$$

```
CFunction(CalcAcc,[A_a,A_b]);

void CalcAcc(
        double b,
        double a,
        double g,
        double V_b,

                    double V_a,
        double *A_a_,
        double *A_b_)
{
    double A_a;
    double A_b;
    A_b = -(3.0*V_a*V_a*sin(b)-8.0*cos(a+b)*g-
            V_a*V_a*sin(2.0*b)+4.0\
*cos(b)*g
*cos(a+b)+8.0*cos(b)*cos(a)*g-
V_a*V_b*sin(2.0*b)-V_b*V_b*sin(2.0*b)/2-\
8.0*cos(a
)*g+2.0*V_a*V_b*sin(b)+V_b*V_b*sin(b))/(-
3.0/2.0+cos(2.0*b)/2);
    A_a = -(V_a*V_a*sin(2.0*b)/2-
4.0*cos(b)*g*cos(a+b)-V_a*V_a*sin(b\
)+8.0*cos
(a)*g-2.0*V_a*V_b*sin(b)-V_b*V_b*sin(b))/(-
3.0/2.0+cos(2.0*b)/2);
    *A_a_ = A_a;
    *A_b_ = A_b;
}
```

Table 6: CFunction() can be used to create a C function to evaluate these accelerations.

In order to ensure that all the equations in our output sequence are preserved we use the undocumented Maple function `optimize/statseq()` with as its first parameter the list of equations to be optimized and as its second parameter the list of output variables. This ensures these outputs are preserved.

`optimize/statseq`([r=2*a,x=r,y=r*sin(t)],[x,y]);

$$[\,r = 2\,a,\, x = r,\, y = r\,\sin(t)\,]$$

4. Conclusions

We have presented a system which allows engineering mechanics problems to be specified in an extremely intuitive and graphical manner through the medium of a variational geometry system while a mathematical formulation of the problem within Maple is made available to the user. We have thus, for this problem domain, automated the error prone process of mathematical modeling which generally precedes mathematical analysis. The maintenance of an intermediate variable table enabled us to keep the complexity of this mathematical model under control.

5. References

1. Light R., Gossard D., (1982) "Modification of geometric models through variational geometry", CAD 14:209:214

2. Fuller N. & Prusinkiewicz P. (1989) "Applications of Euclidean constructions to computer graphics", Visual Computer 5:53-67

3. Todd P. (1986) "An algorithm for determining consistency and manufacturability of dimensioned drawings" CAD86 pp 36 - 41, Butterworths, London

4. Todd P. and Cherry G., (1989) "Symbolic Analysis of Planar Drawings", Lecture Notes in Computer Science Vol 358 (P. Gianni ed.) 344-355

5. Todd P. (1989) "A k-tree generalisation that characterises consistency of dimensioned engineering drawings", SIAM Journal of Discrete Mathematics 2:255-261.

6. Todd P. (1992) "A Constructive Variational Geometry Based Mechanism Design Software Package" ASME DE-Vol 46, Mechanism Design & Synthesis, pp 267 - 273

Authors

Philip Todd has an M.A. in Mathematics from Cambridge University, an M.S. in Applied Mathematics from Georgia Tech and a Ph.D. in Mathematical Biology from Dundee University. He is the founder and V.P. of Research and Development of Saltire Software. He is currently the principal investigator of an NSF SBIR grant researching automatic mathematical modelling of geometric and mechanical problems.

Robin McLeod received his Ph.D. in Mathematics from Dundee University in 1972. His research interests lie in numerical analysis and applied geometry. He is a Senior Scientist at Saltire Software. His current research involves human readability of symbolic mathematics, and readability-directed simplification.

Marcia Harris received her B.S. in Mathematics from UCLA, and her M.S. in Computer Science from Oregon State University. Her research interests include the symbolic representation of robot paths.

Todd & McLeod can be contacted at Saltire Software, PO Box 1565, Beaverton OR 97075, tel. 503-622-4055

Harris can be contacted at 1334 SE 52nd Street, Portland, OR

Acknowledgement

This research was partially funded by the National Science Foundation under grant number ISI-9223482

SIGNAL PROCESSING USING MAPLE V

John I. Molinder
Department of Engineering, Harvey Mudd College, Claremont, California

Introduction

It is common practice to use frequency domain techniques to determine the characteristics of signals and systems. Signals are often characterized by their frequency spectrum and a great deal of insight into the behavior of a linear time-invariant (LTI) system can be gained by examining its frequency response. As a result of the dramatic increase in the speed and decrease in the cost of digital hardware, discrete-time signal processing techniques based on the discrete Fourier transform (DFT) have become more and more prevalent. This paper describes a number of special commands that exploit both the symbolic and numeric computational capabilities of MapleV to provide a signal processing environment that is both flexible and easy to use. Using these commands, the results of processing continuous-time signals with continuous-time systems described by their s-plane transfer functions or discrete-time systems described by their z-plane transfer functions are easily determined. Important concepts such as aliasing and causality can be explored. Although designed for a class in systems engineering, the procedures should be of use to practicing professionals as well. First, some of the special commands will be described followed by an example of showing the response of both continuous-time and discrete-time systems to a pulse input. Listings of the command procedures are given in the appendix. Additional commands are under development and listings are available from the author.

Underlying Concepts

Sampled continuous-time signals and discrete-time signals can be thought of as vectors with each component representing the value of the signal at a particular time instant. Since it is more natural to represent a continuous-time signal as an expression, the first step is to construct a vector with components equal to the values of the expression at uniformly spaced time instants. Maple's Fast Fourier Transform (FFT) command can then be used to compute the spectrum of the sampled signal. The effect of processing the signal with a LTI system (either continuous-time or discrete-time) can now be determined by computing the spectrum of the output. This requires multiplying the spectrum of the input signal by the frequency response of the system. Finally, the output of the system in the time domain can be determined by taking the inverse Fourier transform of the output spectrum using Maple's iFFT command. While all of these operations can be accomplished using the standard Maple command set, the special commands described in this paper greatly reduce the time and tedium involved (not to mention the probability of error) and allow the user to concentrate on the important principles rather than the details. A short description of each of these commands is given below.

Special Commands

Initialization

Before using the special commands listed below their files must be read and numerical values must be assigned to the variables sf (sampling frequency) and p2 (length of the FFT as a power of 2). In the example that follows sf is set to 64 samples per second and p2 to a value of 8 (corresponding to a FFT of length $n=2^8=256$ points).

ffourier(et)

This command computes a vector of length n whose components are the values of the samples of a continuous-time signal (represented by the expression et), determines its spectrum (using Maple's FFT command) and plots the magnitude of the spectrum as a function of frequency in Hertz (corresponding to the sampling frequency). The result is stored in vectors xr and xi (the real and imaginary parts of the spectrum) where it can be accessed by the filtering commands listed below. Thus this command must be executed before using any of the filtering commands.

lpfilter(fh)

The effect of an ideal low-pass filter with cutoff frequency fh is simulated by this command which also plots the magnitude spectrum of the output. As with all the filter commands, the result is stored in vectors xrf and xif (the real and imaginary parts of the output spectrum) where it is available for the inverse Fourier command.

afilter(tf)

This command determines the effect of processing the signal with a continuous-time system specified by its s-plane transfer function tf (tf must be an expression with independent variable s). The magnitude of the output spectrum is plotted and the result is stored in vectors xrf and xif.

zfilter(dtf)

Analogous to the previous command, the effect of processing the signal with a discrete-time system specified by its z-plane transfer function is determined by using this command. Again, the magnitude of the output spectrum is plotted and the result is stored in vectors xrf and xif.

Fourier()

This command uses Maple's iFFT command to compute the inverse Fourier transform of the output spectrum and plots the output in the time domain. It accesses xrf and xif for its input.

vecprod(n,x,y,z)

This command computes a vector z of length n with components that are the products of the corresponding components of vectors x and y. It is used as part of the lpfilter(fh) command.

Example

Consider the response of a system to a pulse. The first step is to create an expression xt that represents the pulse:

$$xt:=u(t-1.5)-u(t-2.5);$$

where u is the alias of the Heaviside function. A plot is shown in Figure 1. To compute the spectrum and display its magnitude the following commands are issued.

$$p2:=8;$$

$$sf:=64;$$

$$ffourier(xt);$$

This will compute the DFT (using Maple's FFT command) with $n=2^8=256$ points and a sampling frequency of 64 samples per second (covering a total time duration of 256/64=4 seconds. The resulting plot of the magnitude spectrum is shown in Figure 2. Note that the points of the DFT have been connected for clarity but the actual frequency resolution is 64/256=0.25 Hz.

This will set all the components of the DFT corresponding to frequencies higher than 10 Hz equal to zero. The resulting spectrum is shown in Figure 3. A plot of the waveform in the time domain is shown in Figure 4 and is obtained by giving the following command.

iffourier();

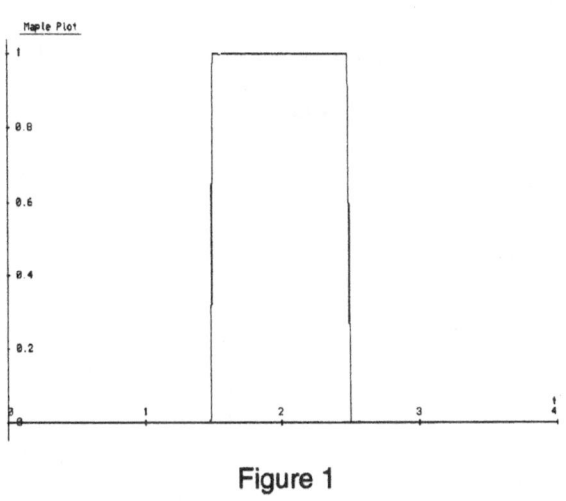

Figure 1

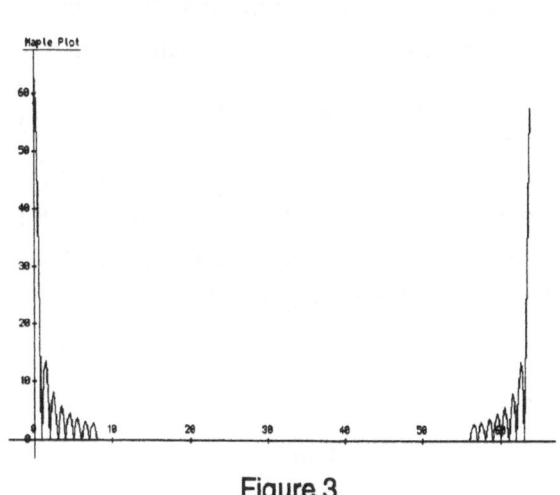

Figure 3

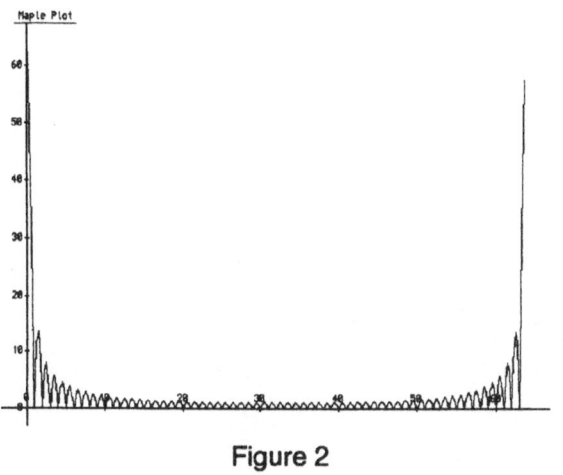

Figure 2

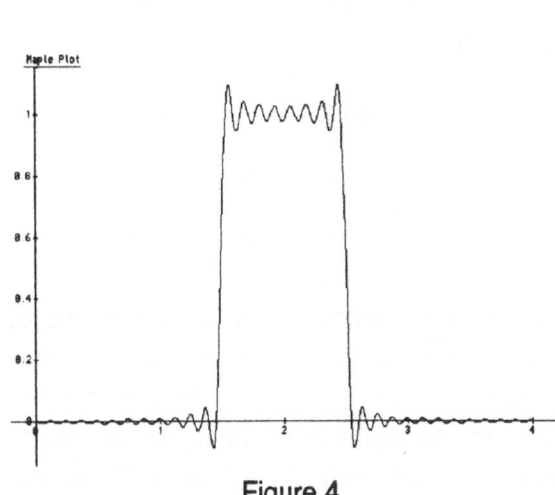

Figure 4

If this waveform is the input to an ideal low-pass filter the spectrum of the output can be determined by giving the command below.

lpfilter(10);

41

Now suppose we want to determine the response of a second-order system with transfer function $H_1(s)$ given below.

$$H_1(s) = \frac{900}{s^2 + 6s + 900}$$

This corresponds to a second-order system with a natural frequency of 30 rad/sec and a per unit damping of 0.1 (highly underdamped). Since the spectrum of the pulse is available it need not be computed again. First the transfer function is entered as an expression as shown below.

H1s:=900/(s^2+6*s+900);

Next the magnitude spectrum of the output (shown in Figure 5) and the output waveform (shown in Figure 6) are computed and plotted by issuing the following commands.

afilter(H1s);

iffourier();

Now consider the transfer function $H_2(s)$ given below.

$$H_2(s) = \frac{900}{s^2 - 6s + 900}$$

Issuing the commands below produces the output waveform shown in Figure 7.

H2s:=900/(s^2-6*s+900);

afilter(H2s);

iffourier();

Notice that the output is not causal (response starts prior to the input). This is to be expected since use of the Fourier transform forces the region of convergence to be to the left of the poles in the right half of the s-plane (to include the imaginary axis) resulting in an anticausal response. The magnitude spectrum is the same as that shown in Figure 5.

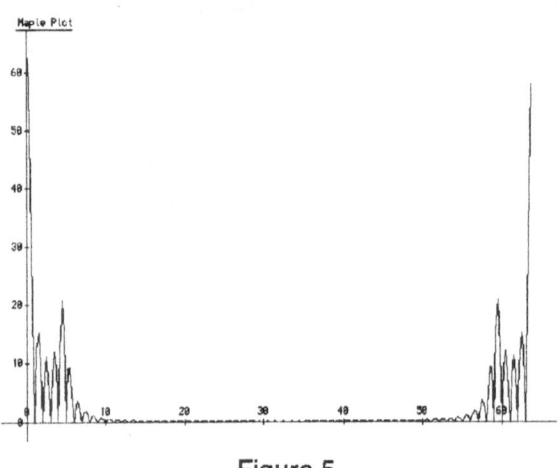

Figure 5

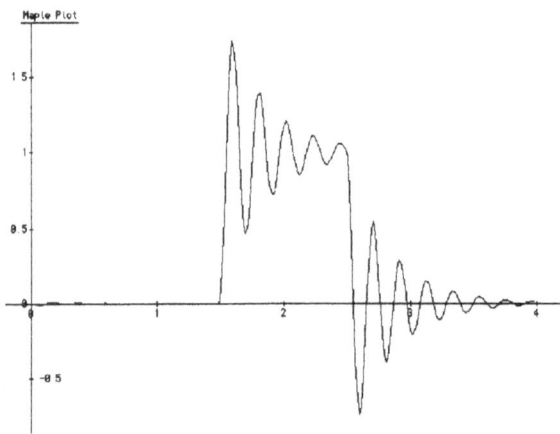

Figure 6

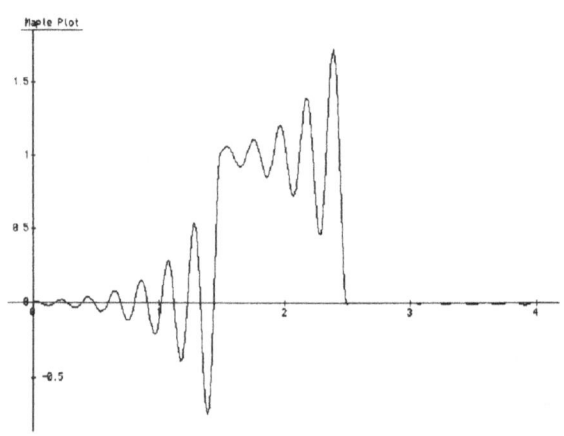

Figure 7

42

Similar sequences of commands produce the outputs shown in Figures 8 and 9 corresponding to the transfer functions listed below.

$$H_3(s) = \frac{10}{s + 10}$$

$$H_4(s) = \frac{10}{s - 10}$$

These represent the output of first-order systems with causal and anticausal responses.

Finally, consider a first-order discrete-time system with the transfer function given below.

$$H_1(z) = \frac{1}{1 - 0.875\, z^{-1}}$$

Issuing the commands below produces the output waveform shown in Figure 10.

H1z:=1/(1-.875/z);

zfilter(H1z);

iffourier();

As expected, the result is similar to the output of the first-order continuous-time system shown in Figure 8.

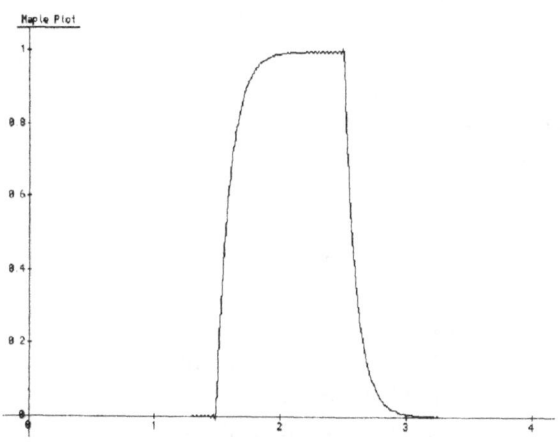

Figure 8

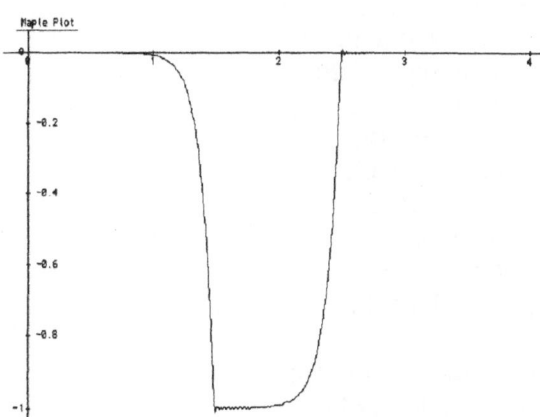

Figure 9

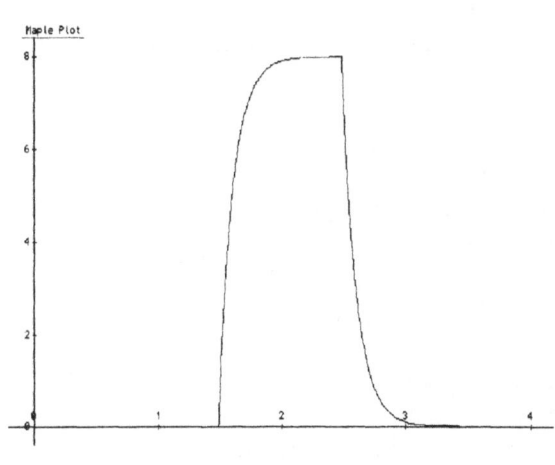

Figure 10

Conclusions

The special commands described in this paper provide a convenient method for determining the properties of signals and systems using MapleV. Combined with Maple's symbolic manipulation capabilities these procedures allow the student or professional to design and simulate a variety of signal processing systems. Similar commands are under development for areas such as modulation and communication link analysis.

Appendix

Listings for several of the special commands are given below.

fourier.mpl

```
# Set up MapleV to handle FFT and unit step
# functions

readlib(FFT):
readlib(Heaviside):
alias(u=Heaviside):

# Load vector manipulation routine

# read `vecprod.mpl`;

# Routine to compute and graph FFT of a
# sampled time waveform
# npts=2^p2=number of points
# sf=sampling frequency
# et=time waveform expression (variable must
# be t)

ffourier:=proc(et)
npts:=2^p2;
en:=subs(t=(n-1)/sf,et);
fcn:=unapply(en,n);
xr:=vector(npts,fcn);
xi:=vector(npts,0);
FFT(p2,xr,xi);
m:=vector(npts,0);
for k to npts do
  m[k]:=sqrt(xr[k]^2+xi[k]^2);
od;
df:=sf/npts;
g:=n->df*(n-1);
xax:=vector(npts,g);

p:=concat(xax,m);
p1:=convert(p,listlist);
plot(p1);
end:
```

lpfilter.mpl

```
# Ideal low-pass filter routine
# npts=2^p2=number of points in FFT
# fh=high frequency cutoff
```

```
# NOTE:  REQUIRES THAT FOURIER.MPL BE
# READ BEFORE USING

lpfilter:=proc(fh)
npts:=2^p2;
mx:=vector(npts,1);
np:=npts/2;
for k from 2 to np+1 do
  if xax[k] > fh then
    mx[k]:=0;
    mx[npts+2-k]:=0;
  fi;
od;
vecprod(npts,xr,mx,xrf);
vecprod(npts,xi,mx,xif);
mf:=vector(npts,0);
for k to npts do
  mf[k]:=sqrt(xrf[k]^2+xif[k]^2);
od;

p:=concat(xax,mf);
p1:=convert(p,listlist);
plot(p1);
end:
```

afilter.mpl

```
# Arbitrary filter routine
# npts=2^p2=number of points in FFT
# sf=sampling frequency
# tf=transfer function (variable must be s)

afilter:=proc(tf)
npts:=2^p2;
np:=npts/2;
pi:=evalf(Pi);
dw:=evalf(2*pi*sf/npts);
ffd:=subs(s=I*(n-1)*dw,tf);
ffdf:=unapply(ffd,n);
mx:=vector(np+1,ffdf);
x:=vector(np+1,0);
xf:=vector(npts,0);
xrf:=vector(npts,0);
xif:=vector(npts,0);
mf:=vector(npts,0);
for k to np+1 do
  x[k]:=xr[k]+I*xi[k];
od;
xf[1]:=evalc(x[1]*mx[1]);
for k from 2 to np+1 do
  xf[k]:=evalc(x[k]*mx[k]);
  xf[npts+2-k]:=evalc(conjugate(xf[k]));
od;
```

44

```
for k to npts do
  xrf[k]:=evalc(Re(xf[k]));
  xif[k]:=evalc(Im(xf[k]));
  mf[k]:=sqrt(xrf[k]^2+xif[k]^2);
od;

p:=concat(xax,mf);
p1:=convert(p,listlist);
plot(p1);
end:
```

zfilter.mpl

```
# Arbitrary discrete filter routine

# npts=2^p2=number of points to be computed
# sf=sampling frequency
# dtf=transfer function (variable must be z)

zfilter:=proc(dtf)
npts:=2^p2;
pi:=evalf(Pi);
dw:=evalf(2*pi/npts);
ffd:=subs(z=exp(I*(n-1)*dw),dtf);
ffdf:=unapply(ffd,n);
mx:=vector(npts,ffdf);
x:=vector(npts,0);
xf:=vector(npts,0);
xrf:=vector(npts,0);
xif:=vector(npts,0);
mdf:=vector(npts,0);
for k to npts do
  x[k]:=xr[k]+I*xi[k];
  xf[k]:=evalc(x[k]*mx[k]);
  xrf[k]:=evalc(Re(xf[k]));
  xif[k]:=evalc(Im(xf[k]));
  mdf[k]:=sqrt(xrf[k]^2+xif[k]^2);
od;
sf1:=sf;
g:=n->(n-1)*sf1/npts;
xax:=vector(npts,g);

p:=concat(xax,mdf);
p1:=convert(p,listlist);
plot(p1);
end:
```

ifourier.mpl

```
# Computes and plots real part of inverse FFT
# npts=2^p2=number of points
```

```
# sf=sampling frequency

iffourier:=proc()
npts:=2^p2;
iFFT(p2,xrf,xif);
dt:=1/sf;
gt:=n->dt*(n-1);
xaxt:=vector(npts,gt);

p:=concat(xaxt,xrf);
p1:=convert(p,listlist);
plot(p1);
end:
```

vecprod.mpl

```
# Read in linear algebra routines for vector ma-
nipulation

with(linalg):

# Compute the product of real vectors x and y
term by term
# n=vector length
# z=resulting vector

vecprod:=proc(n,x,y,z)
  z:=vector(n,1);
  for i to n do
    z[i]:=x[i]*y[i];
  od;
end:
```

Reference

A. V. Oppenheim, A. S. Willsky, and I. T. Young, *Signals and Systems*, Prentice-Hall, Inc., New Jersey, 1983.

John I. Molinder (john_molinder@hmc.edu) received his B.S. degree from the University of Nebraska, his M.S. degree from the Air Force Institute of Technology, and his Ph.D. degree from Caltech, all in Electrical Engineering. Since 1970 he has been a member of the faculty at Harvey Mudd College where he is Professor and Harvey S. Mudd Fellow of Engineering.

IMPLEMENTATION OF A MAPPING PROCEDURE FOR THE DOMAIN DECOMPOSITION FOR PDEs IN MAPLE

Peiyuan Li and Richard L. Peskin
CAIP Center, Rutgers University, Piscataway, NJ

Abstract

Domain decomposition for partial differential equations(PDEs) is implemented using singular perturbation analysis. The programming work is accomplished in a symbolic programming environment, namely Maple.

First we establish the definitions of a domain and boundary conditions in the symbolic programming environment, Maple, then apply the characteristic curve method to transform the reduced equation to a first order system. After a series of transformations, the first order system can be transformed to a third order ODE. If the third order ODE can be solved in Maple, the solutions of the reduced equation are obtained. Thus, an approximate solution of the equation is constructed from a solution of the reduced equation and correction terms corresponding to boundary layers and interior layers. By applying the above methods a set of criteria to determine a stable partition of a domain can be built. Based on the set of criteria, the search for all possible stable partitions is implemented by traversing on the graph of a given domain. Using the similar traversing method on the partitioned domain, each individual subdomain is selected automatically so that it can be distributed to a processor for further processing. The structure of the system implemented in Maple is described in a pseudo-code. In this paper only elliptic equations are discussed.

1 Introduction

The analysis for singular perturbation problem has been successfully used for domain decomposition by several researchers. Russo and Peskin[8] combined the asymptotic analysis and an artificial intelligence technique, plan generating system, to decompose the domains of second order nonlinear ODEs and generate parallel programs for finite difference method for solving the problems. Hedstrom and Howes[1] used asymptotic analysis to determine a domain decomposition method for a singular perturbation convection-diffusion equation with turning points, and also suggested the basis functions to be used in a finite element method. A numerical algorithm based on asymptotic analysis for a 2-dimentional boundary value problem has been implemented by Rodrigue and Reiter[9]. The authors[5] have successfully applied singular perturbation analysis[2, 3, 4] to domain decomposition for partial differential equation(PDEs). In our previous paper[5] the presentation is focused on mathematical analysis and mathematical methods used for the domain decomposition. In this paper we present the implementation of domain decomposition in a symbolic computation environment(Maple).

General elliptic partial differential equations (PDEs) can be specified as follows,

$$A(\mathbf{x}, u) \cdot \nabla u + h(\mathbf{x}, u) = \varepsilon \nabla^2 u, \quad \mathbf{x} \text{ in } \Omega, \quad (1)$$

$$u(\mathbf{x}, \varepsilon) = \phi(\mathbf{x}) \quad \mathbf{x} \text{ on } \Gamma = \partial\Omega.$$

where

$$\mathbf{x} = (x_1, \ldots, x_N),$$

$$\nabla = (\frac{\partial}{\partial x_1}, \ldots, \frac{\partial}{\partial x_N})$$

and

$$A(\mathbf{x}, u) = (a_1(\mathbf{x}, u), \ldots, a_N(\mathbf{x}, u)).$$

Ω is a bounded region in $\mathcal{R}^N$ whose boundary Γ is a smooth $(N-1)$-dimensional manifold.

The associated reduced equation is,

$$A(\mathbf{x}, u) \cdot \nabla u + h(\mathbf{x}, u) = 0, \quad \mathbf{x} \text{ in } \Omega, \quad (2)$$

which is a quasilinear first order PDE. The solution of the reduced equation will be denoted as u_0, u being the solution of the elliptic problem (1).

As a result of singular perturbation analysis, the solution of an elliptic problem can be approximated, under certain conditions, by the solution of the associated reduced equation and correction terms. That is, in a large portion of the domain of the elliptic problem the solution of the problem can be replaced by the solution of the reduced equation; however there exist subdomains, boundary and interior layers, in which the solution of the problem is a combination of the reduced solution and corresponding correction terms. Some conditions, called stable conditions, are required for admissible solutions. These stable conditions are described in detail as theorems in [5] and a set of rules are built from the stable condition requirements. The aim of our system is to search for stable partitions from all possible partitions using previously mentioned rules and traversing on a graph of the domain. After stable partitioning, the domain is divided into a set of subdomains. Finally the system selects each subdomain individually so that each subdomain can be distributed across different processors for parallel computation. During the selecting process the size of the subdomain has to be adjusted properly in order to satisfy requirements from the Schwarz method[10].

2 Representations of Domain and BCs

The first step for implementing, in Maple, domain decomposition for 2-D PDEs is the representation of domain and boundary conditions. Here we do not distinguish the concepts, domain and region rigorously. The strict definitions of domain and region are given in [5]. We use lists to express domains and boundary conditions because lists are accessible data structures in Maple. In Maple, a list can be easily tranformed into a set or an expression sequence. This transformation allows different type manipulations to be applicable to the same object. For instance, for a domain, set operations(union, minus and intersect), and list operation(concatenation) can be used. More importantly, the list, representing a domain, can be regarded as a graph, so that algorithms for graphs can be directly applied to the list(domain).

For a general domain in 2-D, as shown in Figure 1, its boundary consists of several sides. The

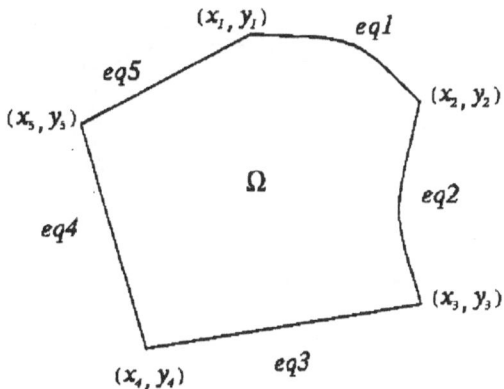

Figure 1: The domain expressed in Maple

list, expressing the domain, is as follows:

$$[[x_1, y_1, eq1],$$
$$[x_2, y_2, eq2],$$
$$[x_3, y_3, eq3],$$
$$[x_4, y_4, eq4],$$
$$[x_5, y_5, eq5]],$$

in which every element is a sublist, corresponding to a side of the domain. The order of elements in a list is the clockwise order of sides of a domain. The first and second elements in a sublist are the x and y coordinates of the start location of a side, and the third element is the equation of curve of the side. Here we do not limit the equation for a side; however implementation requires restriction to equations recognizable by Maple. Both convex and concave domains are permissible in our definition.

The boundary conditions in 2-D are expressed as a list also. For the domain in Figure 1, we have a following list as boundary conditions,

$$[[1,5], 1, f_1(x,y), f_2(x,y), f_3(x,y), f_4(x,y),$$
$$f_5(x,y), f_1(x_1,y_1) = f_5(x_1,y_1)].$$

The first element of the above list is a sublist, which contains the indices of incoming sides. By incoming side, we mean that the characteristic curves for the reduced equation (2) are incoming through this side. The detail description of incoming side can be found in [5]. The first and fifth sides are incoming side in the above list. The second element is the number of the additional boundary conditions. The third to seventh elements, $f_1(x,y), f_2(x,y), f_3(x,y), f_4(x,y), f_5(x,y)$, are given boundary conditions for the first to fifth

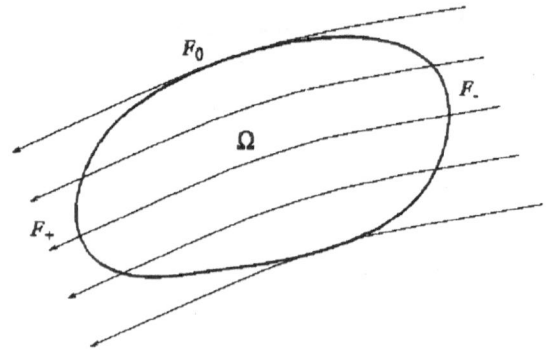

Figure 2: A general domain

sides respectively. The last element, $f_1(x_1, y_1) = f_5(x_1, y_1)$, is the additional boundary condition, which means the values of $f_1(x, y)$ and $f_5(x, y)$ are equal at the point (x_1, y_1). (There is one additional condition, so the second element is the number, 1.)

3 Solving the Reduced Equation

The solution of the associated reduced equation(2) is the main part of the approximated solution of PDE(1). Solving the reduced equation is one of the principal parts of our system implemented in Maple.

Suppose a general domain for PDE is depicted as in Figure 2. The boundary of domain, Γ, consists of three parts, F_-, the incoming side, F_0, the tangent side, F_+, the outgoing side. For the reduced equation, we cannot have a solution satisfying the complete boundary conditions, only a solution which satisfies the given boundary conditions along a part of the boundary, say, the incoming side, F_-. Using the method of characteristics [11] for solving the first order PDE (2), we obtain the first order system,

$$
\begin{aligned}
\frac{dx}{ds} &= a_1(x, y, u) \\
\frac{dy}{ds} &= a_2(x, y, u) \qquad (3) \\
\frac{du}{ds} &= -h(x, y, u)
\end{aligned}
$$

with initial conditions at $s = 0$ given as $x = x(\tau)$, $y = y(\tau)$ and $u = u(\tau)$ with τ as a parameter. The solution of (3) are called the characteristic curves of the equation (2). Here we have assumed that $a_1(x, y, u)$, $a_2(x, y, u)$ and $h(x, y, u)$ are sufficiently

smooth and do not all vanish at the same point. The first order system has a parametric form solution, $x = x(s), y = y(s)$, and $u(s)$. Also the initial condition is referred to as the initial curve C in (x, y, u) space. To solve this initial value problem we pass a characteristic curve through each point of the initial curve C. If these curves generate a surface, this integral surface is the solution of the initial value problem. The conditions, under which a unique solution of the initial value problem for (3) can be obtained, are expressed as,

$$
\begin{aligned}
\Delta(\tau) \equiv\ & \frac{dy}{d\tau} a_1[x(\tau), y(\tau), u(\tau)] \\
& - \frac{dx}{d\tau} a_2[x(\tau), y(\tau), u(\tau)] \neq 0, \qquad (4)
\end{aligned}
$$

on C. The initial conditions are obtained from the boundary conditions along incoming side F_-.

3.1 Obtaining the Initial conditions from Boundary Conditions

If the reduced equation (2) satisfies the following boundary condition along the incoming side F_-,

$$ u_0(x, y) = \phi(x, y), \quad \text{for } (x, y) \text{ in } f(x, y) = 0. $$

For the boundary, $f(x, y) = 0$ we change the form to, $y = g(x)$. Then for the first order system (3), the initial condition can be rewritten as,

$$ s = 0, \ x = \tau, \ y = g(\tau), \ u = \phi(\tau, g(\tau)). $$

As a special case, if the boundary is $x = k$, where k is a constant, the initial condition then is,

$$ s = 0, \ x = k, \ y = \tau, \ u = \phi(k, \tau). $$

Similarly, for the boundary $y = k$, we have the initial condition for the first order system (3),

$$ s = 0, \ x = \tau, \ y = k, \ u = \phi(\tau, k). $$

3.2 Solving the First Order System in Maple

The initial value problem may not have a solution at all or it may have infinitely many solutions; these cases are not treated in the current paper. We only treat the case in which the condition (4) is satisfied, that is, the initial value problem has a unique solution. This first order system is solved in Maple. Using the method of elimination, a third order ordinary differential equation(ODE) is derived from the first order system.

Differentiating the first equations of (3) we obtain the equation,

$$\frac{d^2x}{ds^2} = \mathcal{F}(\frac{dx}{ds}, \frac{dy}{ds}, \frac{du}{ds}, x, y, u), \qquad (5)$$

Upon the substitution of (3) into (5), the equation (5) becomes,

$$\frac{d^2x}{ds^2} = \mathcal{F}(x, y, u), \qquad (6)$$

Differentiating the equation (6) and substituting (3) into it, we find

$$\frac{d^3x}{ds^3} = \mathcal{M}(x, y, u). \qquad (7)$$

Coupling the first equation of (3) and the equation (6) and solving for y, u,

$$y = \mathcal{Y}(\frac{d^2x}{ds^2}, \frac{dx}{ds}),$$
$$u = \mathcal{U}(\frac{d^2x}{ds^2}, \frac{dx}{ds}),$$

Finally, substituting the above equations into the equation (7) we obtain the third order ODE

$$\mathcal{T}(\frac{d^3x}{ds^3}, \frac{d^2x}{ds^2}, \frac{dx}{ds}, x) = 0, \qquad (8)$$

If resulting equation (8) is linear, it can be solved directly. However if it is nonlinear, it may not be directly solved using Maple. Fortunately there is no explicit variable s in the equation (8), so we can introduce the transformation,

$$\frac{dx}{ds} = p,$$
$$\frac{d^2x}{ds^2} = p\frac{dp}{dx}, \qquad (9)$$
$$\frac{d^3x}{ds^3} = p^2\frac{d^2p}{dx^2} + p(\frac{dp}{dx})^2,$$

in which p is a function of x. The equation (8) is transformed into the second order ODE

$$\mathcal{S}(\frac{d^2p}{dx^2}, \frac{dp}{dx}, p, x) = 0. \qquad (10)$$

Attempts to solve the second order ODE symbolically are limited by current capabilities of systems like Maple. If the nonlinear second ODE has the form

$$\mathcal{S}(\frac{d^2p}{dx^2}, \frac{dp}{dx}, p) = 0, \qquad (11)$$

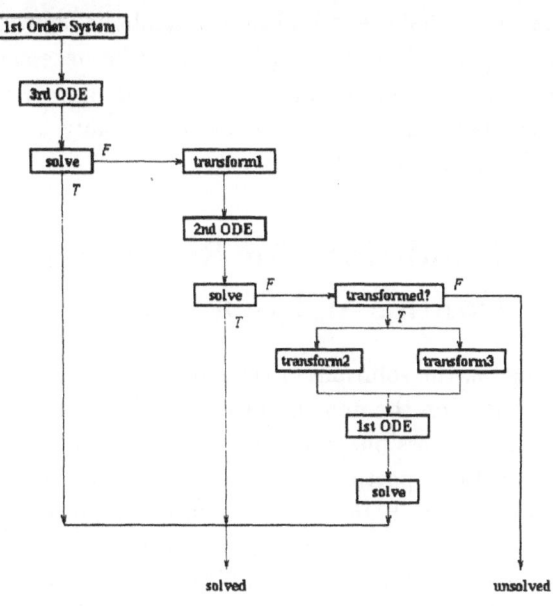

Figure 3: Flowchart for solving the first order system

we use the transformation

$$\frac{dp}{dx} = q,$$
$$\frac{d^2p}{dx^2} = q\frac{dq}{dp}, \qquad (12)$$

where q is the function of p, to reduce the second order ODE(11) to the first order ODE

$$\mathcal{F}(\frac{dq}{dp}, q, p) = 0, \qquad (13)$$

which can be solved using Maple. Another limited form of the second order ODE we consider is,

$$\mathcal{S}(\frac{d^2p}{dx^2}, \frac{dp}{dx}, x) = 0. \qquad (14)$$

Similarly to the above case, we apply the transform

$$\frac{dp}{dx} = r,$$
$$\frac{d^2p}{dx^2} = \frac{dr}{dx}, \qquad (15)$$

and obtain the first order ODE

$$\mathcal{F}(\frac{dr}{dx}, r, x) = 0, \qquad (16)$$

which can be solved using Maple. The flowchart expressing the above procedure is depicted in Figure 3.

The ODEs for which the above special cases apply can be solved using Maple. However we cannot guarantee that we will obtain a symbolic solution to the general ODE like (7). It may be necessary to resort to a numerical solution. Using numerical methods in the singular perturbation analysis will be discussed in a future paper.

4 Conditions for Stable Partitioning and Searching

Using u_0, the solutions of the reduced equation, we can partition the domain into a set of subdomains. Among all possible partitions we are only interested in the stable partitions with which u, the solution of the elliptic PDE is approximated in every subdomains by u_0 and is convergent to u_0 as $\varepsilon \to 0$.

Certain stability conditions required by convergence have to be imposed on such solutions. In [5], three theorems, Theorem 1, 2, 3, for boundary layer and two theorems, Theorem 4, 5, for interior layer are described in detail. Each theorem includes certain stability conditions as assumptions. When all assumptions in one of these thorems are satisfied, there exists a positive constant ε_0 such that for $0 < \varepsilon \le \varepsilon_0$ the equation (1) has a solution $u = u(\mathbf{x}, \varepsilon)$ of class $c^{(2,\alpha)}(\bar{\Omega}_i)$, which is defined as the smoothness of a function in [4]. Moreover, for $\mathbf{x}$ in $\bar{\Omega}_i$ we have that

$$|u(\mathbf{x},\varepsilon) - u_0| \le W_i(\mathbf{x},\varepsilon) + c_i\varepsilon, \qquad (17)$$

where

$$
\begin{aligned}
W_1(\mathbf{x},\varepsilon) &= K_1 \exp[-k_1\varepsilon^{-1}\Sigma_1(\mathbf{x})] + \ldots \\
&\quad + K_{p_1}\exp[-k_1\varepsilon^{-1}\Sigma_{p_1}(\mathbf{x})], \\
W_2(\mathbf{x},\varepsilon) &= L_1\exp[-(m_1\varepsilon^{-1})^{1/2}\Psi_1(\mathbf{x})] + \ldots \\
&\quad + L_{r_1}\exp[-(m_1\varepsilon^{-1})^{1/2}\Psi_{r_1}(\mathbf{x})], \\
W_3 &= W_1 + W_2,
\end{aligned}
$$

and the c_i are known positive constants. Here

$$
K_p = \max_{\Sigma_p^{-1}(0)} |\phi(\mathbf{x}) - u_0(\mathbf{x})|,
$$

$$
L_r = \max_{\Psi_r^{-1}(0)} |\phi(\mathbf{x}) - u_0(\mathbf{x})|,
$$

$0 < k_1 < k$, and $0 < m_1 < m$. The parameters k and m are constants defined in assumptions of theorems[5].

The formula (17) indicates that the solution of (1), u, is approximated by the reduced solution of (2) with the layer corrector $W_i(\mathbf{x},\varepsilon)$. $W_i(\mathbf{x},\varepsilon)$ is the summation of several terms each of which corresponds to a boundary or interior layer. The layer correctors, $W_i(\mathbf{x},\varepsilon)$, give the nature of layers in detail. The layer effect exponentially decays along the inner normal direction of the corresponding boundary or interior singular curve. And the width of layer is of order ε or of order $\varepsilon^{1/2}$ depending on the nature of the boundary or singular curve [5].

Based on the stability conditions, Theorem 1 - 5[5], we build a production system [7], which uses the stability conditions as rules to search for all stable partitions of the given domain. In the production system, the global database consists of symbolic functions, symbolic equations and lists which, as defined in Section 2, express the given domain and subdomains as well as boundary conditions. The search process is implemented by traversing on the graph which includes the given domain and all singular curves in the domain. Moreover, the search process is controlled by the following steps.

1. Along whole boundary every side is distinguished as F_+, F_0 and F_-.

2. On every side of boundary it is verified that there is a theorem among Theorem 1, Theorem 2 and Theorem 3, which is satisfied.

3. The reduced solution u_0 is tested to determine if it is smooth.

4. If the u_0 is nonsmooth, the singular curves must be found.

5. Along the singular curves it is verified if there is a theorem between Theorem 4 and Theorem 5 which is satisfied.

6. The widths of boundary layers and interior layers are determined.

Once the traversing on a given domain is completed, the domain has been stably partitioned into a set of subdomains which include boundary and interior layers and the other subdomains. The next step is to "select" each subdomain individually for further computation.

5 Selecting

Selecting subdomains means, that based on a partitioned domain, each individual subdomain is automatically tagged so that it can be distributed to a processor for further numerical processing. The selecting process is implemented by traversing on a domain, similar to that used for searching. During the selecting process, the size of the subdomain has to be adjusted properly in order to assure satisfaction with requirements from the Schwarz method [10, 9], that is, the subdomain has to overlap with adjacent subdomains. The overlapping of subdomains can accelerate the convergence of iterations in computation for PDEs. The following rules are applied in the selecting process:

1. The selecting starts from the incoming side.

2. When the traversal moves from an incoming side to a boundary side which has a boundary layer, we have to select a birth-boundary layer subdomain. This subdomain is defined to be the region located at the corner between incoming side and boundary layer side.

3. When the traversal moves along a boundary side which has a layer, we can select a boundary layer subdomain.

4. When the traversal moves from a boundary side to another boundary side, both of which have boundary layers, a corner-boundary layer subdomain can be selected. This subdomain is located at the corner between two boundary sides.

5. When traversal moves from an incoming side to a singular curve, an birth-interior layer subdomain can be selected. This subdomain is defined to be the region located at the corner between singular curve and incoming side.

6. When the traversal moves along a singular curve, we can select an interior layer subdomain.

7. When the traversal moves from a singular curve to a boundary side which has a boundary layer, a corner-interior layer subdomain can be selected. This subdomain is located at the corner between singular curve and boundary side.

8. When the traversal is finished on a domain, after removing the subdomains created in the above steps, the remaining part, that is the main subdomain, is selected.

6 Structure of Program

The whole package of domain decomposition in Maple consists of 78 functions. The structure of the package can be represented in the following pseudocode.

```
pertur
{
  _solve the reduced equation(sReduequ)
  _determine the singular curve with the
   solution of reduced equation(sicurve)
  _verify the first assumption of the stable
   conditions for boundary layers(criteB1)
  _verify the second assumption of the
   stable conditions for boundary
   layers(criteB2)
  _verify the third assumption of the
   stable conditions for boundary
   layers(criteB3)
  _determine the widths of boundary layers
   for all sides of domain(Wid, boundco)
  _form the set of sides of all boundary
   layers(formlyr)
  _determine the kind of interior layers
   for interior lines(tyinter)
  _verify the stable conditions for interior
   layers(criteI)
  _determine the width of interior layers(widI)
  _form the set of sides of all interior
   layers(formiyr)
     if no interior layer(sicu=NULL)
       _select subdomains in turn, obtain a set
        of sudomains(wsosubs)
     elif
       _select subdomains in turn, including
        interior layers, obtain a set of
        subdomains(obtsubs)
     fi
}
```

Following is an example which illustrates the process.

7 Example

Consider an elliptic problem.

$$\varepsilon \Delta u = -\frac{\partial u}{\partial x} + u,$$

in the square region Ω,

$$S = (x, y) : F(x, y) < 0 \quad \text{for } F(x, y) = (1 - x^2)(y^2 - 1),$$

and

$$\phi(x, y) = \begin{cases} 0 & \text{side 1:} x = 1, 0 \le y \le 1 \\ 1 & \text{side 2:} x = 1, -1 \le y \le 0 \\ \exp(x - 1) & \text{side 3:} -1 \le x \le 1, y = -1 \\ \exp(-2) & \text{side 4:} x = -1, -1 \le y \le 0 \\ 0 & \text{side 5:} x = -1, 0 \le y \le 1 \\ 0 & \text{side 6:} -1 \le x \le 1, y = 1 \end{cases}$$

The parameters are

$$\varepsilon = 10^{-4}, \qquad \eta = 10^{-4},$$

here η is the decay ratio of layers defined in [5]. The reduced equation is

$$-\frac{\partial u}{\partial x} + u = 0.$$

The reduced solutions satisfying the boundary conditions along side 1 and 2 is

$$u_0 = \begin{cases} 0 & 0 \le y \\ \exp(x-1) & y \le 0 \end{cases}$$

$y = 0$ is a singular curve for the reduced solution. The system travels the region and automatically applies stability criteria, Theorem 3 and 5, to the graph, finally get the stable partition of domain, depicted in Figure 4, and the widths of layers. Width of boundary layer:

$$0, 0, .07963856349, .002378981685,$$

$$.0009210340374, .09210340372$$

Width of interior layer:

$$.04291932052.$$

The set of subdomains selected is shown in Figure 5. The corresponding list of subdomains selected is as follows.

The list of subdomains selected:

[[[.8618448944, 1., y = 1],
[1, 1, x = 1],
[1., .8618448944, y-.8618448944 = 0],
[.8618448944, .8618448944, x-.8618448944 = 0]],

[[-1., .8618448944, x = -1],
[-1, 1, y = 1],
[-.9986184489, 1., -x-.9986184489 = 0],
[-.9986184489, .8618448944, y-.8618448944 = 0]],

[[-.9990789660, .8618448944, -x-.9990789660 = 0],
[-.9990789660, 1., y = 1],
[.9078965963, 1., x-.9078965963 = 0],
[.9078965963, .8618448944, y-.8618448944 = 0]],

[[-.9986184489, 4291932051.0E-11, -y+4291932051.0E-11 = 0],
[-1., 4291932051.0E-11, x = -1],
[-1., .9078965963, y-.9078965963 = 0],
[-.9986184489, .9078965963, -x-.9986184489 = 0]],

[[1., 4291932051.0E-11, -y+4291932051.0E-11 = 0],
[-.9990789660, 4291932051.0E-11, -x-.9990789660 = 0],
[-.9990789660, .9078965963, y-.9078965963 = 0],
[1., .9078965963, x = 1]],

[[1., -.8805421548, x = 1],
[1, -1, y = -1],
[.8805421548, -1., x-.8805421548 = 0],
[.8805421548, -.8805421548, -y-.8805421548 = 0]],

[[-.9964315275, -1., y = -1],
[-1, -1, x = -1],
[-1., -.8805421548, -y-.8805421548 = 0],

[-.9964315275, -.8805421548, -x-.9964315275 = 0]],

[[.9203614365, -.8805421548, x-.9203614365 = 0],
[.9203614365, -1., y = -1],
[-.9976210183, -1., -x-.9976210183 = 0],
[-.9976210183, -.8805421548, -y-.8805421548 = 0]],

[[-.9964315275, -.9203614365, -y-.9203614365 = 0],
[-1., -.9203614365, x = -1],
[-1., -4291932051.0E-11, y+4291932051.0E-11 = 0],
[-.9964315275, -4291932051.0E-11, -x-.9964315275 = 0]],

[[1., -.9203614365, -y-.9203614365 = 0],
[-.9976210183, -.9203614365, -x-.9976210183 = 0],
[-.9976210183, -4291932051.0E-11, y+4291932051.0E-11 = 0],
[1., -4291932051.0E-11, x = 1]],

[[1., -6437898077.0E-11, y+6437898077.0E-11 = 0],
[.9356210192, -6437898077.0E-11, x-.9356210192 = 0],
[.9356210192, 6437898077.0E-11, -y+6437898077.0E-11 = 0],
[1., 6437898077.0E-11, x = 1]],

[[-1., 6437898077.0E-11, -y+6437898077.0E-11 = 0],
[-.9964315275, 6437898077.0E-11, -x-.9964315275 = 0],
[-.9964315275, -6437898077.0E-11, y+6437898077.0E-11 = 0],
[-1., -6437898077.0E-11, x = -1]],

[[.9570806795, -6437898077.0E-11, y+6437898077.0E-11 = 0],
[-.9976210183, -6437898077.0E-11, -x-.9976210183 = 0],
[-.9976210183, 6437898077.0E-11, -y+6437898077.0E-11 = 0],
[.9570806795, 6437898077.0E-11, x-.9570806795 = 0]]]

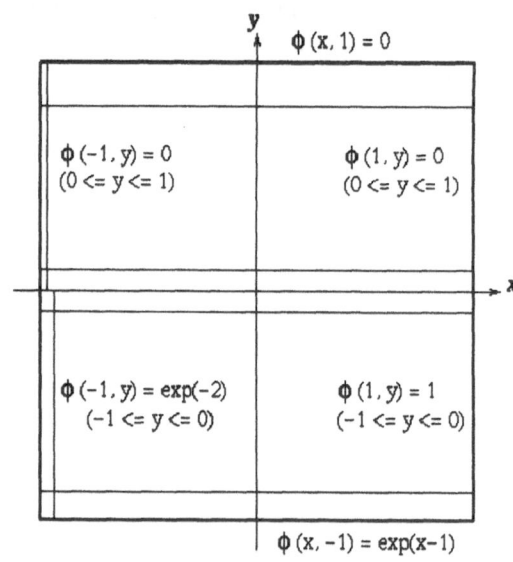

Figure 4: The stable partitioned domain

8 Conclusions

In our present symbolic-numeric system we only handle the partition of simple domains. By including advanced computational geometry functions, the partition of complicated domains can be implemented. Based on the partitioned domain, we identify the data dependency among the set of subdomains. Thus we can create a synchronous parallel

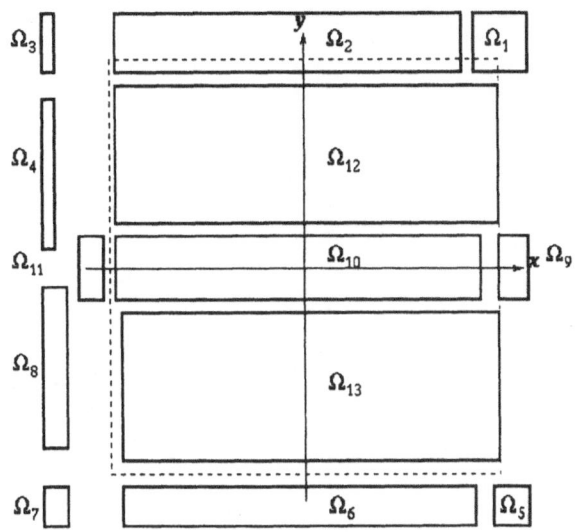

Figure 5: The selected subdomains

algorithm employing the intrinsic data dependency among the set subdomains. Using the set of subdomains, the synchronous parallel algorithm and the proper mapping of communication among the processors we can expect to complete the parallel distributed computation for PDEs with good parallel computing time, speedup and efficiency. Thus, symboic systems like Maple are useful, not only for analysis of singular perturbation PDEs, but also for domain decomposition for parallel computation.

9   Acknowledgment

This study has been supported by the National Science Foundation under grant ECS-9110424 at Rutgers University.

References

[1] G.W. Hedstrom and F.A. Howes, "A Domain Decomposition Method for a Convection Diffusion Equation with Turning Point", *Proceedings of the Second International Symposium on Domain Decomposition Methods*, California, pp38-46, 1988.

[2] F.A. Howes, "Perturbed Boundary Value Problems Whose Reduced Solutions Are Nonsmooth", *Indianna University Mathematics Journal*, Vol.30, No.2, pp267-280, 1981.

[3] F.A. Howes, "Singular Perturbed Semilinear Elliptic Boundary Value Problems", *Comm. Partial Differential Equations*, Vol.4(1), pp1-39, 1979.

[4] F.A. Howes, "Some Singular Perturbed Nonlinear Boundary Value Problems of Elliptic Type", *proceedings of Conference on Nonlinear Partial Differential Equations in Engineering and Applied Science*, University of Rhode Island, pp151-166, 1979.

[5] P. Li and R.L. Peskin, "Domain Decomposition for Singular Perturbation PDEs", *Mathematics and Computers in Simulation*, Vol. 36, 1994.

[6] P. Li and R.L. Peskin, "A New Method in Domain Decomposition for ODEs", *Mathematics and Computers in Simulation*, Vol. 36, 1994.

[7] N.J. Nilsson, *Principles of Artificial Intelligence*, Morgan Kaufmann Publishers, Inc., 1980.

[8] M.F. Russo, and R.L. Peskin, "Automatically Identifying the Asymptotic Behavior of Nonlinear Singularly Perturbed Boundary Value Problems", *Journal of Automated Reasoning*, Vol.8, pp395-419, 1992.

[9] G. Rodrigue, and E. Reiter, "A Domain Decomposition Method for Boundary Layer Problems", *Proceedings of the Second International Symposium on Domain Decomposition Methods*, California, pp226-234, 1988.

[10] H.A. Schwarz, *Gesammelte Mathematische Abhandlungen*, Springer, Berlin, 2, pp133-134, 1890.

[11] E. Zauderer, *Partial Differential Equations of Applied Mathematics*, John Wiley & Sons, Inc., pp35-77, 1983.

Peiyuan Li
(pli@yinayng.rutges.edu,   pli@caip.rutgers.edu, Tel:908-445-3304) received the Ph.D. degree in aerospace engineering from Rutgers University in 1994, the M.S. degree in computer science from Rutgers University in 1992 and the M.S. degree in computational mechanics from Shanghai Jiao Tong University in 1982. His current research interests include parallel algorithm, parallel computing and high performance computing as well as applied software in symbolic and object oriented environments.
Richard L. Peskin
(peskin@caip.rutgers.edu, Tel:908-445-3685)

is Professor of Mechanical and Aerospace Engineering at Rutgers University, and is Director of the CAIP Computational Engineering Systems Lab. His research interests include the interface between symbolic and numeric computing, as well as the development of scientific user data management and interface systems. He has published numerous papers related to computer software and applications, as well as in the fields of computational fluid dynamics, nonlinear dynamics, and combustion.

IIA. MAPLE IN EDUCATION

THE INFLUENCE OF MAPLE ON A LINEAR ALGEBRA COURSE AT THE DELFT UNIVERSITY OF TECHNOLOGY

Harry A.W.M. Kneppers
Department of Pure Mathematics, Delft University of Technology, The Netherlands

1. Introduction

For the past two academic years, we have been using Maple in the linear algebra course for first-year students in the faculty of Systems Engineering, Policy Analysis and Management at the Delft University of Technology. The course consists of three components: four hours per week for lectures, two hours per week for tutorials and another two hours per week for computer laboratory.

In the lectures, no computers are used. Definitions and theorems are illustrated by small examples, which do not require too many computations, so that students can see it "with their own eyes." Premature use of a computer would give an undesired magic touch to the subject.

The tutorials are there to help students digest the things they have learned. To this end, they get small exercises so that they can do it "with their own hands." For example, to get some idea of what a row operation does to a matrix, students should first do it by hand a couple of times, before using `addrow`, `swaprow` and `mulrow`.

The purpose of the computer laboratory is to *practise* linear algebra. During the lab classes, each student has a PC with Maple at his or her disposal. The lecturer is there to answer questions.

2. The use of Maple

We shall now look at some ways in which Maple has proved to be useful in our course.

2.1 Elementary computations

When students are practising linear algebra, a lot of time is needed for elementary computations. By "elementary" we mean: the type of computations that have already been practised before, either in primary school or in secondary school or in an earlier part of our course.

Example 2.1.1

After teaching the students how to multiply matrices, we show them some ways in which matrix algebra differs from the algebra of real numbers. We ask them to compute AB and BA, we ask them to multiply $\begin{bmatrix} 1 & -2 \\ -3 & 6 \end{bmatrix}$ with $\begin{bmatrix} 4 & -2 \\ 2 & -1 \end{bmatrix}$, etcetera. All this can easily be done by hand, provided the matrices are small. Working with matrix powers, however, is different. The following exercise is given to the students in the computer lab class during the second week of the course.

Exercise

(a) Let $A = \begin{bmatrix} 1 & 2 & 3 \\ 4 & 5 & 6 \\ 7 & 8 & 9 \end{bmatrix}$. Compute A^2, A^3, A^4, A^5 and A^{25}.

(b) Sometimes, the powers A^n of a matrix A behave according to a certain pattern. Compute A^2, A^3, A^4 and A^5 and give a formula for A^n if:

(i) $A = \begin{bmatrix} -1 & -6 & -4 \\ -4 & 4 & 1 \\ 7 & -9 & -3 \end{bmatrix}$

(ii) $A = \begin{bmatrix} -3 & -1 & 0 \\ 4 & 7 & -10 \\ 4 & 3 & -3 \end{bmatrix}$

(iii) $A = \begin{bmatrix} -2 & 2 & -5 \\ 12 & -7 & 20 \\ 6 & -4 & 11 \end{bmatrix}$

(iv) $A = \begin{bmatrix} 0 & -10 & -6 \\ -1 & -8 & -5 \\ 2 & 11 & 7 \end{bmatrix}$

(v) $A = \begin{bmatrix} -2 & 14 & 8 \\ -1 & 10 & 6 \\ 1 & -13 & -8 \end{bmatrix}$

(vi) $A = \begin{bmatrix} 0 & 1 & 1 & 1 & 1 \\ 0 & 0 & 1 & 1 & 1 \\ 0 & 0 & 0 & 1 & 1 \\ 0 & 0 & 0 & 0 & 1 \\ 0 & 0 & 0 & 0 & 0 \end{bmatrix}$

(vii) $A = \begin{bmatrix} 1 & 1 & 1 \\ 0 & 1 & 1 \\ 0 & 0 & 1 \end{bmatrix}$

Of course, students are not surprised with the outcome of question (a). But when they answer (b), they usually become enthusiastic. (Without a computer, they would not have been so enthusiastic about this exercise.)

Example 2.1.2
We teach our students how to find the matrix of a linear map relative to certain bases, and we prove that the matrix of f times the matrix of g is equal to the matrix of the composite map $f \circ g$. They see a lot of $\sum$'s but most of them don't know what it is all about. Students now have the opportunity to practise this within a reasonable time. For the following exercise, all they need to know is the definition of the matrix of a linear map.

Exercise
V is the vector space with basis (a_1, a_2, a_3), where

$a_1 = \begin{bmatrix} 2 \\ 3 \\ -1 \\ 1 \end{bmatrix}$, $a_2 = \begin{bmatrix} 3 \\ 0 \\ 1 \\ 2 \end{bmatrix}$ and $a_3 =$

$\begin{bmatrix} -1 \\ -2 \\ 0 \\ 1 \end{bmatrix}$, and W is the vector space

with basis (b_1, b_2), where $b_1 = \begin{bmatrix} 2 \\ 0 \\ 3 \\ 0 \\ 0 \end{bmatrix}$

and $b_2 = \begin{bmatrix} 1 \\ 0 \\ 1 \\ 0 \\ 0 \end{bmatrix}$. f is the linear map

from V to W defined by

$$f\left(\begin{bmatrix} v_1 \\ v_2 \\ v_3 \\ v_4 \end{bmatrix} \right) = \begin{bmatrix} 2v_1 - v_2 - 6v_3 - 3v_4 \\ 0 \\ 3v_1 - v_2 - 8v_3 - 4v_4 \\ 0 \\ 0 \end{bmatrix}$$

(in Maple: f:=v->vector([
2*v[1]-v[2]-6*v[3]-3*v[4],0,
3*v[1]-v[2]-8*v[3]-4*v[4],0,0]);)

(a) Let $v = \begin{bmatrix} 0 \\ 4 \\ -3 \\ 1 \end{bmatrix}$. Write v as a linear combination of (a_1, a_2, a_3) and put the coefficients of this linear combination in the column vector x. This x is called the *column of co-ordinates* of v relative to the basis (a_1, a_2, a_3).

(b) Compute $f(v)$ through the Maple command f(v);. Find the column of co-ordinates y of $f(v)$ relative to (b_1, b_2).

(c) Find the matrix F of f relative to (a_1, a_2, a_3) and (b_1, b_2).

(d) Verify that $Fx = y$.

(e) g is the linear map from W to R^2 defined by

$$g\left(\begin{bmatrix} w_1 \\ w_2 \\ w_3 \\ w_4 \\ w_5 \end{bmatrix} \right) = \begin{bmatrix} 2w_1 - 3w_3 + w_5 \\ -w_1 + 7w_2 + 4w_3 \end{bmatrix}$$

In R^2 we choose the basis (c_1, c_2), where $c_1 = \begin{bmatrix} 1 \\ 0 \end{bmatrix}$ and $c_2 = \begin{bmatrix} 1 \\ 1 \end{bmatrix}$. Find the matrix G of g relative to (b_1, b_2) and (c_1, c_2).

(f) h is the linear map defined by $h(v) = g(f(v))$ (so h is the composite map of g and f). Find the matrix H of h relative to (a_1, a_2, a_3) and (c_1, c_2).

(g) Verify that $H = GF$.

Example 2.1.3

Suppose the students have learned and practised solving systems of linear equations, and we now have come to the issue of finding eigenspaces. Again, lots of systems must be solved. There is no need to do this by hand any longer, so the students may use Maple here. This will save a lot of time which they can use for practising the things we want them to practise, like determining the algebraic and geometric multiplicity and diagonalizing matrices.

Exercise
Given the matrices
$$A = \begin{bmatrix} 83 & 80 & -60 \\ -20 & 3 & 30 \\ 0 & 0 & 43 \end{bmatrix} \text{ and}$$
$$B = \begin{bmatrix} 7 & -20 & -5 & 5 \\ 5 & -13 & -5 & 0 \\ -5 & 10 & 7 & 5 \\ 5 & -10 & -5 & -3 \end{bmatrix}:$$

(a) Use only **det** and **solve** to find the eigenvalues of A and B. Check your answer with **eigenvals**.

(b) Use only **nullspace** to find a basis of each of the eigenspaces of A and of B. Check your answer with **eigenvects** (The output of **eigenvects(A);** consists of "lists" of the form $[a, b, S]$, in which a is an eigenvalue, b is the algebraic multiplicity of a and S is a basis of the corresponding eigenspace).

(c) Is there a basis of R^3 that consists of eigenvectors of A?

(d) Is there a basis of R^4 that consists of eigenvectors of B?

2.2 Making pictures

Example 2.2.1
Pictures may help to understand the condition number of a matrix. With Maple, it does not take much time to draw those pictures. So immediately after the definition, students can get an impression of what the condition number is trying to tell them.

Exercise
A is the matrix $\frac{1}{\sqrt{2}} \begin{bmatrix} 1 & 1 \\ -1 & 1 \end{bmatrix}$. The unit circle in the x, y-plane is transformed by A into the curve

$$C = \{A \begin{bmatrix} \cos(t) \sin(t) \end{bmatrix} (0 \le t \le 2\pi)\}$$

A picture of this curve is obtained by the Maple commands

```
b:=evalm(A&*vector([cos(t),sin(t)]));
```

```
plot([b[1],b[2],t=0..2*Pi]);
```

Find the determinant and the condition number of A and make a picture of the curve C. Do the same in each of the following cases:

(a) $A = \begin{bmatrix} 1 & 2 \\ 3 & 4 \end{bmatrix}$

(b) $A = \begin{bmatrix} 100 & 100 \\ -100 & 100 \end{bmatrix}$

(c) $A = \begin{bmatrix} 100 & 200 \\ 300 & 400 \end{bmatrix}$

It appears that the determinant tells us something about the measure by which the area of the unit circle is enlarged, whereas the condition number tells us something about the measure by which the shape of the unit circle is "flattened."

Example 2.2.2

One of the most important applications of linear algebra is solving systems of linear differential equations. We would like the students to make phase portraits, but since we are not teaching them calculus, we don't want them to spend too much time on it.

Exercise
The solution of an initial value problem is regarded as the parametric equation of a curve. In this exercise, we are going to make pictures of such curves. In

each of the questions, we are dealing with the initial value problem

$$\begin{cases} u' = Au \\ u(0) = u_0 \end{cases}$$

(a) $A = \begin{bmatrix} -1 & -1 \\ 1 & -1 \end{bmatrix}$ and $u_0 = \begin{bmatrix} 1 \\ 1 \end{bmatrix}$.

Find the solution u and give the commands
`curve1:=[u[1],u[2],t=0..10];`
and
`plot(curve1);`
to make a picture of u.
Repeat this question, but this time with $u_0 = \begin{bmatrix} 1 \\ 2 \end{bmatrix}$. Assign the name `curve2` to this curve. In the same way, make a `curve3` with $u_0 = \begin{bmatrix} 1 \\ 3 \end{bmatrix}$ and a `curve4` with $u_0 = \begin{bmatrix} 1 \\ 4 \end{bmatrix}$. Draw the four curves in one picture with the command
`plot({curve1,curve2,curve3,curve4});`.

(b) The same question, but now with $A = \begin{bmatrix} 0 & -1 \\ 1 & 0 \end{bmatrix}$.

(c) Again the same question, with $A = \begin{bmatrix} 2 & 0 \\ 0 & -\frac{1}{2} \end{bmatrix}$ and for the initial conditions we take respectively $u_0 = \begin{bmatrix} \frac{1}{10} \\ 10 \end{bmatrix}$, $u_0 = \begin{bmatrix} \frac{1}{10} \\ 5 \end{bmatrix}$, $u_0 = \begin{bmatrix} \frac{1}{10} \\ 0 \end{bmatrix}$ and $u_0 = \begin{bmatrix} \frac{1}{10} \\ -5 \end{bmatrix}$. This time, the interval from 0 to 1 is a better range for t. Why was this change not needed in questions (a) and (b)?

2.3 Proving abstract statements

If we give our students an abstract statement and ask them for a proof, they usually have two problems. Firstly, "abstract" means "difficult" or "mysterious" to them. They often don't see what an abstract proof has to do with down-to-earth computations. Secondly, they don't know how to start. The following two exercises are examples of how to help them with both problems. Both of them

would be rather discouraging without a computer, but in a computerlab class, they are manageable for the students.

Example 2.3.1

Exercise

V is the vector space spanned by (a_1, a_2) and W is the vector space spanned by (b_1, b_2, b_3), where $a_1 = \begin{bmatrix} 2 \\ 0 \\ -3 \\ 1 \end{bmatrix}$, $a_2 = \begin{bmatrix} 1 \\ 4 \\ -1 \\ 5 \end{bmatrix}$, $b_1 = \begin{bmatrix} 2 \\ 1 \\ 1 \\ -1 \end{bmatrix}$, $b_2 = \begin{bmatrix} -4 \\ -37 \\ 8 \\ 32 \end{bmatrix}$

and $b_3 = \begin{bmatrix} 32 \\ -54 \\ 36 \\ 44 \end{bmatrix}$.

(a) Verify that both a_1 and a_2 are perpendicular to each of b_1, b_2 and b_3. (The dotproduct of two vectors v and w is computed by Maple through the command `dotprod(v,w);`).

(b) Verify that each vector from V is perpendicular to each vector from W. Hint: Take $v = c_1 a_1 + c_2 a_2$ and $w = d_1 b_1 + d_2 b_2 + d_3 b_3$ (don't assign values to c_1, c_2, d_1, d_2 and d_3! If necessary, give the commands `c1:='c1';` etc.) and show that v is perpendicular to w.

(c) Now the general statement. V is the vector space spanned by $(a_1, a_2, ..., a_n)$ and W the vector space spanned by $(b_1, b_2, ..., b_m)$. It is given that a_i is perpendicular to b_j for $i = 1, ..., n$ and $j = 1, ..., m$. Prove that each vector from V is perpendicular to each vector from W.

Example 2.3.2

Exercise

(a) A is the matrix $\begin{bmatrix} 16 & -14 & -23 \\ 9 & -7 & -15 \\ 4 & -4 & -5 \end{bmatrix}$

$$\text{and } B = \begin{bmatrix} 17 & -49 & 70 \\ 6 & -20 & 34 \\ 0 & -2 & 7 \end{bmatrix}.$$

    (i) Find a matrix S and a matrix T such that $S^{-1}AS = T^{-1}BT = \Lambda$, where Λ is a diagonal matrix.

    (ii) Compute $M = ST^{-1}$.

    (iii) Verify that $B = M^{-1}AM$.

(b) A is an $n \times n$ matrix with n distinct eigenvalues and B is an $n \times n$ matrix with the same eigenvalues as A. Prove that B is similar to A.

3. Teaching materials

In our view, most books with titles such as "Linear Algebra with a Computer" make too much use of a computer (see, for example, "Experiments in Computational Matrix Algebra" by D. R. Hill (Random House,1988) or "Linear Algebra with Maple V" by E. Johnson (Wadsworth, 1993)). Definitions are often surrounded by lots of computer commands and sometimes the reader is expected to type complete programs. This approach focuses too much on the computer. So we decided to use a textbook on Linear Algebra without a computer. We have chosen G. Strang's book "Linear Algebra and its Applications" (third edition, Harcourt Brace Jovanovich, San Diego 1988). Then we wrote a supplement for the lab classes. The exercises in the above examples are from this supplement. Writing the supplement, we faced the following three difficulties.

First, we were tempted to make the matrices too big. For example, we asked for the rank of the matrix

$$\begin{bmatrix} 1-t & -4 & 7-t & -6 \\ 2 & t^2-2t+1 & -4 & t^2-2t+3 \\ t-6 & -6 & t+3 & -9 \\ 1 & 2t-4 & 7-3t & 3t-6 \\ 3t-12 & 2t-12 & 6 & 3t-18 \end{bmatrix}$$

for each value of t. It turned out that the students needed more time than expected to type all the entries and to detect typing errors.

Second, we did not want the students to use all the available Maple functions. In example 2.1.3,
we did not want them to use **eigenvals** immediately, since this exercise was meant to practise the computing of eigenvalues. So we often had to indicate which Maple functions they were allowed to use.

Third, we did not want to spend too much time on teaching them Maple. So we decided to introduce each function at the moment it was needed. Furthermore, we provided a list of all the Maple functions that are used in our course. This turned out to be a good policy: there was hardly any need to give the students an introduction to Maple. They just used it.

4. The results

Comparing the experience of the past two years with the teaching of linear algebra without a lab class, we make the following observations.

4.1 Students' motivation

On the average, the motivation of the students is better. Of course, there always is the usual group of students who are motivated anyway, and also there are those who just want to pass their exams with minimal efforts. But talking about students who are somewhere between these extremes (which is the largest group) we noticed that they do much better than the norm for this group of students.

4.2 Understanding the theory

We teach definitions, we give examples and we prove theorems, and our students sit and listen (and, hopefully, make notes). But a great deal of our theory cannot be practised by the students themselves, for the simple reason that there is not enough time. With computer algebra, we can overcome this problem. For example, observing the way students handle the exercise in example 2.1.2, it turns out that the recipe for finding the matrix of a linear map is not as straightforward as we expected it to be for the students. After completing this exercise in the computer lab class, they have a much better understanding of what this matrix is all about.

4.3 Tests

Every other week, at the end of a lab class, we give the students a 20-minute test. With Maple available, we can ask them the things we want to ask. A typical example is a test in which several 3 by 3 matrices are presented, and students are asked to diagonalize them, if possible. Thus we can see if they have learned how to handle eigenvalues and eigenvectors, without going through the time-consuming process of solving lots of systems of linear equations.

5. Conclusion

It hardly takes time to teach Maple to the students, whereas using Maple saves a lot of time. With the assistance of Maple, we observe an improvement in the motivation of the students. Furthermore, they get a better understanding of the theory and we can test them more effectively.

Harry Kneppers studied Mathematics at the Leiden State University (The Netherlands) and recieved his Ph.D. in Mathematics from the Free University (Amsterdam, The Netherlands) in 1982. At present, he is a lecturer at the Delft University of Technology. His address is as follows:

Harry A.W.M. Kneppers
Department of Mathematics
Faculty of Technical Mathematics and
Informatics
Delft University of Technology
P.O. Box 5031
2600 GA Delft
The Netherlands

E-mail: H.A.W.M.Kneppers@twi.tudelft.nl

CAN MAPLE HELP IN TEACHING CALCULUS WITH LIMITED TIME AND STAFF?

Scott R. Fulton
Department of Mathematics and Computer Science, Clarkson University,
Potsdam, NY

Abstract

This paper describes our experiences incorportating Maple into a calculus course with limited time and staff. Our approach to teaching the basics of Maple and its use in lectures and take-home projects are described. Student reactions are summarized and compared to those at another institution. We conclude that our low-key approach can have a positive impact on the course without radical changes in the course format or substantial increases in contact hours or staffing.

1 Introduction

Computer algebra systems such as Maple have enormous potential for transforming the teaching and learning of Calculus. Well-established programs, such as that at Rensselaer Polytechnic Institute [1] have made it clear that Maple can be used effectively if the necessary commitment, facilities, time, and personnel are there. But what if any of these factors are lacking? Can Maple still be used effectively? This paper describes our experiences incorporating Maple into a calculus course with adequate facilities but a distinct lack of time and personnel. We hope this information will prove useful to others interested in using Maple under less than ideal conditions.

The paper is organized as follows. Section 2 describes the institutional setting and constraints. The format of the calculus course and the ways in which Maple were used are described in section 3. Section 4 presents results from a survey of student reactions to using Maple. Our conclusions and thoughts for the future appear in section 5.

2 Background

In the 1993–94 academic year we introduced a significant component of Maple into the standard Calculus course at Clarkson University. This section describes the background for this experiment: the facilities, time and staffing constraints, and prior uses of Maple at the University.

2.1 Facilities

Clarkson University has a wide range of computer facilities available for students. A generous site license from Waterloo Maple Software allows the full Academic version of Maple (currently Maple V release 2) to be available on all machines which are capable of running it. This includes:

- Student PCs: Each student entering the University is issued a personal computer. For 1993–94, this computer was an IBM PS/2 Model 25SX (20MHz 80386SX processor) with 4MB RAM, an 80MB hard disk, and a color VGA montior. Maple was pre-installed on each student PC.

- Networked PCs: Various PCs in dorm common areas, computer labs, classrooms, and lecture halls are available to students (when not in use in classes). Many of these are capable of running Maple, accessed from a central file server.

- Workstations: Over 125 IBM RS/6000 color workstations are available for student use. Many of these are concentrated in three computer classroom/labs with 20–30 workstations each; all run Maple.

2.2 Constraints

The standard freshman calculus course at Clarkson (taken primarily by engineering and science majors) meets for two hours of lecture and one hour of recitation per week. This austere schedule poses a challenge for teaching and learning calculus, even with a conventional approach. Adding a new aspect to the course (such as using Maple) must be balanced against the prospect of having to cut something else out due to lack of time. In Fall 1993, there were three lecture sections (150–160 students each) and 12 recitations (about 38–40 students each), with a total of about 460 students. Total staffing was one professor and three and a half Teaching Assistants. In Spring 1994 there were about 15% fewer students, the same number of sections, one professor, and four TAs. Undergraduate tutors were available for scheduled group and drop-in tutoring and help sessions. Student consultants were available in the computer labs in the evenings and on weekends.

2.3 History

The efforts to incorporate Maple into calculus described in this paper build on the efforts of others in previous years. In Fall 1991, an experimental section of Calculus II (for students with advanced placement credit) met regularly in a PC-equipped classroom/lab. Also, an experimental course combining calculus and physics used Maple regularly. Anecdotal evidence suggests that students were not completely satisfied with either approach; however, the experience gained by the instructors helped guide later experiments.

In Fall 1992, all recitations of Calculus I were held in a workstation-equipped classroom/lab. The Teaching Assistants showed students how to use Maple to solve problems related to their homework. Although it did not generate a student revolt like at the University of Pennsylvania [2], this experiment was not a complete success. Survey results indicated that many students felt that they did not learn enough about Maple to use it effectively on their own, and that the recitation time would be better spent going over homework and answering questions, rather than dealing with Maple. Some compared this approach to a typing class: "The TA wrote a command on the board and we typed it in but didn't understand anything."

3 Implementation

Maple (or any other computer algebra system) can serve at least two distinct purposes in a calculus course:

- As a learning tool: Maple has the potential to help students better understand the concepts of calculus by making it easy to explore an idea graphically and numerically as well as analytically.

- As a problem-solving tool: Maple allows students to solve problems more difficult and/or realistic than the standard "textbook" problems, allowing students to focus on the concepts involved rather than the details of the calculations.

While we have tried to incorporate both of these aspects of Maple into the course, our focus is on the first: using Maple to help teach and learn calculus.

3.1 Recitations

Our first goal was to teach the students enough about Maple to allow them to use it effectively on their own. To do so, the first three weeks of recitations were held in a workstation-equipped classroom/lab. Experience from the previous year showed that standing in front and writing on the board was a poor way to explain Maple (the students cannot see the board well from behind the computer monitors), so we prepared tutorial handouts to lead students through the basics of Maple.

The tutorial handout for the first week led students through starting Maple, getting help, using Maple as a calculator, basic Maple commands and syntax, and some basic commands for algebra. All algebra was limited to expressions (not functions) for this first week. For this first week only, we had two TAs in each recitation to help deal with the initial confusion of logging in, lost account names and passwords, etc.

The tutorial handout for the second week explained more commands for algebra, introduced lists (which arise in solving equations) and the Maple notation for functions, and showed how to generate simple plots. Most students appeared to be able to work through these tutorials during the recitation; some optional exercises were included for working at home, but were ignored by most students. Having students work in pairs worked well: this appeared to help overcome the typing problems and general confusion experienced the previous year. It was also inevitable, since there were more students per recitation than workstations in the lab.

The third week of recitation (still in the computer lab) was devoted to working through a "Practice Project" intended to give students an idea of what to expect in the Maple Projects they would later do on their own (see below). This project led students through some simple transformations of functions (translations and dilations) and then used these ideas in solving a simple applied problem (finding the equation of a curve describing the shape of a tunnel). Most students did the first part without trouble, while many were not able to complete the last part. This work was not handed in.

All remaining recitations were held in a regular classroom and devoted to reviewing material from lecture, going over homework problems, and answering questions. Thus, after the third week, students were on their own as far as using Maple. The book *Maple V Flight Manual* [3] was recommended as a useful supplement, but it appears that few students bought it.

3.2   Maple Projects

Students completed three take-home Maple Projects during each semester. The goals of the projects were two-fold: to reinforce the calculus concepts being learned through lecture and homework, and to show how Maple can be used as a tool to solve "real-life" problems. Each project consisted of a handout with explanation and instructions and an answer sheet to be completed and turned in. Some of the projects were entirely new, while others were based on ideas from other sources, notably [4]. Students had one week to complete each project, and were allowed to work in pairs if desired (less than half did so). Most students used Maple on their own PCs, rather than go to one of the networked workstations or PCs.

Given the large enrollment and limited staff, we tried to design the projects so the instructions would be crystal clear (to avoid unnecessary confusion) and to make the grading possible in a reasonable amount of time. Initially, most of the commands needed were given explicitly in the instructions (with explanation); later, students were expected to figure out how to accomplish a given task in Maple on their own. Since the projects

were intended as learning exercises, grading was intentionally lenient: anyone making an honest effort would get a majority of the credit available. Each project counted 5% of the course grade, and included an optional extra credit section.

The three projects for the Fall consisted of:

- *Project 1: The Derivative.*
  This led students through the concept of the derivative as the slope of the line tangent to a curve, using both graphics and numerical calculations in an attempt to beat the concept to death. An application section considered a problem involving a rocket under constant thrust, using the derivative to compute the acceleration from the velocity. Maple was used to estimate the time at which an astronaut would black out due to the effects of acceleration. An extra credit section pointed out that the acceleration increases rapidly as the fuel runs low, and asked students how to adjust the assumptions to avoid crushing the astronaut.

- *Project 2: Optimization.*
  An initial section had the students use Maple to find the derivative of a function, set it equal to zero, and solve for the critical points. Graphs were used to see what was happening, and Maple's *fsolve* command was introduced. An application section used Maple to solve the classic textbook min-max problem of moving a beam around a rectangular corner in a hallway, and then illustrated the power of Maple by considering the case where the thickness of the beam was not negligible (i.e., a 4 × 4). An extra credit section generalized the problem by allowing the beam to be tipped in the vertical.

- *Project 3: The Definite Integral.*
  This used commands from Maple's *student* package to illustrate areas under a curve (graphically and numerically). Students then used an integral to solve an applied probability problem (what percentage of sheet steel can be expected not to meet the specifications?); this integral could not be evaluated using antiderivatives. An extra credit section explored the accuracy of answers given by various numerical approximations (*leftsum*, *rightsum*, *middlesum*, *trapezoid*, and *simpson*).

The three projects for the Spring consisted of:

- *Project 1: Volumes of Solids of Revolution.*
  An initial section led the students through defining and plotting a solid of revolution (like a football), and computing its volume by the disk method. In the next section, students repeated this for a doughnut using the washer method, and then worked out the Maple commands for the shell method on their own. An extra credit section had students compute the volume of frosting needed to cover the top of the doughnut (maple frosting, of course).

- *Project 2: Numerical Integration.*
  After defining the "exact" expression for the period of a pendulum as an integral, this project used Maple to investigate the accuracy of the linear approximation to the pendulum and the accuracy of the trapezoidal approximation to the integral. Students also found the length required for a given period.

- *Project 3: Convergence of Infinite Series.*
  After introducing the Maple commands to generate (and plot) partial sums of a series, students used Maple to investigate whether a given series would converge, and estimated its sum. They then computed approximations to π using four different series and compared the number of terms required to give three decimal place accuracy.

Grades on the Maple Projects were uniformly high, which is not suprising given the simple nature of the projects and the intentionally lenient grading policy. It appears that most students were able to complete most of each project without trouble. While few students expressed great enthusiasm for the projects, few complaints were heard. Finally, despite the opportunity given to work with a partner, it is likely that some cheating also occurred (this was suggested by several students); given the current lack of staff for the course, it is difficult to see what could be done to prevent or even detect it.

3.3 Lectures

Maple was used every day in lecture, typically for 5–10 minutes of a 50 minute lecture. The primary use was to help illustrate the concepts being taught. Some examples include:

- On the first day of class, simple animations of secant lines approaching a tangent line and boxes filling up the area under a curve were used to illustrate the dynamic nature of the fundamental ideas of calculus.

- In discussing limits (e.g., $\lim_{x\to 0} \sin(x)/x = 1$), it proved useful to have Maple generate tables of numerical values and graphs (these seem more believable when produced on the spot, rather than copied from the instructor's notes).

- Using Maple helped bring Newton's method to life.

- Using Maple for graphs was especially helpful, since it allows one to draw accurate graphs that everyone can see.

- In discussing infinite series, the fact that the harmonic series diverges was made more believable by using Maple to comute partial sums of a large number of terms (e.g., $10^{10,000}$ terms).

In addition to helping communicate the concepts being taught, using Maple in lecture helped the students take the program seriously, and provided them with examples of its use. Also, it helped break up the monotony of the lecture format, which may have helped enhance learning and improve students' perceptions of the course.

We started the year using the DOS version of Maple on a 386 PC. This arrangement proved frustrating, since the characters produced were barely readable from most of the lecture hall. Several students suggested trying the PC-windows version; this proved to be great, and was used the rest of the year. With the larger fonts available, text was clearly visible from everywhere in the room, and the black-on-white format was much clearer. Also, the ability to point and click on a graph to find coordinates of a point was helpful.

3.4 Homework

Students turned in homework to be graded approximately once per week. None of the problems specified using Maple (with the exception of the Maple projects). However, students were encouraged to use Maple to check their answers (not to do the work!), and they were shown how to do so in many cases in lecture. It appears that about 30–40% of the students did use Maple this way (see section 4 below).

3.5 Exams

During the fall semester, each of the three hour exams contained questions which involved Maple, in each case worth a total of 10% of the exam. The purpose of these questions was two-fold: to encourage students to learn Maple, and to evaluate whether or not they had. The first exam had two questions testing basic Maple syntax, the second had two questions about how to express certain operations in Maple, and the third asked which of five sequences of Maple commands would solve a given max-min problem. With only one exception, average scores on the Maple questions were substantially higher than on the non-Maple questions; nevertheless, many students were upset that Maple appeared on the exams, and there was great relief when it was announced that the final exam would not involve Maple.

During the spring semester, no Maple appeared on the exams. In retrospect, this was a mistake, as several students pointed out: it allowed students who "worked with a partner" on the Maple projects to avoid any real involvement (and learning) with Maple.

3.6 Notes

To keep our focus on learning calculus—rather than struggling with computers—we intentionally stuck to the basics of Maple. In particular, we did not teach or use:

- programming constructs (if/then, loops, procedures, etc.)

- advanced graphics

- printing Maple sessions or graphics

Implicit plots and animation were used a few times in lecture, but the students were not asked to use them.

During the Spring semester, no additional instruction in Maple was given (except for optional review sessions and help for the few students who were not enrolled in the course in the Fall). However, most of the students (the engineering majors) were also enrolled in an introductory engineering course in computers which used Maple for 30–60% of the time.

4 Evaluation

There was relatively little negative feedback about Maple during the year. In six opportunities for written feedback about the course, few students expressed an opinion about Maple (although those who did were often negative). In talking with students and TAs, it appeared that student reactions to Maple were generally favorable.

A survey about Maple conducted in lecture during the last week of class each semester provided more quantitative feedback. The responses are shown in Table I for the Fall semester (299 of 444 students responding) and Table II for the Spring semester (177 of 361 students responding). Note that questions 1 and 10 were changed between semesters. For comparison, many of the statements on the survey were taken directly from a similar survey used at Rensselaer Polytechnic Institute (RPI) and reported in [1]. The columns labeled "RPI" give the percentages of students at RPI who agreed with the statement in Fall 1992 (Table I) and Spring 1993 (Table II). It should be noted that the calculus course at RPI has substantially more time (3 or 4 lectures, 1 recitation, 1 computer lab per week) and resources (31 TAs for about 1000 students) than at Clarkson (2 lectures, 1 recitation, no computer lab, with 3.5–4 TAs for about 400 students). Finally, a simple survey about Maple was done at Clarkson in Fall 1992; only one of the questions was repeated here (#17), as the others did not seem applicable. In 1992 the responses for #17 were 1%, 10%, 28%, 28%, and 31% (strongly agree to strongly disagree), showing that students in the present year feel significantly more comfortable with using Maple on their own.

Students were also given the opportunity to write their own comments on the surveys. While their written comments were few, many of these were negative. A common thread was the request for more help in learning Maple: either a handbook, a separate computer lab, or a separate course in Maple. Indeed, in the Spring—when many students were enrolled in the engineering computing course—several students wrote that we should "make Maple a separate course" and have the calculus course "stick to learning calculus".

	strongly agree	agree	uncertain or no opinion	disagree	strongly disagree	RPI
1. The first three weeks of recitation in the computer lab taught me enough about Maple to be able to use it effectively in this course.	5	25	9	44	15	—
2. The use of Maple in lectures helped me to better understand the concepts of calculus.	11	39	21	20	8	—
3. The Maple projects were interesting.	6	31	26	25	12	50
4. The Maple projects were reasonable in length and difficulty.	12	52	20	12	4	54
5. Working on the Maple projects with a partner was helpful and worthwhile.	37	38	17	3	3	86
6. Working the Maple projects helped me to better understand the concepts of calculus.	7	29	27	26	10	—
7. Attending weekly computer labs (in addition to recitations) would have been better than take-home Maple projects.	25	23	18	23	10	—
8. I used Maple often to check my homework or to help understand material from class.	10	26	12	24	27	—
9. I prefer to use Maple on my personal computer rather than on one of the networked computers (workstation or PC) in the computer lab rooms.	31	30	26	6	5	—
10. The exam questions involving Maple were reasonable and fair.	10	33	18	24	14	—
11. Maple enables me to solve some problems that would be almost impossible by hand calculations.	32	46	11	6	4	87
12. By using Maple I am able to see new approaches for solving some problems.	5	29	31	25	8	57
13. Maple was easy to apply to problems of various kinds.	6	23	25	34	10	41
14. The use of Maple in this course has improved my problem-solving skills.	2	14	31	35	16	42
15. The use of Maple revealed aspects of calculus that I hadn't thought about before.	3	21	35	29	11	49
16. The use of Maple was satisfactorily interwoven with the rest of the course.	11	46	20	17	6	71
17. I have learned Maple well enough to feel confident using it on my own.	6	30	19	29	15	—
18. My knowledge of Maple will probably help me in other science and engineering courses.	12	32	26	17	11	70
19. I enjoyed the course.	14	36	23	16	9	76
20. I would rather have had a calculus course that did not use Maple.	21	15	23	23	16	46

Table I: Responses (percent) to Maple Survey (Calculus I, Fall 1993)

	strongly agree	agree	uncertain or no opinion	disagree	strongly disagree	RPI
1. I learned enough about Maple last fall in Calculus I to be able to use it effectively in this course.	13	44	10	26	6	—
2. The use of Maple in lectures helped me to better understand the concepts of calculus.	12	38	23	19	7	—
3. The Maple projects were interesting.	4	30	31	25	10	64
4. The Maple projects were reasonable in length and difficulty.	8	54	15	16	6	64
5. Working on the Maple projects with a partner was helpful and worthwhile.	42	36	15	3	2	89
6. Working the Maple projects helped me to better understand the concepts of calculus.	4	37	23	23	12	—
7. Attending weekly computer labs (in addition to recitations) would have been better than take-home Maple projects.	15	20	19	23	23	—
8. I used Maple often to check my homework or to help understand material from class.	12	33	14	24	17	—
9. I prefer to use Maple on my personal computer rather than on one of the networked computers (workstation or PC) in the computer lab rooms.	20	28	23	13	15	—
10. I sometimes use Maple now for learning or solving problems, even when it is not required for a course.	6	30	21	22	20	—
11. Maple enables me to solve some problems that would be almost impossible by hand calculations.	38	47	10	2	2	91
12. By using Maple I am able to see new approaches for solving some problems.	4	29	39	21	6	69
13. Maple was easy to apply to problems of various kinds.	3	41	26	22	7	59
14. The use of Maple in this course has improved my problem-solving skills.	2	21	37	31	7	57
15. The use of Maple revealed aspects of calculus that I hadn't thought about before.	5	28	35	25	6	62
16. The use of Maple was satisfactorily interwoven with the rest of the course.	14	53	13	15	5	82
17. I have learned Maple well enough to feel confident using it on my own.	14	49	12	14	9	—
18. My knowledge of Maple will probably help me in other science and engineering courses.	28	41	12	10	7	74
19. I enjoyed the course.	12	44	25	10	8	80
20. I would rather have had a calculus course that did not use Maple.	10	10	24	32	22	36

Table II: Responses (percent) to Maple Survey (Calculus II, Spring 1994)

5 Conclusions and Thoughts for the Future

Overall, our experience with introducing Maple in our calculus course was positive. The student reactions, while not overwhelmingly positive, indicate that many students found the use of Maple at least acceptable, and in some cases helpful. On the other hand, some students appear to view Maple as an add-on to the course, and as such it provides a convenient focal point for complaints. Some positive aspects:

- Providing tutorial handouts allowed recitations in the computer lab to function well.

- Allowing students to work together on Maple proved to be a good idea.

- Using Maple in lectures opened up new ways to illustrate concepts, made numerical results more believable, and contributed to a more interesting style of presentation.

- Having Maple on the student PCs (rather than in a computer lab to which students must make a separate trip) made it much more palatable.

Some possible improvements for the future:

- Identify a specific subset of Maple commands to be learned and used in the course, and provide a handout listing and describing them.

- Require students to purchase an appropriate printed manual for Maple, such [3].

- Find more realistic applications to include in the Maple projects.

Has using Maple increased students' understanding of calculus? It may be impossible to answer this question. Nevertheless, it appears that even when used in the relatively low-key way described here, Maple can have a positive effect on the course. Students have been exposed to another way of approaching calculus and have seen the power of a modern tool for doing mathematics; this should help equip them for subsequent courses and careers.

Acknowledgements: If this work has been successful, it is in part due to the efforts of those who came before me and those who helped. Professor A. George Davis helped bring Maple to Clarkson and was the first to use it in calculus here; Professor A. S. Fokas continued that effort. The suggestions of Professor Michael Felland were also of great help. The TAs for the course (Benjamin Brown, Joel Helms, Taras Lakoba, Frank Michielsen, and Anwar Saleh) did much of the front-line work with the students. The Educational Resource Center staff (especially Thomas Wright and Robert Barringer) provided invaluable technical support. Finally, the people at Waterloo Maple Software made this work possible by their generous license which allowed us to put Maple on all student computers.

References

[1] W. E. Boyce and J. G. Ecker, "The Computer-Oriented Calculus Course at Rensselaer Polytechnic Institute," presented at the Maple Summer Workshop and Symposium, Ann Arbor, MI, August 1993; accepted for publication in *The College Mathematics Journal.*

[2] T. J. DeLoughry, T. J., "A Revolt Over Software." *Chronicle of Higher Education,* 11/24/93, p. A17.

[3] W. Ellis, E. Johnson, E. Lodi, and D. Schwalbe, *Maple V Flight Manual.* Pacific Grove, California: Brooks/Cole, 1992.

[4] M. H. Holmes, J. G. Ecker, W. E. Boyce. and W. L. Siegmann, *Exploring Calculus with Maple.* Reading, Massachusetts: Addison-Wesley, 1993.

Scott Fulton earned a B.A. degree from Kalamazoo College and M.S. and Ph.D. degrees from Colorado State University. Since 1986 he has been at Clarkson University, where he is currently Associate Professor in the Department of Mathematics and Computer Science. Besides teaching, he conducts research in computational mathematics, including spectral, multigrid, and semi-implicit methods, and applications to atmospheric modeling and numerical weather prediction. He may be contacted at:

Dept. of Mathematics and Computer Science
Clarkson University
Box 5815
Potsdam, NY 13699–5815
email: fulton@sun.mcs.clarkson.edu

ENLIVENING THE MATHEMATICS CURRICULUM WITH MAPLE

G. F. Fitz-Gerald and W. P. Healy
Department of Mathematics, Royal Melbourne Institute of Technology, Melbourne,
Victoria, Australia

Abstract

We describe the use of Maple in the undergraduate mathematics curriculum at a large Australian University, the Royal Melbourne Institute of Technology (RMIT). Maple has been integrated into (a) the first-year programme of a specialized Mathematics Degree Course for mathematics majors and (b) some higher-year service mathematics subjects for Engineering and Applied Science students. Maple sessions are supported by high-quality teaching materials published as part of the RMIT Lecture Notes in Mathematics Series. We believe the introduction of Maple has enlivened the teaching and learning of mathematics at RMIT. Some discussion is given of the changes in both teaching methods and subject content brought about by the use of Maple. The results of a student questionnaire and an analysis of student performance in one of the higher-year engineering mathematics subjects are also presented.

1 Introduction

The mathematics department at RMIT introduced Maple into the teaching programme in 1992. A subsidised licence was obtained on the adoption of the textbook [1] in the large-enrollment first-year service calculus subject. This provided the opportunity to install Maple (Version 4.2) on a centralised computer network supporting 17 Macintosh SE computers. Subsequently we upgraded to Version 5 and more recently to Version 5 Release 2.

The availability of such a small number of machines meant that any experiment in the use of Maple would involve only small-enrollment subjects. For this reason the introduction of Maple to the teaching programme was at first restricted to our mathematics degree course. The first-year quota for this course is approximately sixty students. (In contrast, the first-year service calculus subject has an enrollment of approximately eight hundred students.) Four classes of fifteen first-year mathematics degree students became the "guinea pigs" for our introduction of Maple to enliven the teaching of a traditional mathematics programme at RMIT.

Our perception of the effect of introducing Maple-based material into this level of the degree programme is that traditional topics that had become unpopular with students were now approached in a more understanding way. The topic of Riemann sums in integration theory is a case in point. Encouraged by student response to the use of Maple in the first-year mathematics degree programme, we have begun to introduce similar material into some of the larger-enrollment service subjects. Over the past few years the content of some subjects had to be modified because of the lack of algebraic manipulation skills and trigonometrical and geometrical knowledge on the part of many students entering tertiary education. This has meant that some of the more interesting, but more analytically demanding, aspects of the subjects could only be touched upon. Maple now provides the opportunity to include varied and realistic examples on these more difficult topics. Students can concentrate on the parameters of any particular model rather than on the intricacies of the methods involved. (We remark parenthetically that this may force the mathematical community to confront the possibility that an education in mathematics should provide more than simply a kit bag of tools applied to a range of contrived problems.)

In Section 2 below we give details of the integration of Maple into the first year of the mathematics degree course and discuss the educational philosophy that motivated this teaching initiative. In Section 3 we describe the use of Maple in higher-year service mathematics subjects and present an evaluation based on a student questionnaire and an analysis of student performance in one of these subjects. Planned new developments are outlined in Section 4 and our conclusions are summarised in Section 5.

2 Mathematics Degree

2.1 Structure of the Programme

As was mentioned in Section 1, Maple was introduced into the teaching of the first-year mathematics degree programme in 1992. This was achieved by developing a set of Laboratory Sheets structured around a pre-existing mathematics laboratory subject in which students performed "pen and paper" calculations. The first year of the mathematics degree programme is a traditional one that covers the areas of analysis, linear algebra and applied mathematics (mathematical modelling). Presentation of the material has always been in the form of small classes that are typically a combination of lecture and tutorial. However, in support of the theoretical presentations, a formal component of the course involves students attending the mathematics laboratory sessions. In these, staff assistance can be obtained while the students attempt problems related to work recently covered in lectures.

The five core subjects and one laboratory subject in the first year of the mathematics degree course are given in Table 1. All of these are year-long subjects taken over two 13-week semesters. In addition to the core subjects, there are also four hours per week of non-mathematics elective subjects. The mathematics laboratory subject MA192 was designed to develop the students' problem-solving skills with problems being drawn from the four core mathematics subjects (MA111, MA212, MA532 and MA533). Only one of the two contact hours for MA192 is now devoted to solving problems by the traditional pen and paper methods. The other contact hour is taken up by a weekly Maple Laboratory Session.

2.2 Maple Laboratory Sheets

The introduction of Maple was implemented by an interested and enthusiastic team of writers who were familiar with the entire course and developed material that would support the overall aims of the programme. Initially it required the weekly preparation of a hand-out, in the form of a laboratory sheet along the lines suggested in [2], on some topic of the course. As the Maple sessions ran in parallel with the traditional laboratory sessions, each week students were solving similar problems using both pen and paper calculations and Maple. The opportunity to explore algebraically more difficult problems than a typical modern student could handle without recourse to a symbolic manipulation package was not lost on the developers of the Maple laboratory material. In Table 2, a list of all the first-year topics, grouped by subject area, for which Maple laboratory sessions have been developed is presented.

A handbook of laboratory sheets for use by students in the Maple sessions has now been prepared in two parts with one part for each semester. The handbook has been published by the Department of Mathematics as part of the RMIT Lecture Notes in Mathematics Series [3,4]. (These handbooks reflect the high quality of the lecture note series in general and provide excellent resources to the student at a modest cost.) Each laboratory sheet includes:

- a statement of the aims of the session;

- a review of the mathematical background for the session;

- worked examples;

- student exercises;

- the syntax of the Maple commands to be used;

- a report sheet.

A shortened version of the laboratory sheet for the session on Polar Coordinates is given as an example in Appendix A. (The Maple command syntax and report sheet are omitted in this version.)

Table 1: Core subjects in first year.

Subject Code	Subject Name	Hours/Week
MA111	Analysis 1	3
MA212	Linear Algebra	2
MA532	Mathematical Methods and Models A	3
MA533	Mathematical Methods and Models B	2
CS600	Digital Programming 1	4
MA192	Mathematics Laboratory 1	2

2.3 Maple Laboratory Sessions

The four weekly one-hour Maple classes, each with about fifteen students, are supervised by an academic staff member. The staff member is either a lecturer or tutor in the first year of the mathematics degree programme. The rôle of the supervisor is to:

- explain the aim of the session to students;

- respond to difficulties that students may have on the use of Maple, the Macintosh SE system or the mathematics underlying the laboratory session;

- observe student progress and offer assistance to individual students.

In order to prepare for the laboratory sessions, students are encouraged to read the laboratory sheets beforehand and, if necessary, to review the relevant mathematical topics. They are expected to go through the worked examples using Maple during the session and to submit their solutions to the exercises on the report sheet at the end of the session. The report sheets are marked and returned to the students at the beginning of the next session.

The Maple laboratory sessions contribute to the assessment of the subject MA192, for which the only grades available are pass and fail. In order to obtain a pass in MA192 students are first of all required to attend classes in the subject. They must then obtain a passing grade in at least 75% of the traditional pen and paper problem-solving sessions and at least 75% of the Maple laboratory sessions taken over the whole academic year.

2.4 Educational Philosophy

The aims for student learning or educational 'philosophy' behind the introduction of Maple to the first-year mathematics degree course are as follows.

1. The Maple sessions are intended to support (rather than completely replace) the traditional pen and paper laboratory sessions.

2. The use of Maple promotes students' understanding by engaging them actively with the subject. In each Maple session, the students review and use mathematics that they have met at least twice before in the preceding two or three weeks — once in lectures and once in the pen and paper laboratory sessions. The mathematical content is therefore reinforced by being dealt with again in a different context.

3. Maple's graphics capabilities enable the students to plot graphs of functions and to visualise solutions of differential equations very easily. In the past, although most students were capable of deriving the formal solutions of differential equations, for example, they often had a poor understanding of the behaviour of the solutions and their physical implications.

4. Maple's computational power and algebra 'engine' enable students to solve problems more efficiently and to tackle problems that are more difficult algebraically than those that modern students are capable of dealing with otherwise. This allows the possibility of dealing with more realistic problems.

Table 2: List of first-year topics available as Maple laboratories.

Subject	Topics
Maple	Introduction
	Variables and functions
Analysis	Inequalities
	Composition of functions
	Limits
	Differentiation
	Riemann sums
	Non-linear equations
	Integration
	Polar coordinates
	Functions of two real variables
	Sequences and series
Linear Algebra	Spanning sets
	Vectors
	Matrices
	Linear transformations
	Inner product spaces
	Eigenvalues and eigenvectors; diagonalization
	Applications of diagonalization of matrices
Mathematical Methods and Models	Differential equations
	Particle motion
	Phase plots
	Resonance

5. It is easy to create individual Maple assignments for students, although the need for doing this has not actually arisen in MA192.

6. The students are taught how to use a symbolic manipulation package. That mathematics students should be aware of and familiar with mathematical software packages is a desirable educational end in itself, quite apart from the use of such packages as teaching aids or as tools for the students during their degree programmes.

3 Service Subjects

3.1 Comparison with Mathematics Degree

Based on the success of the Maple material with the mathematics degree students, a similar programme is being developed for some of the large-enrollment service subjects. Material has already been prepared for courses in MA003 Differential Equations A, MA004 Differential Equations B, MA015 Residues and Integral Transforms and MA019 Finite Element Methods. We will discuss here only the subjects MA003 and MA004. These courses have no time allowance for the formal presentation of Maple material and the large numbers preclude the formal laboratory sessions adopted in the mathematics degree course. The same style of laboratory session is nevertheless followed. The current availability of Maple on only a small number of machines for a large number of students is clearly a problem from the point of view of conducting supervised laboratory sessions.

3.2 Differential Equations A

Maple laboratory sessions have been developed for the four main topics in the subject MA003 Differential Equations A. These cover the areas of

- Laplace transforms;
- Series solutions of differential equations;
- Fourier series;
- Partial differential equations.

MA003 is taken by the majority of students in the Faculty of Engineering and the Department of Applied Physics at RMIT. The enrollment is 550 students taught by several staff in classes ranging in size from 40 to 120. Because of the numbers involved, no attempt has been made so far to make the Maple sessions a compulsory part of the MA003 programme. Once a wider access to Maple is achieved this situation will be redressed.

3.3 Differential Equations B

Maple laboratory sessions have been developed for several topics in the subject MA004 Differential Equations B. The aim of MA004 is to extend the basic techniques required in the solution of boundary-value problems in mathematical physics and engineering. MA004 also includes a brief introduction to systems of non-linear differential equations and stability.

MA004 is a compulsory subject in the third year of the Bachelor of Engineering course in Electrical Engineering at RMIT and is taught as one single class. The class normally has an enrollment of 60 students with a majority of these being Electrical Engineering students. The remaining are usually Applied Physics students who take MA004 as an elective subject in their final year.

The lecturer demonstrates the use of Maple during class using a portable computer and a screen projector. MA004 students are given free access to the Macintosh SE Laboratory when it is not being used for classes. As these students are in the third year of an engineering or physics degree course, they have much more experience in computer programming and the use of computer software packages than first-year mathematics students.

Maple laboratory sessions for MA004 cover the following topics:

- Introduction to Maple;
- Generalized Fourier series;
- Partial differential equations;
- Application of Bessel functions;
- Phase plots.

For each session, laboratory sheets are provided for student use. These have the same format as the sheets used in MA192. The students are expected to go through the worked examples in the laboratory in their own time and are required to submit their solutions to the exercises on the report sheet. Students are required to submit three assignments which, together, contribute 20% of the total assessment for the subject. All three assignments require the use of Maple.

3.3.1 Evaluation

Student Questionnaire

A questionnaire was handed out to the MA004 class near the end of the course in 1993. The students were encouraged to complete the questionnaire but were told that doing so was voluntary. Eleven students answered the questionnaire. Even though this represents only about one sixth of the class, there was nevertheless a spectrum of opinion expressed in the answers.

The questionnaire consisted of the following three items. (Note that DEB stands for Differential Equations B.)

1. What aspects of the use of Maple have helped your understanding of topics in DEB?

2. Please comment on what aspects of the use of Maple you feel have hampered your development of understanding of topics in DEB.

3. What, in your opinion, are the **three most important things** you have learned from using Maple in DEB?

A broad categorization of the respondents is that nine were 'positive', i.e. they found Maple useful in helping to learn the subject, and the remaining two were 'negative', i.e. they did not find Maple useful. The answers of a typical 'positive' respondent (Student A) and of a 'negative' respondent (Student B) to the three questions are reproduced in full below.

Student A

1. Visualizing the series and graphical solution of the boundary-value problems; understanding the steps used in deriving the Fourier coefficients.

2. The process of producing the program output is not necessarily time consuming. However, getting these printed (when many people are on the system) can be. One suggestion—someone from the Mathematics Department needs to assist the computer staff with Maple printing.

3. (a) Using another or alternative computer package to 'Mathematica'.

   (b) To visualize particular series expansions.

   (c) Being able to solve 'long-winded' boundary-value problems.

Student B

1. I found that Maple was unable to enlighten me in the understanding of topics in DEB. I believe it is a waste of our time using it.

2. The friendliness towards the user was awful.

3. (a) Familiarization with the keyboard, monitor and disk drives.

   (b) Ergonomics in my study environment.

   (c) One more program that I can write on my résumé.

The single most common aspect of Maple mentioned by the students was its two- and three-dimensional graph-plotting capabilities and the way in which this enabled them to visualize solutions of differential equations. This is illustrated in the answer to Question 1 by Student A and in the comment by another student that 'seeing in three dimensions the graphical output of heat equations' was one of the three most important things learned from using Maple. Seven of the eleven respondents to the questionnaire mentioned this aspect of Maple as a positive feature. Four of the respondents did not answer Question 2 and two others made only brief comments to the effect that the package was not very 'user friendly' (see also the response of Student B given above). Others, however, had deeper specific criticisms in response to Question 2. For example,

> Perhaps try to tie the Maple commands more in with the lectures, but keep the assignments and examples.

and

> Sometimes the Maple exercises are so easy to do because you just had to copy the worked examples (maybe change a few things), so at times like that we weren't really thinking about the problem on our own.

Both of these students were positive about using Maple, however. Their answers to Question 1 were

> Being able to verify theorems from lectures; also being able to change variables and compute new answers quickly and easily.

and

> The preparation for the assignments where we had to look through the lecture notes to understand the solution process of the assignment. The explanation-type questions were good because they tested our knowledge and understanding of DEB concepts.

respectively.

Analysis of Student Performance

The assignment requirements were not only to produce numerical answers or graphical output but also to give a physical interpretation or commentary on the results obtained. Most students did the assignments very well. Approximately 84% of the students achieved a total mark of 15 or greater out of 20 for the three assignments combined. Only 5% of students achieved a mark of less than 10 out of 20. Many students sought advice from the lecturer, particularly for Assignment 1. The chief request was for clarification of technical details of Maple commands. Many students obtained their own copy of Maple for use on their own computers. A few students went to extraordinary lengths in producing graphical output for the assignments and this was accompanied by commentaries and interpretations that indicated an excellent understanding of the subject matter. These students were also among the top performers in the examination.

The pass rate for MA004 in 1993 was 80%. This compares with pass rates of 74% and 70% in 1992 and 1991 respectively. Of course not too much can be read into one year's figures but it is at least clear that the introduction of Maple has not had a detrimental effect on student learning in this subject. On the contrary, there is evidence from the answers to the examination questions that some common misconceptions held by students in the past have to a large extent been eliminated by the use of Maple. For example, egregious errors in the manipulation of infinite series solutions of partial differential equations were almost entirely absent. It is significant that an understanding of such manipulations was necessary in all three assignments in order to be able to use Maple to produce the required solutions and graphs. This argument is supported by the comment of one student, already quoted above, that 'the preparation for the assignments where we had to look through the lecture notes to understand the solution process of the assignment' was an aspect of the use of Maple that helped the understanding of topics in this subject.

4 Planned Developments

4.1 Evaluation

The integration of Maple into our undergraduate programmes is still in its early stages of development. Its effectiveness in terms of students' performance and understanding needs to be further evaluated. It will be useful to carry out evaluations of other subjects along the lines of that already discussed for the subject MA004.

4.2 New Laboratory

The large numbers enrolled in service courses is a major difficulty with attempting to introduce Maple. However the department has recently acquired new accommodation that includes a laboratory with approximately fifty 486 machines each running Maple under the Windows environment. This should also have an impact on perceptions of "user friendliness" of Maple. We look forward to the opportunity to exploit this in introducing Maple in service mathematics where the provision of more realistic examples has always been a problem. Furthermore, we intend to schedule supervised classes in this new laboratory for all students.

The fact that some students have their own copy also assists the introduction of Maple to these courses. We are happier suggesting students consider such a purchase once a full version is readily available on the RMIT campus.

5 Conclusion

Maple has been successfully introduced into a variety of mathematics programmes covering a wide range of topics offered at RMIT. Supporting material in the form of laboratory sheets provides a useful student resource in understanding the mathematics and the implementation of the Maple material.

The introduction of Maple to the mathematics teaching programme has helped to put into focus the mathematically important elements of the mathematics curriculum. There is an opportunity to delineate precisely the core mathematical ideas needed to be presented in a course without the added distractions of routine methods getting in the way of understanding any new material being presented. Too often students are labouring in their grasp of new topics simply because of their lack of technical expertise in material covered at an earlier stage (often before they arrive at the tertiary level).

Students participating in the programme have enjoyed the experience and have responded in positive ways. Many use Maple to solve problems in other subjects they are currently studying. This develops further the use of Maple in the entire mathematics curriculum at RMIT.

References

[1] Stewart, J. *Calculus*, 2nd ed., Belmont California, Brookes/Cole, 1991.

[2] Bauldry, W. C., & Fielder, J. R. *Calculus Laboratories with Maple*, Belmont California, Brookes/Cole, 1991.

[3] Connell, H. J., Fitz-Gerald, G. F., Healy, W. P. and Luo, Y. S. *Maple Laboratory Sessions Part 1*, RMIT Lecture Note in Mathematics, Melbourne, Mathematics Department RMIT, 1993.

[4] Connell, H. J., Fitz-Gerald, G. F., Healy, W. P. and Luo, Y. S. *Maple Laboratory Sessions Part 2*, RMIT Lecture Note in Mathematics, Melbourne, Mathematics Department RMIT, 1993.

Appendix A — Example Laboratory Session
1 Aim

In this laboratory session you will use the commands

plot and **fsolve**

to plot curves defined by polar equations and to find points of intersection of such curves.

2 Background

If r and θ are polar coordinates in the plane, a curve is defined by the equation $r = f(\theta)$, where f is a given function, or, more generally, by $g(r, \theta) = 0$ where g is a given function. The curve consists of all points that have at least one set of polar coordinates (r, θ) that satisfy the equation. [Note that (r, θ), $(-r, \theta + \pi)$ and $(r, \theta + 2n\pi)$, where n is any integer, all represent the same point.] Certain symmetries of a curve may be inferred from its polar equation. In particular,

(a) if the equation is invariant under the substitution $(r, \theta) \rightarrow (r, -\theta)$, then the curve is symmetric about the initial line;

(b) if the equation is invariant under the substitution $(r, \theta) \rightarrow (r, \pi - \theta)$, then the curve is symmetric about the straight line $\theta = \frac{1}{2}\pi$ — this is the straight line through the origin and perpendicular to the initial line;

(c) if the equation is invariant under the substitution $(r, \theta) \rightarrow (-r, \theta)$ or the substitution $(r, \theta) \rightarrow (r, \theta + \pi)$, then the curve is symmetric about the origin.

Points of intersection of two curves defined by polar equations are obtained by solving the equations simultaneously for r and θ. However, this method may not give *all* points of intersection, since some points of intersection may have to be represented by two different sets of polar coordinates in order to satisfy the two equations. In these cases, it is often necessary to transform the polar equation of one of the curves. In general, it is useful to sketch the two curves on the same diagram to ensure that all points of intersection have been located.

3 Graphs of Polar Equations

When Maple's **plot** command is used to obtain the graph of a polar equation, the curve must be given parametrically in the form $[r(t), \theta(t), t = t_1 .. t_2]$ where t is a real parameter and $t_1 \leq t \leq t_2$. The option **coords = polar** must also be specified. *In this laboratory session, we will always take the parameter t to be the coordinate θ itself.* The polar equation will therefore be passed to the **plot** routine as $[r(t), t, t = t_1 .. t_2]$.

Example 1 *Use the* **plot** *command to sketch the curve defined by the polar equation* $r = 2\cos\theta$ *where* $0 \leq \theta \leq \pi$.

We obtain the plot by following the procedure outlined above with

$$r(t) = 2 * \cos(t).$$

- plot($[2*\cos(t), t, t = 0 .. \mathrm{Pi}]$, coords = polar);

We know from lectures that the equation $r = 2a\cos\theta$ represents a circle that passes through the origin, has radius a and has a diameter along the initial line. (Note also that since $\cos\theta$ is even in θ, the graph must be symmetric about the initial line.) The curve produced by Maple, however, is distorted and does not look like a circle. To rectify this, we can specify horizontal and vertical ranges (for the Cartesian coordinates x and y) to ensure that the aspect ratio of the graph is the same as that of the screen. The aspect ratio is the ratio of the width to the height and for Macintosh screen this is about 1.5. In the following **plot** command, the horizontal and vertical ranges are $-3\,..\,3$ and $-2\,..\,2$ respectively. This gives an aspect ratio of $6/4$ or 1.5 and the curve obtained does look like a circle when the plot window is 'maximized' to occupy the entire screen. We also use the 'title' option to give the graph a name that will appear in the plot window.

- plot($[2*\cos(t), t, t = 0\,..\,\text{Pi}], -3\,..\,3, -2\,..\,2,$
    $\text{coords} = \text{polar},$
    $\text{title} = \text{'Example 1} - \text{Circle'});$

We can plot more than one curve in the same window by passing a *set* of polar equations to the **plot** command, as in the next example.

Example 2 *Sketch the cardioid*

$$r = 1 - \cos\theta$$

and the parabola

$$r = 1/(1 + \cos\theta)$$

in the same window.

We use the same horizontal and vertical ranges as in Example 1 and take the domain of θ to be $(-\pi, \pi)$. Note that on the parabola, $r \to \infty$ as $\theta \to \pm\pi$.

- plot($\{[1 - \cos(t), t, t = -\text{Pi}\,..\,\text{Pi}],$
    $[1/(1 + \cos(t)), t, t = -\text{Pi}\,..\,\text{Pi}]\},$
    $-3\,..\,3, -2\,..\,2, \text{coords} = \text{polar},$
    $\text{title} = \text{'Example 2} - \text{Cardioid}$
    $\text{and Parabola'});$

It appears from the plots that the only points of

intersection of the cardioid and the parabola are $(1, \pm\frac{1}{2}\pi)$ and that the two curves touch at these points.

4 Exercises 1

1. For each of the following polar equations,

    (i) use the **plot** command to sketch the curve;

    (ii) state the symmetries, if any, the curve possesses;

    (iii) explain how the symmetries can be inferred from the polar equation.

    (a) $r = 3 - 2\cos\theta$ (limaçon without a loop)

    (b) $r = 1 + 2\sin\theta$ (limaçon with a loop)

2. (a) Sketch the 4-leaved rose $r = 4\sin 2\theta$ and the circle $r = 4$ in the same window.

    (b) Find all points of intersection of the two curves.

5 Points of Intersection

Although sketching the graphs of two polar equations sometimes enables us to find the points of intersection of the two curves, it is often necessary to solve the polar equations simultaneously to find such points. Maple's **fsolve** command allows us to solve the equations numerically. This is illustrated in the following example.

Example 3 *Find all points of intersection of the spiral $r = 0.2\theta$ and the ellipse $r = 1/(2 - \cos\theta)$ where $0 \le \theta \le 2\pi$.*

We first plot the two curves.

- plot($\{[0.2*t, t, t = 0\,..\,2*\text{Pi}], [1/(2 - \cos(t)),$
    $t, t = 0\,..\,2*\text{Pi}]\},$
    $-1.5\,..\,1.5, -1\,..\,1, \text{coords} = \text{polar},$
    $\text{title} = \text{'Example 3} - \text{Spiral and}$
    $\text{Ellipse'});$

It is evident from the graphs that there is just one point of intersection. This occurs in the second quadrant, in which $\frac{1}{2}\pi < \theta < \pi$. We use **fsolve** to

---

79

obtain the solution of $0.2\theta = 1/(2 - \cos\theta)$ in this interval.

- fsolve$(0.2*t = 1/(2 - \cos(t)), t, Pi/2 .. Pi)$;

    2.039428915

The corresponding value of r is then obtained by substitution. The right-hand side of either equation can be used but here it is easier to use $r = 0.2\theta$.

- $0.2*$";

    .4078857830

Polar coordinates for the point of intersection are therefore $(0.408, 2.039)$ correct to 3 decimal places.

6 Exercises 2

(a) Sketch the 4-leaved rose $r = \cos 2\theta$ and the cardioid $r = 1 - \cos\theta$ in the same window.

(b) Your graphs should show that there are seven points of intersection. Polar coordinates for three of these points can be read directly from the graphs. What are they?

(c) Of the other four points of intersection, two are in the first quadrant and the other two are reflections of these in the initial line. It is therefore sufficient to determine the polar coordinates of the two points of intersection in the first quadrant. One of these two points appears to have a θ coordinate in the interval $(0, \frac{1}{4}\pi)$.

   (i) Plot $r = \cos 2\theta$ and $r = 1 - \cos\theta$ for $0 < \theta < \frac{1}{4}\pi$ only.

   (ii) Using **fsolve**, find polar coordinates for this point of intersection correct to 3 decimal places.

(d) The other point of intersection in the first quadrant appears to have a θ coordinate in the interval $(\frac{1}{4}\pi, \frac{1}{2}\pi)$. However, using **fsolve** to try to find a solution of $\cos 2\theta = 1 - \cos\theta$ in this interval produces a null result.

   (i) Explain this null result by plotting $r = \cos 2\theta$ and $r = 1 - \cos\theta$ for $\frac{1}{4}\pi < \theta < \frac{1}{2}\pi$ only. (Note that $\cos 2\theta$ is negative on this interval.)

   (ii) Note that the four-leaved rose is symmetric about the origin and can also be represented by the polar equation $-r = \cos 2\theta$, although its petals will now be traced out in a different order as θ increases from 0 to 2π. Plot $r = -\cos 2\theta$ and $r = 1 - \cos\theta$ for $\frac{1}{4}\pi < \theta < \frac{1}{2}\pi$ only and then use **fsolve** to find polar coordinates for this point of intersection correct to 3 decimal places.

G. F. Fitz-Gerald (*rmagf@minyos.xx.rmit.edu.au*) is a Senior Lecturer in the Mathematics Department at RMIT. He holds a B. Sc.(Hons.) and Ph. D. degrees in Applied Mathematics from the University of Western Australia. His main research interests include water gravity waves and the use of mathematical methods to solve applied problems arising in applications. Such applications have recently included bluff-body aerodynamic flows, electromagnetic wave propagation and aero-acoustics. He has been particularly interested in the teaching of service mathematics and is the coordinator of all higher year service mathematics at RMIT. The introduction and use of symbolic computation packages (such as Maple) into the teaching programme at RMIT is another of his interests.

W. P. Healy (*rmalh@minyos.xx.rmit.edu.au*) is a Senior Lecturer in the Department of Mathematics at RMIT. He received a B. Sc.(Hons.) degree in Applied Mathematics from North-East London Polytechnic in 1972, a Ph. D. from University College London in 1975 and a Graduate Diploma in Computing Studies from the University of Melbourne in 1991. Prior to joining RMIT as a Lecturer in 1982 he was a Postdoctoral Research Fellow for four years and a Lecturer for four years and a Lecturer for two years at the Australian National University, Canberra. His research interests are chiefly in Quantum Electrodynamics. He is the author of a book "Non-Relativistic Quantum Electrodynamics" (Academic Press, London, 1982) and has published more than twenty research papers. His teaching interests are in undergraduate Pure and Applied Mathematics, Engineering Mathematics and pedagogical uses of symbolic manipulation packages.

IIB. MAPLE IN SCIENCE

ANALYTICAL APPROACHES TO SOLVING COUPLED NONLINEAR SCHROEDINGER EQUATIONS USING MAPLE V

M.F. Mahmood and T. L. Gill
Computational Science and Engineering Research Center, Howard University,
Washington, DC

Abstract

The coupled nonlinear Schroedinger (CNLS) equations involving highly oscillatory terms attracts great current interest owing to both its fundamental meaning and potential applications in optical logic devices. We have used Maple V in solving the problem of coupled nonlinear Schroedinger equations. A paper describing the main features of this work will soon be published.

Introduction

A significant amount of work has been done in the past in the investigation of nonlinear processes in optical fibers.[1,2] Propagation of soliton pulses as information bits in optical communication systems has attracted a great deal of interest since it was shown experimentally that they can propagate in single—mode optical fibers.[3]

Solitons are nonlinear pulses that propagate without dispersion in the anomalous dispersion regime of single—mode optical fibers. The combined effects of group—velocity dispersion and self—phase modulation lead to the formation of optical solitons in these fibers and can be described by a nonlinear Schroedinger (NLS) equation.[1] Single—mode optical fibers support two distinct modes of polarization[4] because of linear birefringence with weak intermodal dispersion and can be coupled together through the Kerr effect. The coupling between the two polarization modes in a birefringent single—mode optical fiber can be regarded as an important effect, since the coupling between the modefields is possible over long propagation distances.

The interaction between the two optical modes in a birefringent single—mode fiber may be described by a system of coupled nonlinear Schroedinger (CNLS) equations and it gives rise to a large number of schemes for making all—optical switches.

In this paper, we present a general formalism for the study of soliton propagation in birefringent single—mode optical fibers in the framework of a system of CNLS equations with highly oscillatory terms due to nonlinear polarization and used Maple V specially in evaluating integrals wherever it was possible. We follow an adiabatic approach using a time—averaged variational formulation[5] to describe nonlinear intermode interactions of soliton pulses in the two polarizations. In our analysis, the CNLS equations are reduced to a Lagrangian problem when solitons are considered as effective particles. Using NLS solitons as trial functions in an averaged Lagrangian formulation, a set of ordinary differential equations (ODE's) for various soliton parameters are derived.

Variational analysis of CNLS equations

The pulse propagation in a nonlinear birefringent optical fiber with two distinct modes operating in the anomalous group—velocity dispersion regime is governed by a system of CNLS equations[6]

$$i(\frac{\partial u}{\partial z} + \delta\frac{\partial u}{\partial t}) + \frac{1}{2}\frac{\partial^2 u}{\partial t^2} + (|u|^2 + \epsilon|v|^2)u$$

$$+ \frac{\epsilon}{2}v^2 u^* \exp(-iR\delta z) = 0, \quad (1a)$$

$$i(\frac{\partial v}{\partial z} - \delta\frac{\partial v}{\partial t}) + \frac{1}{2}\frac{\partial^2 v}{\partial t^2}$$

$$+ (|v|^2 + \epsilon|u|^2)v$$

$$+ \frac{\epsilon}{2}u^2 v^* \exp(iR\delta z) = 0, \quad (1b)$$

$u(z,t) = v(z,t)$ and $v = v(z,t)$ are the slowly varying wave envelopes of the slow and fast polarized modes, respectively, and δ is one—half the difference in the group velocities of the two modes. We assume that the fiber is lossless. The highly oscillatory terms in equations (1a) — (1b) cannot be neglected for weakly birefringent fibers since this leads to polarization instability [6] − [7]. These equations without the highly oscillatory terms were studied earlier with the initial conditions

$$u(0,t) = \frac{A}{\sqrt{2}} \text{ secht,} \qquad (2a)$$

$$v(0,t) = \frac{A}{\sqrt{2}} \text{ secht.} \qquad (2b)$$

Equations (1a) and (1b), when solved with $\delta = 0$ and the oscillatory terms neglected, give an exact solution, $u = v$, which is the solution of the NLS equation:

$$i\frac{\partial \psi}{\partial z} + \frac{1}{2}\frac{\partial^2 \psi}{\partial t^2} + |\psi|^2 \psi = 0, \qquad (3)$$

where $\psi = \sqrt{1 + \epsilon}\, u = \sqrt{1 + \epsilon}\, v$. Equation (3) is exactly integrable by the inverse scattering transform (IST) method and, when applied to the Cauchy problem with the initial condition

$$u(0,t) = v(0,t)$$

$$= \frac{1}{\sqrt{1 + \epsilon}} \psi(0,t)$$

$$= \frac{K}{\sqrt{1 + \epsilon}} \text{ secht,} \qquad (4)$$

the soliton

$$u(z,t) = v(z,t) = \frac{1}{\sqrt{1 + \epsilon}} \psi(z,t)$$

$$= \frac{2\zeta}{\sqrt{1 + \epsilon}} \exp(2i\zeta^2 z)\text{sech}(2\zeta t) \quad (5)$$

evolves for $\frac{1}{2} < K < \frac{3}{2}$ with an amplitude ζ related to the pulse intensity K by the relation

$$\zeta = K - \frac{1}{2} = \sqrt{\frac{1 + \epsilon}{2}}\, A - \frac{1}{2}. \qquad (6)$$

We assume that the above results are also valid in the case of coupled NLS equations for the symmetric two−component input pulse.

We reformulate the problem of coupled NLS equations (1a) − (1b) using a variational principle in Lagrangian form,

$$\delta \int_{-\infty}^{\infty} \int_{-\infty}^{\infty} L\,dz\,dt = 0. \qquad (7)$$

The Lagrangian, L is given by

$$L = \frac{i}{2}(u\,\frac{\partial u^*}{\partial z} - u^*\,\frac{\partial u}{\partial z})$$

$$+\,\frac{i}{2}(v\,\frac{\partial v^*}{\partial z} - v^*\,\frac{\partial v}{\partial z})$$

$$+\,\frac{i\delta}{2}(u\,\frac{\partial u^*}{\partial t} - u^*\,\frac{\partial u}{\partial t})$$

$$+\,\frac{i\delta}{2}(v^*\,\frac{\partial v}{\partial t} - v\,\frac{\partial v^*}{\partial t})$$

$$+\,\frac{1}{2}(|\frac{\partial u}{\partial t}|^2 - |u|^4)$$

$$+\,\frac{1}{2}(|\frac{\partial v}{\partial t}|^2 - |v|^4)$$

$$-\,\epsilon|u|^2\,|v|^2 - \frac{\epsilon}{4}u^2v^{*2}$$

$$\exp(iR\delta z) - \frac{\epsilon}{4}u^{*2}v^2\exp(-iR\delta z), \qquad (8)$$

where $*$ represents complex conjugation. A time—averaged Lagrangian may be defined as

$$<L> = \int L\mathrm{d}t. \qquad (9)$$

We apply a reduced variational principle in which $\epsilon|u|^2\,|v|^2$ is treated as a perturbation and consider the following soliton—form ansatz for the polarization components:

$$u = \frac{2\zeta_1}{\sqrt{1+\epsilon}}\,\text{sech}[2\zeta_1(t-\xi_1)]$$

$$\exp[2i\mu_1(t-\xi_1) + i\nu_1], \qquad (10a)$$

$$v = \frac{2\zeta_2}{\sqrt{1+\epsilon}}\,\text{sech}[2\zeta_2(t-\xi_2)]$$

$$\exp[2i\mu_2(t-\xi_2) + i\nu_2]. \qquad (10b)$$

Inserting the unperturbed solitonic components (10) into equation (8) and using (9) in conjunction with Maple V software, we obtain

$$<L> = -\frac{8\zeta_1\mu_1}{(1+\epsilon)}\frac{\partial \xi_1}{\partial z}$$

$$+\,\frac{4\zeta_1}{(1+\epsilon)}\frac{\partial \nu_1}{\partial z} - \frac{8\zeta_2\mu_2}{(1+\epsilon)}\frac{\partial \xi_2}{\partial z}$$

$$+\,\frac{4\zeta_2}{(1+\epsilon)}\frac{\partial \nu_2}{\partial z} + \frac{8\delta\zeta_1\,\mu_1}{(1+\epsilon)}$$

$$-\,\frac{8\delta\zeta_2\,\mu_2}{(1+\epsilon)} + \frac{8\zeta_1\,\mu_1^2}{(1+\epsilon)}$$

$$+\frac{8}{3}\frac{\zeta_1^3}{(1+\epsilon)}+\frac{8\zeta_2\,\mu_2^2}{(1+\epsilon)}$$

$$+\frac{8}{3}\frac{\zeta_2^3}{(1+\epsilon)}-\frac{16}{3}\frac{\zeta_1^3}{(1+\epsilon)^2}$$

$$-\frac{16}{3}\frac{\zeta_2^3}{(1+\epsilon)^2}-\frac{1}{(1+\epsilon)}$$

$$<L_{uv}^1>-\frac{e^{iR\delta z}}{(1+\epsilon)}$$

$$<L_{uv}^2>-\frac{e^{-iR\delta z}}{(1+\epsilon)}<L_{uv}^3>, \qquad (11)$$

where

$$<L_{uv}^1>=\frac{8\,\epsilon\,\zeta_1\zeta_2^2}{(1+\epsilon)}\int_{-\infty}^{\infty}\operatorname{sech}^2 x_1$$

$$\operatorname{sech}^2 x_2 dx_1,$$

$$<L_{uv}^2>=\frac{2\,\epsilon\,\zeta_1\zeta_2^2}{(1+\epsilon)}\int_{-\infty}^{\infty}\operatorname{sech}^2 x_1\,\operatorname{sech}^2 x_2$$

$$\exp[\frac{2i(\mu_1-\mu_2)}{\zeta_1}x_1-4i\mu_2(\xi_1-\xi_2)$$

$$+2i(\nu_1-\nu_2)]dx_1$$

$$<L_{uv}^3>=\frac{2\,\epsilon\,\zeta_1\zeta_2^2}{(1+\epsilon)}\int_{-\infty}^{\infty}\operatorname{sech}^2 x_1\,\operatorname{sech}^2 x_2$$

$$\exp[\frac{-2i(\mu_1-\mu_2)}{\zeta_1}x_1+4i\mu_2(\xi_1-\xi_2)$$

$$-2i(\nu_1-\nu_2)]dx_1$$

and $x_r = 2\zeta_r(t-\xi_r)$, $r = 1,2$.

Using the Euler–Lagrange equations, a coupled system of eight ODE's for the evolution of u and v soliton parameters ζ_r, ξ_r, μ_r, and ν_r ($r = 1,2$) are obtained.

$$\zeta_r\frac{d\xi_r}{dz}+(-1)^r\,\delta\zeta_r-2\mu_r\,\zeta_r$$

$$+\frac{1}{8}e^{iR\delta z}\frac{\partial}{\partial\mu_r}<L_{uv}^2>$$

$$+\frac{1}{8}e^{-iR\delta z}\frac{\partial}{\partial\mu_r}<L_{uv}^3>=0, \quad (12)$$

$$-8\frac{d}{dz}(\mu_r\zeta_r)+\frac{\partial}{\partial\xi_r}<L_{uv}^1>$$

$$+e^{iR\delta z}\frac{\partial}{\partial\xi_r}<L_{uv}^2>$$

$$+e^{-iR\delta z}\frac{\partial}{\partial\xi_r}<L_{uv}^3>=0, \qquad (13)$$

$$\frac{d\zeta_r}{dz}+\frac{1}{4}e^{iR\delta z}\frac{\partial}{\partial\nu_r}<L_{uv}^2>$$

$$+ \frac{1}{4} e^{-iR\delta z} \frac{\partial}{\partial \nu_r} <L_{uv}^3> = 0, \quad (14)$$

$$\frac{d\nu_r}{dz} = 2\mu_r^2 + 2\zeta_r^2 \left(\frac{1 - \epsilon}{1 + \epsilon} \right)$$

$$- \frac{\mu_r}{4\zeta_r} e^{iR\delta z} \frac{\partial}{\partial \mu_r} <L_{uv}^2>$$

$$- \frac{\mu_r}{4\zeta_r} e^{-iR\delta z} \frac{\partial}{\partial \mu_r} <L_{uv}^3>$$

$$+ \frac{1}{4} \frac{\partial}{\partial \zeta_r} <L_{uv}^1> + \frac{1}{4} e^{iR\delta z} \frac{\partial}{\partial \zeta_r}$$

$$<L_{uv}^2> + \frac{1}{4} e^{-iR\delta z} \frac{\partial}{\partial \zeta_r} <L_{uv}^3>, \quad (15)$$

In equation (12), the birefringence parameter δ appears in a symmetrical way, and this suggests the study of the perturbation—induced dynamics of the soliton solution. These ODE's give a criterion for the two pulses to form a bound state. The nonlinearity of the medium makes it possible to lock the u and v polarization components together above threshold amplitude through their effective interaction. This threshold varies with the birefringence parameter, δ

and is calculated from soliton phenomenology. The formula for the threshold amplitude, A_{thr}, is obtained as a function of birefringence and is given by

$$A_{thr}(\delta) = [2(1 + \epsilon)]^{-1/2}$$

$$+ 0 \cdot 26\epsilon^{-1/2} \delta. \quad (16)$$

Our calculated values of the threshold amplitude by formula (16) was found to depend nonlinearly on birefringence.

In conclusion, we found Maple V to be quite useful, making our lives easier by evaluating most of the integrals in our analytical approach to solving coupled nonlinear Schroedinger equations.

REFERENCES

1. V.E. Zakharov and A.B. Shabat, Sov. Phys. JETP 34, 62(1972).

2. A. Hasegawa and F. Tappert, Appl. Phys. Lett. 23, 142(1973).

3. L.F. Mollenauer, R.H. Stolen and J.P. Gordon, Phys. Rev. Lett. 45, 1095(1980).

4. D. Gloge, Appl. Opt. 10, 2252(1971).

5. D. Anderson, M. Lisak, and T. Reichel,

   Phys. Rev. A $\underline{38}$, 1618(1988).

6. C.R. Menyuk, J. Opt. Soc. Am. B$\underline{5}$,

   392(1988).

7. K.J. Blow, N.J. Doran, and D. Wood,

   Opt. Lett. $\underline{12}$, 202(1986).

8. S. Trillo, S. Wabnitz, R.H. Stolen, G.

   Assanto, C.T. Seaton and G.I. Stegeman,

   Appl. Phys. Lett. $\underline{49}$, 1224(1986).

Dr. Mohammad F. Mahmood received his Ph.D. degree from Howard University, Washington, D.C. in 1988. His current research interests include nonlinear wave propagation in optical fibers, applications to all—optical switching devices and lasers, soliton theory and computer architecture. He is currently a Senior Research Scientist at Computational Science and Engineering Research Center, Howard University, Washington, D.C.

Dr. Tepper L. Gill received his Ph.D. degree from Wayne State University in 1974. His current research interests include simulation modeling of nonlinear systems, relativistic quantum theory, high speed networking and computing, and soliton theory. He is currently Professor of Electrical Engineering & Mathematics and Director of Computational Science and Engineering Research Center, Howard University, Washington, D.C.

SOLUTION OF THE SOLAR CELL TRANSPORT EQUATIONS USING MAPLE

Randy T. Dorn and R. J. Soukup
Department of Electrical Engineering, University of Nebraska-Lincoln, Lincoln, NE

Abstract

The solar cell transport equations and the associated boundary conditions can cause many symbolic algebra computer programs difficulties. However, it is shown in this paper that Maple can be used to find the transport equation solutions for both a dark and an illuminated solar cell. The correct procedures must be used and some precautions taken but once learned, these techniques will be useful in many applications.

Introduction

Symbolic math packages such as Maple are of great interest to the scientist or engineer because of the potential to free them from the drudgery of tedious algebraic manipulations. Often equations can be very long and complicated and it becomes a chore just to keep track of the negative signs. One particular area where Maple has been used to great advantage is the solution of the transport equation for a solar cell.

In its simplest form the transport equation is a second order linear ordinary differential equation with well known general solutions. Maple can easily solve this equation and even apply simple boundary conditions. However, solar cells can have unusual boundary conditions that often cannot be easily applied. In either case, once a solution is found, it may be several pages in length and difficult to put into an intelligible form.

The commands required to solve a differential equation, apply the boundary conditions, and arrange the solution in the simplest form are not always obvious. Often it may seem easier to abandon Maple and find the solution by hand than to spend the time necessary to discover a command sequence that gives the desired results.

This paper will show how to make Maple find the transport equation solutions for both a dark and an illuminated solar cell without resorting to hand calculations. Several common boundary conditions are explored and the precautions necessary to get intelligible answers are explained. The tricks and techniques used to solve these problems should be useful in many other applications.

Solar Cell Model

A solar cell is a pn junction that has been optimized to convert light into electricity. The most general models begin with the carrier concentration continuity equations and these are usually simplified further into the minority carrier diffusion equations. These equations have closed form symbolic solutions and can be considered the starting point in most device analysis.[1]

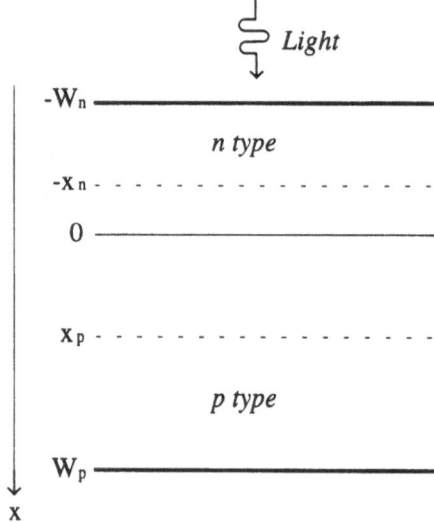

Figure 1. Solar Cell configuration

The configuration is shown in figure 1. Light is incident on the surface at -Wn on a cell that is $W_p + W_n$ thick. The pn junction is at $x = 0$ and the depletion region is $x_p + x_n$ thick. The steady state minority carrier diffusion equations and associated boundary conditions are:

p region $(xp < x < Wp)$

$$\frac{d^2 \Delta n_{p(x)}}{d x^2} - \frac{\Delta n_{p(x)}}{Ln^2} = -\frac{Go\, e^{-\alpha x}}{Dn}$$

$$\Delta n_{p(xp)} = \Delta n_{po}\left(e^{\left(\frac{V}{Vt}\right)} - 1\right)$$

$$\lim_{x \to \infty} \Delta n_{p(x)} = 0$$

n region $(-Wn < x < -xn)$

$$\frac{d^2 \Delta p_{n(x)}}{d x^2} - \frac{\Delta p_{n(x)}}{Lp^2} = -\frac{Go\, e^{-\alpha x}}{Dp}$$

$$\Delta p_{n(-xn)} = \Delta p_{no}\left(e^{\left(\frac{V}{Vt}\right)} - 1\right)$$

$$\frac{d\, \Delta p_{n(x)}}{d x}\bigg|_{x = -Wn} = \frac{Sp}{Dp}\Delta p_{n(-Wn)}$$

Ln is the diffusion length and Dn is the diffusion coefficient for a minority electron and both are assumed constant. Δn_p denotes the number of excess electrons and $Go\, e^{-\alpha x}$ is the generation due to light. The boundary condition at xp is common for device analysis and is often referred to as the law of the junction. For the boundary condition at Wp we will assume that the p side of the solar cell is much thicker than the average electron diffusion length so that Wp is effectively at infinity. Therefore the number of excess electrons at infinity should be zero. This is the long base diode approximation and simplifies the solution.[2]

The minority carrier diffusion equation for the n region is identical to that in the p region except that we deal with minority holes and $\Delta n_p \to \Delta p_n$, $Ln \to Lp$, and $Dn \to Dp$. The boundary condition at -xn is comparable to that at xp. The boundary condition at -Wn relates Δp_n and its derivative by Sp, the surface recombination velocity.[3] Note that the depletion region ($-xn < x < xp$) is not discussed here.

Dark Cell Solution

p region

The equations for a dark cell are identical to those of a regular pn junction. Since there is no light, $G = 0$ and the minority diffusion equation in the p region reduces to:

f1:=diff(dnp(x),x$2)-dnp(x)/Ln^2=0;

$$f1 := \left(\frac{\partial^2}{\partial x^2}\, dnp(x)\right) - \frac{dnp(x)}{Ln^2} = 0$$

Since it is inconvenient to use a "Δ" in Maple, a "d" will be used in its place. The boundary condition at the edge of the depletion region (xp) is given by:

bc1:=dnp(xp)=dnpo*(exp(V/Vt)-1);

$$bc1 := dnp(xp) = dnpo\left(e^{\left(\frac{V}{Vt}\right)} - 1\right)$$

The most reliable way to handle an infinite boundary in Maple is to assume first it is finite and then take the limit after the differential equation is solved.

bc2:=dnp(Wp)=0;

$$bc2 := dnp(Wp) = 0$$

To solve f1 with the listed boundary conditions we use the dsolve command. The simplify command reduces the result to a common denominator and combines the exponentials.

f2:=simplify(dsolve({f1,bc1,bc2},dnp(x)));

$$f2 := dnp(x) =$$

$$\frac{dnpo\left(e^{\left(\frac{V}{Vt}\right)} - 1\right)\left(e^{\left(-\frac{Wp-x}{Ln}\right)} - e^{\left(\frac{Wp-x}{Ln}\right)}\right)}{e^{\left(-\frac{Wp-xp}{Ln}\right)} - e^{\left(\frac{Wp-xp}{Ln}\right)}}$$

Before the limit can be taken we must have a single expression with no = sign. To do this we take the right hand side of f2 and give it the name dnpx.

dnpx:=rhs(f2):

In addition the assume command must be used to inform Maple of the sign of Ln. This is necessary so that Maple can determine if exp(-Wp/Ln) is 0 or undefined. The limit will return unevaluated if assume is not included.

assume(Ln>0):

dnpx:=simplify(limit(dnpx,Wp=infinity));

$$dnpx := dnpo \left(e^{\left(\frac{V}{Vt}\right)} - 1 \right) e^{\left(-\frac{x-xp}{Ln\sim}\right)}$$

Ln~ is Maple's notation that this result was computed using an assumption on Ln. It is now a simple matter to compute the current density due to electrons and evaluate it at x = xp:

Jnx:=q*Dn*diff(dnpx,x):

Jnxp:=simplify(subs(x=xp,Jnx));

$$Jnxp := -\frac{q\,Dn\,dnpo \left(e^{\left(\frac{V}{Vt}\right)} - 1 \right)}{Ln\sim}$$

---

n region

The minority carrier diffusion equation for the n region of a dark cell (G = 0) is:

f3:=diff(dpn(x),x$2)-dpn(x)/Lp^2=0;

$$f3 := \left(\frac{\partial^2}{\partial x^2} dpn(x) \right) - \frac{dpn(x)}{Lp^2} = 0$$

The boundary condition at the edge of the depletion region (-xn) is comparable to bc1. However, there is little sense in inserting such a long constant so soon in the calculations. We can use Cdpn for now and substitute the actual expression {dpno*[exp(V/Vt)-1]} later. Another advantage is that this constant will not be converted into trig form in later manipulations so that the final expression will be simpler.

bc3:= dpn(-xn)=Cdpn;

$$bc3 := dpn(-xn) = Cdpn$$

The boundary condition at -Wn is given by:

bc4:=D(dpn)(-Wn)=Sp/Dp*dpn(-Wn);

$$bc4 := D(dpn)(-Wn) = \frac{Sp\,dpn(-Wn)}{Dp}$$

Maple cannot completely evaluate this boundary condition in one step and additional commands will be necessary.

f4:=dsolve({f3,bc3,bc4},dpn(x));

$$f4 := dpn(x) = \frac{\left(Cdpn\, e^{\left(\frac{Wn}{Lp}\right)} Dp + Sp\,dpn(-Wn)\, e^{\left(\frac{xn}{Lp}\right)} Lp \right) e^{\left(\frac{x}{Lp}\right)}}{\left(e^{\left(-\frac{Wn}{Lp}\right)} e^{\left(\frac{xn}{Lp}\right)} + e^{\left(-\frac{xn}{Lp}\right)} e^{\left(\frac{Wn}{Lp}\right)} \right) Dp}$$
$$+ \frac{\left(-e^{\left(-\frac{xn}{Lp}\right)} Sp\,dpn(-Wn)\, Lp + Cdpn\,Dp\, e^{\left(-\frac{Wn}{Lp}\right)} \right) e^{\left(-\frac{x}{Lp}\right)}}{\left(e^{\left(-\frac{Wn}{Lp}\right)} e^{\left(\frac{xn}{Lp}\right)} + e^{\left(-\frac{xn}{Lp}\right)} e^{\left(\frac{Wn}{Lp}\right)} \right) Dp}$$

Note that dpn(x) contains a dpn(-Wn). We can evaluate dpn(-Wn) by substituting -Wn for x and solving for dpn(-Wn). If dpn(-Wn) is then defined to be that expression, further references to f4 will include the evaluated dpn(-Wn).

dpn(-Wn):=solve(subs(x=-Wn,f4),dpn(-Wn)):

In its present form f4 is a very long and complicated expression and will not be shown here. To simplify f4 we use convert to put it into trig form and then we can use combine to gather the trig terms. It is also necessary to use simplify after each of these commands to get the simplest expression.

f4a:=simplify(convert(f4,trig)):

f4b:=simplify(combine(f4a,trig)):

dpnx:=rhs(f4b);

$$dpnx := \frac{Cdpn\left(Dp\,\cosh\!\left(\dfrac{Wn+x}{Lp}\right) + Sp\,Lp\,\sinh\!\left(\dfrac{Wn+x}{Lp}\right)\right)}{Dp\,\cosh\!\left(\dfrac{Wn-xn}{Lp}\right) + Sp\,Lp\,\sinh\!\left(\dfrac{Wn-xn}{Lp}\right)}$$

If we had not taken the precaution of substituting Cdpn for dpno*[exp(V/Vt)-1] Maple would have also converted it to trig form and created a very long expression.

Now we can compute the current density due to holes and evaluate it at -xn.

Jpx:=-q*Dp*diff(dpnx,x):

Jpxn:=simplify(subs(x=-xn,Jpx));

$$Jpxn := -\frac{q\,Dp\,Cdpn\left(Dp\,\sinh\!\left(\dfrac{Wn-xn}{Lp}\right) + Sp\,\cosh\!\left(\dfrac{Wn-xn}{Lp}\right)Lp\right)}{Lp\left(Dp\,\cosh\!\left(\dfrac{Wn-xn}{Lp}\right) + Sp\,Lp\,\sinh\!\left(\dfrac{Wn-xn}{Lp}\right)\right)}$$

Now it is a simple matter to substitute the mathematical expression for Cdpn.

Jpxn:=subs(Cdpn=dpno*(exp(V/Vt)-1),Jpxn);

$$Jpxn := -\frac{q\,Dp\,dpno\left(e^{\left(\frac{V}{Vt}\right)}-1\right)\left(Dp\,\sinh\!\left(\dfrac{Wn-xn}{Lp}\right) + Sp\,\cosh\!\left(\dfrac{Wn-xn}{Lp}\right)Lp\right)}{Lp\left(Dp\,\cosh\!\left(\dfrac{Wn-xn}{Lp}\right) + Sp\,Lp\,\sinh\!\left(\dfrac{Wn-xn}{Lp}\right)\right)}$$

Jp(-xn) is more complicated than Jn(xp) because of the different boundary conditions. However, it is easy to show that the expressions are comparable if we make the comparable assumption that Wn is essentially at infinity.

assume(Lp>0):

limit(Jpxn,Wn=infinity);

$$-\frac{dpno\,Dp\,q\left(e^{\left(\frac{V}{Vt}\right)}-1\right)}{Lp\!\sim}$$

This simpler form of Jp(-xn) added to Jn(xp) is known as the ideal diode equation.[2]

Illuminated Cell Solution

The equations for an illuminated cell include a generation term which makes the solutions more complicated. Maple can still find the solution using the same procedure as for the dark cell but it will be very long and difficult to comprehend. This does not mean that Maple is no longer useful, only that some experience and intuition are necessary.

p region

The minority diffusion equation in the p region for the illuminated cell is:

f5:=diff(dnp(x),x$2)-dnp(x)/Ln^2= -Go*exp(-a*x)/Dp;

$$f5 := \left(\frac{\partial^2}{\partial x^2} dnp(x) \right) - \frac{dnp(x)}{Ln^2} = -\frac{Go\, e^{(-a\,x)}}{Dp}$$

The boundary conditions are the same as for the dark cell and will not be repeated here.

From our experience with the dark cell we can guess that the solution is of the form:

dnpx:=C1*sinh((x-xp)/Ln)+C2*cosh((x-xp)/Ln)+ C3*exp(-a*x);

$$dnpx :=$$

$$C1\, \sinh\left(\frac{x - xp}{Ln} \right) + C2\, \cosh\left(\frac{x - xp}{Ln} \right) + C3\, e^{(-a\,x)}$$

If this is actually a solution it should solve the original differential equation.

simplify(subs(dnp(x)=dnpx,f5));

$$\frac{C3\, e^{(-a\,x)} \left(a^2 Ln^2 - 1 \right)}{Ln^2} = -\frac{Go\, e^{(-a\,x)}}{Dp}$$

It is a good idea to give the constants a different name at this time. This will keep the other constants a manageable size. dnpx is a solution of f5 if C3 = C3a is:

C3a:=simplify(solve(",C3));

$$C3a := -\frac{Go\, Ln^2}{Dp \left(a^2 Ln^2 - 1 \right)}$$

The first boundary condition is evaluated as follows:

simplify(subs(x=xp,dnpx)=Cdnp);

$$C2 + C3\, e^{(-a\,xp)} = Cdnp$$

C2a:=solve(",C2);

$$C2a := -C3\, e^{(-a\,xp)} + Cdnp$$

For the second boundary condition:

simplify(subs(x=Wp,dnpx)=0);

$$C1\, \sinh\left(\frac{Wp - xp}{Ln} \right) + C2\, \cosh\left(\frac{Wp - xp}{Ln} \right)$$
$$+ C3\, e^{(-a\,Wp)} = 0$$

C1a:=simplify(solve(",C1));

$$C1a := -\frac{C2\, \cosh\left(\dfrac{Wp - xp}{Ln} \right) + C3\, e^{(-a\,Wp)}}{\sinh\left(\dfrac{Wp - xp}{Ln} \right)}$$

C1a is the only term with a Wp. Since we are interested in the solution when Wp is infinite it is enough to evaluate C1a at Wp = infinity. This time it is necessary to make an assumption about both Ln and a.

assume(Ln>0): assume(a>0):

limit(C1a,Wp=infinity);

$$-C2$$

Clearly C1 = -C2 = -C2a, C2 = C2a, and C3 = C3a. Although dnpx is long it is still understandable because it is in separate parts. If we had let Maple find the solution using dsolve it would not be so clear.

The current density due to electrons evaluated at x = xp is:

Jnx:=q*Dn*diff(dnpx,x):

94

a:='a': Ln:='Ln':

Jnxp:=expand(subs(x=xp,Jnx));

$$Jnxp := \frac{q\,Dn\,C1}{Ln} - \frac{q\,Dn\,C3\,a}{e^{(a\,xp)}}$$

We can substitute for the constants, but the result is an expression that simplify is unable to put in the shortest form. Instead of using simplify, collect is used to gather Go, and then the op command is used to split the expression into separate parts. These parts are simplified individually and then the equation is reassembled.

C1:=-C2a: C3:=C3a: Cdnp:=dnpo*(exp(V/Vt)-1):

J1:=collect(Jnxp,Go):

Jnxp:=simplify(op(1,J1))+J1-op(1,J1);

$$Jnxp :=$$
$$\frac{Go\,q\,Dn\,Ln\,e^{(-a\,xp)}}{(Ln\,a+1)\,Dp} - \frac{q\,Dn\,dnpo\left(e^{\left(\frac{V}{Vt}\right)}-1\right)}{Ln}$$

n region

The minority carrier diffusion equation for the n region of an illuminated cell is:

f6:=diff(dpn(x),x$2)-dpn(x)/Lp^2=
-Go*exp(-a*x)/Dp;

$$f6 := \left(\frac{\partial^2}{\partial x^2}\,dpn(x)\right) - \frac{dpn(x)}{Lp^2} = -\frac{Go\,e^{(-a\,x)}}{Dp}$$

The boundary conditions are the same as for the dark cell. We can guess that the solution is of the form:

dpnx:=C4*sinh((x+xn)/Lp)+C5*cosh((x+xn)/Lp)+
C6*exp(-a*x);

$$dpnx :=$$
$$C4\,\sinh\left(\frac{x+xn}{Lp}\right) + C5\,\cosh\left(\frac{x+xn}{Lp}\right) + C6\,e^{(-a\,x)}$$

The procedure is the same as for the p region so not as many details will be shown.

If this is actually a solution it should solve the original differential equation.

subs(dpn(x)=dpnx,f6):

C6a:=simplify(solve(",C6));

$$C6a := -\frac{Go\,Lp^2}{Dp\left(a^2\,Lp^2-1\right)}$$

The first boundary condition is evaluated as follows:

subs(x=-xn,dpnx=Cdpn): C5a:=solve(",C5);

$$C5a := -C6\,e^{(a\,xn)} + Cdpn$$

For the second boundary condition:

subs(x=-Wn,diff(dpnx,x))=
Sp/Dp*subs(x=-Wn,dpnx):

C4a:=simplify(solve(",C4));

$$C4a :=$$
$$\frac{C5\,\sinh\left(\frac{Wn-xn}{Lp}\right)Dp + C6\,a\,e^{(a\,Wn)}Lp\,Dp}{\cosh\left(\frac{Wn-xn}{Lp}\right)Dp + Sp\,\sinh\left(\frac{Wn-xn}{Lp}\right)Lp}$$
$$+\frac{Sp\,C5\,\cosh\left(\frac{Wn-xn}{Lp}\right)Lp + Sp\,C6\,e^{(a\,Wn)}Lp}{\cosh\left(\frac{Wn-xn}{Lp}\right)Dp + Sp\,\sinh\left(\frac{Wn-xn}{Lp}\right)Lp}$$

Often Sp is not known and we are interested in the solution when Sp = 0 and Sp = infinity.

C4Sp0:=limit(C4a,Sp=0);

$$C4Sp0 := \frac{C5\,\sinh\left(\frac{Wn-xn}{Lp}\right) + C6\,a\,e^{(a\,Wn)}Lp}{\cosh\left(\frac{Wn-xn}{Lp}\right)}$$

C4Spi:=limit(C4a,Sp=infinity);

$$C4Spi := \frac{C5\,\cosh\left(\frac{Wn-xn}{Lp}\right) + C6\,e^{(a\,Wn)}}{\sinh\left(\frac{Wn-xn}{Lp}\right)}$$

The current density due to holes evaluated at x = -xn is:

Jpx:=-q*Dp*diff(dpnx,x):

Jpxn:=expand(subs(x=-xn,Jpx));

$$Jpxn := -\frac{q\,Dp\,C4}{Lp} + q\,Dp\,C6\,a\,e^{(a\,xn)}$$

For the special case of Sp = 0, Jp(-xn) can be found as follows:

C4:=C4Sp0:   C5:=C5a: Jpa:=collect(Jpxn,C6):

C6:=C6a: Cdpn:=dpno*(exp(V/Vt)-1):

`Jp(-xn,Sp=0)`=Jpa;

The result will not be displayed here because of the length but it is in an intelligible form because of the precautions we have taken. A similar procedure can be used to find Jp when Sp is still unevaluated or the other special case of Sp = infinity.

Conclusion

Two differential equations were solved with three different boundary conditions. We used Maple's "dsolve" to find the solution to the homogenous differential equations f1 and f3. However, additional commands were required to complete the application of the boundary conditions. We were able to insert the mathematical form of the boundary condition for bc1 at the very beginning because of the simplicity of bc2. If we had tried to do the same with bc3 Maple would have converted the exponential term into trig form and the solution would have become very complicated. Since we took special precautions the final solution contained both exponential and trigonometric functions.

Maple's dsolve could also have been used to find a solution for the nonhomogeneous differential equations f5 and f6. However, dsolve typically uses exponentials exclusively and the solution would not have contained the mixed trig and exponential forms that we desire. This would cause the solution to be very long and complicated. By making an appropriate guess of the solution and then solving for the constants we were able to keep the mathematics manageable.

Very few commands were needed to find these solutions. However, many commands are useless unless they are combined with others. A good example of this is the necessity of using "assume" before taking a limit. Other commands that work together are convert and combine. In turn these commands should be followed with simplify, expand or collect in some sequence. Often it is necessary to try several possibilities before the simplest form is reached.

References

1. Pierret, Robert F., *Semiconductor Fundamentals 2nd edition, Addison-Wesley Publishing Co.* 1988 88-92

2. Neudeck, Gerold W., *The PN Junction Diode 2nd edition, Addison-Wesley Publishing Co.* 1988 54-65

3. Fonash, Stephen J., *Solar Cell Device Physics, Academic Press* 1981 154-166

Biographies

Randy T. Dorn received his B.S. in Electrical Engineering from the University of Nebraska-Lincoln in 1992. He is currently a graduate student at the same institution. Interests include solar cells, electronic devices and computer assisted symbolic computations. Email: rmrtd@engvms.unl.edu

Dr. R. J. Soukup received his B.S., M.S., and Ph.D. degrees all in Electrical Engineering, all from the University of Minnesota. He worked for Sperry Rand/Univac (now Unisys) for three years before joining the faculty of the University of Iowa as an Assistant Professor in 1972. In 1976 he moved to the University of Nebraska as an Associate Professor. He became Chairman of the Department of Electrical Engineering there in 1978, a position he still holds, and was promoted to Professor in 1980. He is a Fellow of the IEEE, a member of ASEE, a member of the AVS, a member of and a past President of NEEDHA and President of CSEEDHA. His research is in solar cells. Email: eerdrjs@engvms.unl.edu

KRAMERS AND WANNIER V-MATRICES FOR THE PARTITION FUNCTIONS OF THE ISING MODEL

P. Frempong-Mireku and K.J.K. Moriarty
Department of Mathematics, Statistics and Computing Science,
Dalhousie University, Halifax, Nova Scotia, Canada

Abstract

In this paper we show how to use maple to construct the two dimensional (2-d) and the three dimensional (3-d) Kramers and Wannier V-matrices which are used to compute the partition function (PF) of the Ising model by the perturbation method. The Maple functions which we have built to deal with these problems are *d2cvmat*, *d2lnvmat*, *d2upvmat*, *d2stmat*, *d3cvmat*, *d3lnvmat*, and *d3upvmat*.

1 Introduction

We report on symbolic construction of the V-matrices of Kramers and Wannier [1] in both 2-d and 3-d. These matrices are used in statistical physics to compute the PF of the Ising model by the perturbation method. In 1925 Ising carried out an exact calculation of the PF in a one-dimensional lattice. His work showed that there was no phase transition to a ferromagnetic ordered state at any temperature. He also failed to predict phase transition in 2-d and 3-d as well. However, later works by Onsager [2], Montroll [3], Kramers and Wannier and others have revealed phase transition in the 2-d and 3-d models. In this paper, we follow the steps of Kramers and Wannier, Askin and Lamb [5] and Oguchi [6] to create the V-matrices that are used to compute the PF of the Ising model on square and cubic lattice. Each of the two V-matrices has the advantage of reducing the size of the original matrix by half, yet it produces a PF equal to the original matrix. Another advantage of the matrix approach is that it provides a solution to the long range order and also the propagation of order throughout the lattice. The theoretical background is presented in the next section.

2   2-D Kramers and Wannier V-matrix.

The screw-like construction introduced by Kramers and Wannier [1] is adopted here in considering the theory of a 2-d Ising model on a square lattice. Assume that the lattice sites are regularly distributed along a continuous line, twisting its way in a screw-wise fashion over the surface of a torus. Suppose that it consists of m pitches of n spins. Let the configuration on each pitch be denoted by $(\sigma_1, \sigma_2, ..., \sigma_n)$. The coordinates $\sigma_1, \sigma_2, ..., \sigma_{mn}$ take values $+1$ or -1. Since the interaction energies between neighboring pairs take the form $V(++) = V(--) = -\frac{1}{2}J$ and $V(+-) = \frac{1}{2}J$, the total energy can be written as

$$E = -\frac{1}{2}J \sum_{<i,j>} \sigma_i \sigma_j.$$

The sum is over all pairs (i, j) which are nearest neighbors and J is the coupling constant. Let n be the number of spins that make up one pitch of the screw. By adding the nth spin, one notices that the nth spin interacts with the 0th and the $(n-1)$th spins only, because we are considering only nearest neighboring interaction. Thus the two interaction energies are as follows:

$$-K\sigma_n\sigma_{n-1}, \quad -K\sigma_n\sigma_0,$$

where $K = J/2kT$ and k is Boltzmann's constant. Let the probability of the arrangement $\sigma_{n-1}, \sigma_{n-2}, ..., \sigma_0$ be

$$A(\sigma_{n-1}, \sigma_{n-2}, ..., \sigma_0),$$

and the one including σ_n at the next position be given by

$$P(\sigma_n, \sigma_{n-1}, ..., \sigma_0).$$

Now according to Boltzmann's theorem, the probability for any particular arrangement is proportional to the $\exp(-E/kT)$, where E is the total energy of the system. If this theorem is applied, one may obtain

$$\lambda P(\sigma_n, ..., \sigma_0) = A(\sigma_{n-1}, ..., \sigma_0) \exp[K\sigma_n$$
$$(\sigma_{n-1} + \sigma_0)]. \qquad (1)$$

Summing $P(\sigma_n, ..., \sigma_0)$ over σ_0 produces

$$\sum_{\sigma_0} P(\sigma_n, ..., \sigma_0) = \lambda A(\sigma_n, ..., \sigma_1).$$

The resulting situation is identical to the one described by $A(\sigma_{n-1}, \sigma_{n-2}, ..., \sigma_0)$, provided the screw is very long. The difference being that σ_1 takes the place of σ_0 and σ_2 of σ_1 and so forth. Hence one gets

$$\sum_{\sigma_0} exp[K\sigma_n(\sigma_{n-1} + \sigma_0]A(\sigma_{n-1}, ..., \sigma_0) =$$
$$\lambda A(\sigma_n, ..., \sigma_1), \qquad (2)$$

where λ is the PF per individual spin regardless of the order of the matrix in a zero magnetic field. For the proof of the latter statement see [1]. The equation (2) can be written compactly as

$$\mathcal{M}\mathbf{A}_q = \lambda_q \mathbf{A}_q.$$

The matrix $\mathcal{M}$ is of dimension $2^n \times 2^n$ and is not symmetrical. Thus it is necessary to write the left-handed matrix equation as

$$\mathbf{B}_p\mathcal{M} = \lambda_p \mathbf{B}_p.$$

The matrix $\mathcal{M}$ can be written as a square array by writing the 2^n configurations $\alpha = (\sigma_n...\sigma_1)$ in some definite order. For example, for any configuration $+ + + - - + + +$, replace $+$ by 0 and $-$ by 1. This example gives 00011000 as a base two number. Thus the example just given has an order of 24. Also it is convenient to separate the configurations into two classes. Those with $\sigma_n = 1$ are the ones with orders numbers $0, 1, ..., 2^{n-1} - 1$, and those with $\sigma_n = -1$ have the order numbers $2^n - 1, 2^n - 2, ..., 2^{n-1}$. The reason for this arrangement is that if α and $\bar{\alpha}$ belong to corresponding places in the two classes then their order numbers add up to 2^{n-1} and are conjugate to each other in the sense that by

reversing all the signs of one, the other can be obtained. Define the parameters

$$K = \frac{J}{2k}, \quad \alpha = e^{2K}, \quad \beta = e^{-2K}.$$

Now we are in the position to construct our program to compute the matrix $\mathcal{M}$. The name is taken from 2-d complete V-matrix abbreviated *d2cvmat*. The matrix $\mathcal{M}$ is therefore constructed thus:

```
d2cvmat:= proc(n) local c1,c2, a, b,
                u1, u2,f0,f2,n1 u0;
if nargs=1 then
   a:=exp(2*K): b:=exp(-2*K):
else
    a:=args[2]; b:=args[3];
fi;
 n1:=2^n/2;
 u0:=matrix(n1,n1,[]):
   for i to n1 do
     for j to n1 do
       if j=2*i-1 then u0[i,j]:=a
       elif j=2*i then u0[i,j]:=1
       else            u0[i,j]:=0
       fi;
     od;
  od;
u2:=matrix(n1,n1,[]):
 for i to n1 do
   for j to n1 do
     if j=2*(n1-i)+1 then u2[i,j]:=b
     elif j=2*(n1-i)+2 then u2[i,j]:=1
       else            u2[i,j]:=0
     fi;
   od;
od;
cvm:=augment(stack(u0,u2),
stack(u2,u0));
  end;
```

Testing our program for $n = 2$, the matrix $\mathcal{M}$ has the form

```
>M:=d2cvmat(2);
```

$$M := \begin{bmatrix} e^{2K} & 1 & 0 & 0 \\ 0 & 0 & e^{-2K} & 1 \\ 0 & 0 & e^{2K} & 1 \\ e^{-2K} & 1 & 0 & 0 \end{bmatrix}$$

Overriding the default arguments defined above with w, t; one obtains

```
> M1:=d2cvmat(3,w,t);
```

$$M1 := \begin{bmatrix} w & 1 & 0 & 0 & 0 & 0 & 0 & 0 \\ 0 & 0 & w & 1 & 0 & 0 & 0 & 0 \\ 0 & 0 & 0 & 0 & 0 & 0 & t & 1 \\ 0 & 0 & 0 & 0 & t & 1 & 0 & 0 \\ 0 & 0 & 0 & 0 & w & 1 & 0 & 0 \\ 0 & 0 & 0 & 0 & 0 & 0 & w & 1 \\ 0 & 0 & t & 1 & 0 & 0 & 0 & 0 \\ t & 1 & 0 & 0 & 0 & 0 & 0 & 0 \end{bmatrix}$$

The V-matrix is the one used in the PF computation not the matrix $\mathcal{M}$. One can build the V-matrix directly or apply a transformation to accomplish it. The latter case is adopted. The transformation is defined as

$$\mathbf{H} = \begin{vmatrix} \mathbf{I} & \mathbf{I} \\ \mathbf{I} & -\mathbf{I} \end{vmatrix}, \qquad (3)$$

where $\mathbf{I}$ is the identity matrix of order 2^{n-1}. It is given by the program below.

```
d2stmat:=proc(n) local H1,H2,H,n1;
n1:=2^n; #the order of the matrix
H1:=array(identity,1..n1/2,1..n1/2);
H2:=evalm(-array(identity,1..n1/2,
1..n1/2));
H:=copyinto(H2,stack(augment(H1,H1),
augment(H1,H1)),n1/2+1,n1/2+1);
end:
```

For $n = 3$, the transformation looks like this:

```
> H:=d2stmat(3);
```

$$H := \begin{bmatrix} 1 & 0 & 0 & 0 & 1 & 0 & 0 & 0 \\ 0 & 1 & 0 & 0 & 0 & 1 & 0 & 0 \\ 0 & 0 & 1 & 0 & 0 & 0 & 1 & 0 \\ 0 & 0 & 0 & 1 & 0 & 0 & 0 & 1 \\ 1 & 0 & 0 & 0 & -1 & 0 & 0 & 0 \\ 0 & 1 & 0 & 0 & 0 & -1 & 0 & 0 \\ 0 & 0 & 1 & 0 & 0 & 0 & -1 & 0 \\ 0 & 0 & 0 & 1 & 0 & 0 & 0 & -1 \end{bmatrix}$$

Then performing the following operations the two matrices in the upper left and in the lower right are defined to be the V-matrices in the two classes, $\sigma_n = 1$ and $\sigma_n = -1$.

```
> V:=evalm(d2stmat(3)&*d2cvmat(3,r,s)
&*inverse(d2stmat(3)));
```

$$V := \begin{bmatrix} r & 1 & 0 & 0 & 0 & 0 & 0 & 0 \\ 0 & 0 & r & 1 & 0 & 0 & 0 & 0 \\ 0 & 0 & s & 1 & 0 & 0 & 0 & 0 \\ s & 1 & 0 & 0 & 0 & 0 & 0 & 0 \\ 0 & 0 & 0 & 0 & r & 1 & 0 & 0 \\ 0 & 0 & 0 & 0 & 0 & 0 & r & 1 \\ 0 & 0 & 0 & 0 & 0 & 0 & -s & -1 \\ 0 & 0 & 0 & 0 & -s & -1 & 0 & 0 \end{bmatrix}$$

A clearer picture of a typical 2-d V-matrix in the class $\sigma_n = 1$ has the following form for $n = 4$.

```
> uV:=d2upvmat(4,s,t);
```

$$uV := \begin{bmatrix} s & 1 & 0 & 0 & 0 & 0 & 0 & 0 \\ 0 & 0 & s & 1 & 0 & 0 & 0 & 0 \\ 0 & 0 & 0 & 0 & s & 1 & 0 & 0 \\ 0 & 0 & 0 & 0 & 0 & 0 & s & 1 \\ 0 & 0 & 0 & 0 & 0 & 0 & t & 1 \\ 0 & 0 & 0 & 0 & t & 1 & 0 & 0 \\ 0 & 0 & t & 1 & 0 & 0 & 0 & 0 \\ t & 1 & 0 & 0 & 0 & 0 & 0 & 0 \end{bmatrix}$$

Three constants are extracted from the V-matrix to compute the PF. For $n = 3$, these constants are:

```
> u0:=evalm(U_0);
```

$$u0 := \begin{bmatrix} 1 & 0 & 0 & 0 \\ 0 & 0 & 1 & 0 \\ 0 & 0 & 0 & 0 \\ 0 & 0 & 0 & 0 \end{bmatrix}$$

```
> u1:=evalm(U_1);
```

$$u1 := \begin{bmatrix} 0 & 1 & 0 & 0 \\ 0 & 0 & 0 & 1 \\ 0 & 0 & 0 & 1 \\ 0 & 1 & 0 & 0 \end{bmatrix}$$

```
> u2:=evalm(U_2);
```

$$u2 := \begin{bmatrix} 0 & 0 & 0 & 0 \\ 0 & 0 & 0 & 0 \\ 0 & 0 & 1 & 0 \\ 1 & 0 & 0 & 0 \end{bmatrix}$$

The next section discusses the 3-d computation of the V-matrix.

3 3-D Kramers and Wannier V-matrix.

The 3-d V-matrix is much more complex than the 2-d V-matrix. To build the 3-d ferromagnet, one may have to follow the advice of Oguchi [2]. He showed that the same approach of the screw method can be applied to evaluate the PF of the Ising model on a cubic lattice. This approach adds spins one by one. For the simple cubic lattice, it can be divided into many layers. Then starting from an arbitrary first position, add a spin beyond the one just placed previously. This construction is continued until a full line is arranged. The next line is then arranged in the same sequence until the whole arrangement is completed. Then moving to the first position in the next layer the same process is repeated. Consider the mth layer, if the kth spin is added next, then since the interactions are restricted to only nearest neighbors, only the $(k-1)$th, the jth placed immediately beside it in a preceding line and $0th$ spin placed just under the kth spin in the $(m-1)$th layer interact with it. Assume that each spin has two orientations $\sigma_i = \pm 1$. The interaction energies are $\mp \frac{1}{2} J$ for parallel and antiparallel spins respectively. Thus the three interactions stated above are:

$$-K\sigma_k\sigma_{k-1}, \quad -K\sigma_k\sigma_j, \quad -K\sigma_k\sigma_0.$$

Again K is defined as $K = J/2kT$. Let the arrangement $\sigma_{k-1}, ..., \sigma_{j-1}, ..., \sigma_0$ have the probability $A(\sigma_{k-1}, ..., \sigma_{j-1}, ..., \sigma_0)$. The probability $A(\sigma_k, ..., \sigma_j, ..., \sigma_1)$ is the one in which σ_1 occupies the place of σ_0, σ_2 takes the place of σ_1 etc. Then by Boltzmann's theorem, the probability of any particular arrangement of spins is proportional to $exp(-E/kT)$ because every arrangement has weight 1. Hence one obtains

$$\sum_{\sigma_0=\pm 1} exp[K\sigma_k(\sigma_{k-1} + \sigma_j + \sigma_0)]A(\sigma_{k-1},$$
$$..., \sigma_{j-1}, ..., \sigma_0) = \lambda A(\sigma_k, ..., \sigma_j, ..., \sigma_1). \quad (4)$$

The λ in the above equation is the result of Boltzmann's exponential being proportional to probabilities. It becomes the eigenvalue of the equation. The construction of the matrix $\mathcal{M}$ in 3-d is different from that of 2-d. The 3-d matrix gives block matrices instead of the single elements in the 2-d. The 3-d V-matrix is constructed using the equation (4). The order follows the same pattern as shown for the 2-d. There are two parts of the matrix $\mathcal{M}$, as noted in the 2-d case. The Maple construction of the upper V-matrix in 3-d is shown below. The construction of the lower V-matrix in 3-d is done similarly.

```
d3upvmat:= proc(n) local P,Q,a,b,R,
                        S,j1,nb,k,V;
if nargs=1 then
   a:=exp(K): b:=exp(-K):
 else
   a:=args[2]; b:=args[3];
fi;
 n1:=2^n/2;  k:=round((n-3)/2);
   if n<3 then
    ERROR('n must be 3 or greater');
   fi;
   if n >=3  then
        j1:=n-2-k;
        nb:=2^k;
 fi;
#compute the needed matrices.
P:=matrix(2^(j1-1),2^j1,[]):
#the constant matrix U0
   for i to 2^(j1-1) do
     for j to 2^j1 do
      if j=2*i-1 then P[i,j]:=a^3:
      elif j=2*i then   P[i,j]:=a:
      else              P[i,j]:=0:
     fi:
    od:
  od:
Q:=matrix(2^(j1-1),2^j1,[]):
 for i to 2^(j1-1) do
   for j to 2^j1 do
     if  j=2*i-1 then Q[i,j]:=a:
     elif j=2*i then   Q[i,j]:=b:
     else  Q[i,j]:=0:
     fi:
   od:
 od:
R:=matrix(2^(j1-1),2^j1,[]):
 for i to 2^(j1-1) do
  for j to 2^j1 do
   if j=2^j1-1-2*(i-1) then R[i,j]:=b^3:
   elif  j=2^j1-2*(i-1) then R[i,j]:=b:
   else                R[i,j]:=0:
   fi:
   od:
od:
S:=matrix(2^(j1-1),2^j1,[]):
 for i to 2^(j1-1) do
   for j to 2^j1 do
     if j=2^j1-1-2*(i-1) then
              S[i,j]:=b:
       elif j=2^j1-2*(i-1) then
                S[i,j]:=a:
       else
              S[i,j]:=0:
     fi:
   od:
 od:
V:=matrix(n1,n1,0);
for i from 1 to nb do
V:=copyinto(P,V,(2^(j1-1))*2*(i-1)+
            1, 2^j1*2*(i-1)+1);
```

```
V:=copyinto(Q,V,2^(j1-1)*(2*i-1)+1,
            2^j1*(2*i-1)+1);
V:=copyinto(R,V,n1/2+2^(j1-1)*2*(i-1)+
            1,n1-2^j1*(2*i-1)+1);
V:=copyinto(S,V,n1/2+2^(j1-1)*(2*i-1)+
            1,n1-2^j1*2*i+1);
od;

end;
```

Set $\alpha = e^K$ and $\eta = e^{-K}$; these are used as default parameters for the 3-d V-matrices. Again as in 2-d case, one can override these elements by substituting any suitable elements. For convenience we have used the symbols a and b. Let P, Q, R, S, be the block matrices in the example below each of dimension 1×2. Then a typical member of the upper V-matrix in 3-d is shown below as

```
>d3upvmat(4,a,b);
```

$$\begin{bmatrix} a^3 & a & 0 & 0 & 0 & 0 & 0 & 0 \\ 0 & 0 & a & b & 0 & 0 & 0 & 0 \\ 0 & 0 & 0 & 0 & a^3 & a & 0 & 0 \\ 0 & 0 & 0 & 0 & 0 & 0 & a & b \\ 0 & 0 & 0 & 0 & 0 & 0 & b^3 & b \\ 0 & 0 & 0 & 0 & b & a & 0 & 0 \\ 0 & 0 & b^3 & b & 0 & 0 & 0 & 0 \\ b & a & 0 & 0 & 0 & 0 & 0 & 0 \end{bmatrix}$$

A typical member of the lower 3-d V-matrix is also shown below as

```
>d3lnvmat(4,a,b);
```

$$\begin{bmatrix} a^3 & a & 0 & 0 & 0 & 0 & 0 & 0 \\ 0 & 0 & a & b & 0 & 0 & 0 & 0 \\ 0 & 0 & 0 & 0 & a^3 & a & 0 & 0 \\ 0 & 0 & 0 & 0 & 0 & 0 & a & b \\ 0 & 0 & 0 & 0 & 0 & 0 & -b^3 & -b \\ 0 & 0 & 0 & 0 & -b & -a & 0 & 0 \\ 0 & 0 & -b^3 & -b & 0 & 0 & 0 & 0 \\ -b & -a & 0 & 0 & 0 & 0 & 0 & 0 \end{bmatrix}$$

As in the 2-d case, the constant matrices needed for one to compute the PF can be easily extracted.

As an application in this paper, we demonstrate that an order-disorder transition takes place between states of finite long range order and those with no long range [5]. Due to space limits matrices of small size are used.

The proof, using matrix theory, consists of proving the fact that for sufficiently high temperature, the maximum characteristic value of the matrix $\mathcal{M}(K)$ no longer degenerates. For sufficiently high temperature, we have

$$\lim_{T \to \infty} \frac{J}{2kT} \to 0.$$

Hence $K = J/2kT \to 0$. so

$$e^K = \alpha = 1 = \beta = e^{-K}.$$

Now using our function *d2upvmat* and for $n = 3$, one can generate the matrix

```
> V+(0) := d2upvmat(3,1,1);
```

$$V + (0) := \begin{bmatrix} 1 & 1 & 0 & 0 \\ 0 & 0 & 1 & 1 \\ 0 & 0 & 1 & 1 \\ 1 & 1 & 0 & 0 \end{bmatrix}$$

In the same way using our function *d2lnvmat*, and for $n = 3$, the matrix

```
> V-(0):=d2lnvmat(3,1,1);
```

$$V - (0) := \begin{bmatrix} 1 & 1 & 0 & 0 \\ 0 & 0 & 1 & 1 \\ 0 & 0 & -1 & -1 \\ -1 & -1 & 0 & 0 \end{bmatrix}$$

can be generated. Here, we are dealing with a finite problem for which the matrices are of order 2^r where r is a positive integer. Thus raising the matrices $V + (0)$ and $V - (0)$ to the second power one obtains

```
> evalm(V+(0)^2);
```

$$(V + (0))^2 := \begin{bmatrix} 1 & 1 & 1 & 1 \\ 1 & 1 & 1 & 1 \\ 1 & 1 & 1 & 1 \\ 1 & 1 & 1 & 1 \end{bmatrix}$$

and

```
> evalm(V-(0)^2);
```

$$V - (0)^2 := \begin{bmatrix} 1 & 1 & 1 & 1 \\ -1 & -1 & -1 & -1 \\ 1 & 1 & 1 & 1 \\ -1 & -1 & -1 & -1 \end{bmatrix}$$

This shows that the matrices are of order 2^2. Computing the characteristic values of $V + (0)^2$ with the help of Maple's eigenvals function one gets

```
   for j to n1 do
     if j=2*i-1 then u0[i,j]:=a
       elif j=2*i then u0[i,j]:=1
       else          u0[i,j]:=0
     fi;
    od;
 od;
  u2:=matrix(n1,n1,[]):
 for i to n1 do
    for j to n1 do
     if j=2*(n1-i)+1 then
              u2[i,j]:=-b
         elif j=2*(n1-i)+2 then
              u2[i,j]:=-1
    else      u2[i,j]:=0
   fi;
   od;
 od;
     f0:=delrows(u0,n1/2+1..n1);
     f2:=delrows(u2,1..n1/2);
       stack(f0,f2);
     end;
#Program computes the the upper
#V matrix of Kramers and Wannier.
d2upvmat:= proc(n) local c1,c2,
#a,b,f0,f2,u0,u1,u2,n1;
if nargs=1 then
    a:=exp(2*K): b:=exp(-2*K):
   else
     a:=args[2]; b:=args[3];
fi;
  n1:=2^n/2;
u0:=matrix(n1,n1,[]):
 for i to n1 do
   for j to n1 do
    if j=2*i-1 then u0[i,j]:=a
    elif j=2*i then u0[i,j]:=1
    else          u0[i,j]:=0
   fi;
  od;
 od;
u2:=matrix(n1,n1,[]):
for i to n1 do
 for j to n1 do
  if j=2*(n1-i)+1 then
           u2[i,j]:=b;
   elif j=2*(n1-i)+2 then
            u2[i,j]:=1;
  else     u2[i,j]:=0;
 fi;
od;od;
    f0:=delrows(u0,n1/2+1..n1);
    f2:=delrows(u2,1..n1/2);
   stack(f0,f2);
```

```
  end;
#Program computes the complete
#V matrix of Kramers and Wannier
d3lnvmat:= proc(n) local P,Q,a,b
                    ,R,S,j1,nb,k,V;
if nargs=1 then
    a:=exp(K): b:=exp(-K):
  else
    a:=args[2]; b:=args[3];
fi;
n1:=2^n/2;   k:=round((n-3)/2);
 if n<3 then
  ERROR('n must be 3 or greater');
  fi;
  if n >=3  then
       j1:=n-2-k;
       nb:=2^k;
 fi;
#compute needed matrices P,Q,R,S.
P:=matrix(2^(j1-1),2^j1,[]):
#the constant matrix U0
 for i to 2^(j1-1) do
   for j to 2^j1 do
    if j=2*i-1 then P[i,j]:=a^3:
      elif j=2*i then  P[i,j]:=a:
      else            P[i,j]:=0:
    fi:
   od:
 od:
Q:=matrix(2^(j1-1),2^j1,[]):
  for i to 2^(j1-1) do
   for j to 2^j1 do
    if j=2*i-1 then Q[i,j]:=a:
    elif j=2*i then Q[i,j]:=b:
    else            Q[i,j]:=0:
      fi:
     od:
    od:
R:=matrix(2^(j1-1),2^j1,[]):
for i to 2^(j1-1) do
  for j to 2^j1 do
   if  j=2^j1-1-2*(i-1) then
            R[i,j]:=b^3:
     elif j=2^j1-2*(i-1) then
              R[i,j]:=b:
       else
              R[i,j]:=0:
     fi:
   od:
  od:
S:=matrix(2^(j1-1),2^j1,[]):
 for i to 2^(j1-1) do
   for j to 2^j1 do
    if j=2^j1-1-2*(i-1) then
```

```
eigenvals(V+(0)^2);
```

$$0, 0, 0, 4$$

Similarly the characteristic values of $V - (0)^2$ are

```
> eigenvals(V-(0)^2)
```

$$0, 0, 0, 0$$

One observes that the maximum characteristic value of $V + (0)$ is 2 and the rest vanish, while the characteristic values of $V - (0)$ are all zero. This is because the trace of $(V + (0)^2)$ is 2^2 and that of $(V - (0)^2)$ vanishes. From these computations, one gets the following generalization. For any upper V-matrix $W + (0)$ and lower V-matrix $W - (0)$ of an arbitrary size, there exists a positive integer r such that, since the rank of $(W + (0))^r$ and $(W - (0))^r$ are one, the characteristic equations reduce to

$$\lambda^{2^r} - (trace\mathbf{W} \pm (0)^r)\lambda^{2^r - 1} = 0.$$

The trace of $(W + (0))^r$ is 2^r, hence the characteristic values of $(W + (0))^r$ are $2^r, ..., 0$, giving the maximun characteristic value of $W + (0)$ to be 2, and all the rest are 0. However, the trace of $(W - (0))^r$ is 0. Hence all the characteristic values vanish and so are the characteristic values of $W - (0)$. This shows that for sufficiently high temperature the maximum characteristic value of $\mathcal{M}$ is non-degenerate, hence it has no long range order. The programs are not many so they are given in the appendix below.

4 Conclusion

Taking advantage of Maple's built-in functions in the linear algebra package it has enabled us to describe the V-matrix which is used in statistical physics to compute the PF of the Ising model by the perturbation method. The V-matrix described here is interesting in its own right. It can be used in the classroom as a tool in teaching the screw-like formulation of the Ising model problem formulated by Kramers and Wannier. The computation of the PF is in progress and it may be given in another paper. We are confident that in the future we look to a major role for Maple in statistical physics.

Acknowledgements

We like to express our appreciation to Dr. Thomas Trappenberg of Dalhousie University, for reading through the paper and for his important comments.

Biographies

Peter Frempong-Mireku is a Doctoral student at Dalhousie University, Halifax, Nova Scotia, Canada. He is working on Symbolic Solution of the Potts-Ising model. He obtained his MSc in Mathematics at the University of Saskatchewan. He also holds MSc degree from Kaluga Pedagogical Institute in Russia, where he had his BSc degee in Mathematics. He has taught in the University of Ghana for three years. His e-mail address is pfm@cs.dal.ca.

Kevin J. M. Moriarty, graduated from Imperial College of Science and Technology, University of London, in England. He has taught and done research at Imperial College, the Deutsches Elektronen Synchrotron (DESY) in Hamburg, Germany, the International Center for Theoretical Physics (ICTP), in Trieste Italy, CERN in Geneva, the Technion-Israel Institute of Tech. in Haifa, Israel, Brookhaven National Laboratory (BHL) in Long Island, NY., Colorado State University in Fort Collins, the Institute of Advanced Study in Princeton. He was first Lecturer and then reader in Applied Maths. at Royal Holloway College of the University of London. In 1983, he joined Dalhousie University, where he is professor of computing Science. In 1990, he was appointed director of the computing science at Dalhousie University. In 1988 he started the Scotia High End Computing Ltd., which signed a partnership with Alliance Computer Systems of Littlton Massachusetts, to develop the Scotia Programming Environment and Facility (SPEFY). He is on the editorial board of *Scientific Computing*.

Appendix

```
#Program computes the
#lower negative V matrix
d2lnvmat:= proc(n) local c1,c2,a,
           b,u0,u1,u2,f0,f2,n1;
if nargs=1 then
   a:=exp(2*K): b:=exp(-2*K):
   else
     a:=args[2]; b:=args[3];
fi;
      n1:=2^n/2;
  u0:=matrix(n1,n1,[]):
  for i to n1 do
```

```
                    S[i,j]:=b:
        elif j=2^j1-2*(i-1) then
                    S[i,j]:=a:
        else
                    S[i,j]:=0:
      fi:
    od:
  od:
od:
V:=matrix(n1,n1,0);
for i from 1 to nb do
V:=copyinto(P,V,(2^(j1-1))*2*
        (i-1)+1,2^j1*2*(i-1)+1);
V:=copyinto(Q,V,2^(j1-1)*(2*i-1)+
                1,2^j1*(2*i-1)+1);
V:=copyinto(evalm(-R),V,n1/2+2^
(j1-1)*2*(i-1)+1,n1-2^j1*(2*i-1)+1);
V:=copyinto(evalm(-S),V,n1/2+2^
(j1-1)*(2*i-1)+1,n1-2^j1*2*i+1);
od;
end;
```

References

[1] Kramers, H. A. and Wannier, G. H., Phys. Rev. **60**, 252 (1941).

[2] Onsager, L., Phy. Rev. **65**, 117 (1944).

[3] Montroll, E. W., J. Chem. Phys. **9** 706 (1941).

[4] Montroll, E. W., J. Chem. Phys. **9** 711 (1941).

[5] Ashin, J. and Lamb W. E., Phy. Rev. **64**, 252(1941).

[6] Oguchi, T., J. Phys. Soc. Japan, **5** 75 (1950).

IIIA. MAPLE IN ABSTRACT ALGEBRA

FRAC: A MAPLE PACKAGE FOR COMPUTING IN THE RATIONAL FUNCTION FIELD K(X)

Cesar Alonso[1], Jaime Gutierrez[2] and Tomas Recio[2]

Departamento de Matemáticas, Universidad de Oviedo, Asturias, Spain[1]

Departamento de Matemáticas, Universidad de Cantabria, Santander, Spain[2]

Abstract:

*In this paper we present the programs package FRAC (= **F**unciones **RAC**ionales) which is designed for performing computations in the rational function field. The main objects in FRAC are rational functions over the field of rational numbers, but extensions to other computable fields can be done in a "natural" way. The key tool is using functional decomposition algorithms. We motivate the interest to work with rational function decomposition by presenting applications to computer science, engineering (CAD), pure mathematics or robotics. We also present some simple examples in order to illustrate the use of FRAC. Finally, we include the synopsis of the main procedures of FRAC.*

1 Introduction

The problem of determining algorithmically if an element f of a class of functions can be represented as a composition of two "simpler" functions g and h in the same class, i.e., such that $f = g(h)$, has attracted interest for a long time, see Ritt(1922) from a theoretical point of view or Zippel(1991) from the computational one. This general problem is called the Functional Decomposition Problem. Although not every function can be decomposed in this manner, when such a decomposition does exits many problems become significantly simpler. We present the particular case when the class of the functions are rational functions over an arbitrary

field $\mathbb{K}$. We want to start motivating this problem through some applications:

1.1 Evaluating

Suppose that we want to evaluate a function f at some points. If f can be decomposed, then its value at different points can be computed with $O(\sqrt{n})$ multiplications rather than $O(n)$ multiplications as is required in the general case. In the particular case when the function f is a rational function, we can check that it decomposes using the FRAC procedure **decomposition**. Suppose that we want to evaluate the rational function $f =$:

$$\frac{x^{35} + 5\,x^{28} + 10\,x^{21} + 10\,x^{14} + 5\,x^7 + 1 - 3\,x^{19} - 3\,x^{12}}{x^{12}\,(x^7 + 1 + x^3)}$$

Applying our FRAC code **decomposition** to the function f, we get:

decomposition$(f)=$

$$\left[\frac{3\,x^4 - 1}{x^4(x-1)}, \frac{-x^3}{1+x^7}\right]$$

time 60.35 words 1726942.

Then, $f = g(h)$, where $h = \dfrac{-x^3}{1+x^7}$ and $g = \dfrac{3\,x^4 - 1}{x^4(x-1)}$.

1.2 Polynomial Solving

The equation $f(x) = 0$ can be numerically solved more efficiently if f is decomposable. In the particular case when f is a polynomial, it is easier then to determine if the zeroes of f can be expressed in terms of radicals. The polynomial decomposition algorithms (see Kozen and Landau(1989)) are now supported by all major symbolic algebra systems.

107

For instance, Maple uses the procedure **compoly**, which is based on an algorithm by Gutierrez et al. (1989), in order to compute symbolically the roots of a polynomial. The following polynomial is irreducible over the rational number field:

$$f = x^{12} - 4x^8 - 5x^6 + 4x^4 + 10x^2 + 3,$$

but the polynomial is decomposable, as can be seen using the **compoly** Maple comand:

compoly$(f, x) =$

$$[x^6 - 4x^4 - 5x^3 + 4x^2 + 10x + 3, x = x^2]$$

time 5.05 words 111084.

On the other hand, applying our procedures **decomposition** or **decompol** to the polynomial f, we get a complete decomposition of f, i.e all given components of f are now indecomposable:

decompol$(f, x) =$

$$[x^2 - 5x + 3, x^3 - 2x, x^2]$$

time 1.16 words 23137.

1.3 Computing Subfields of $\mathbb{K}(x)$

Given two rational functions $G(x)$ and $H(x)$, we are interested in finding explicitly the lattice of all subfields related to $\mathbb{K}(G(x))$ and $\mathbb{K}(H(x))$, i.e. the union field, the intersection field and all intermediate subfields $\mathbb{F}$, with $\mathbb{K}(G) \subset \mathbb{F} \subset \mathbb{K}(x)$. As it is mentioned in Helmke (1990), this is a classical issue in Algebra. The following diagram illustrates this:

$$\mathbb{K}(x)$$
$$|$$
$$\mathbb{K}(G(x), H(x))$$
$$\diagup \quad \diagdown$$
$$\mathbb{K}(G(x)) \qquad \mathbb{K}(H(x))$$
$$\diagdown \qquad \diagup$$
$$\mathbb{K}(G(x) \cap \mathbb{K}(H(x))$$
$$|$$
$$\mathbb{K}$$

In order to construct symbolically this lattice, we use the FRAC procedure **interm**. Moreover, if we want to order the subfields, we must use the procedure **intermorden**. In the particular case when the field $\mathbb{K}(G)$ contains a non-constant polynomial, we can use the procedure **interpol** (see Alonso (1994)). Of course, this code is much faster than the one above, as can be checked in the last section of this paper. In this example, we want to compute all intermediate order subfields $\mathbb{F}$, with $\mathbb{Q}(f) \subset \mathbb{F} \subset \mathbb{Q}(x)$, where f is a rational function with rational number coefficients $f = \frac{x^8 + 1}{x^4}$. We apply the procedure **intermorden** to f:

intermorden$(f) =$

$$[\quad [x^2, \frac{-x^2}{1 + x^4}, \frac{x^2}{1 - x^4}, x^4], \quad [\frac{x}{1 - x^2}, \frac{-x^2}{1 + x^4}],$$
$$[\frac{-x}{1 + x^2}, \frac{-x^2}{1 + x^4}], \quad [\frac{-x^2}{1 + x^4}], \quad [\frac{x^2}{1 - x^4}], \quad [x^4] \quad]$$

time 13.28 words 21346.

So, we can note that there exist six proper subfields; see in the next section for details about the inclusion order relation.

1.4 Simplifying Inverse Kinematics

The symbolic solution of the inverse kinematic problem requires solving a set of joint determining equations for the correponding joint variables. Univariate determining equations (after triangulations) for a revolute joint variable-angle θ, is a kind of polynomial equation $f(\theta) \equiv 0$. The particular type of function $f(\theta)$ is called a $sine - cosine$ polynomial or sc-polynomial (see Kovacs & Hommel (1992)):

$$f(\theta) \equiv \sum_{i+j=0}^{m} f_{ij} s^i c^j$$

where s and c stand for $sin(\theta)$, $cos(\theta)$ and $m \in \mathbb{N}$. We can see $f(\theta)$ as a bivariate polynomial modulo the ideal $s^2 + c^2 - 1$. The most familiar conversion method for sc-polynomial is the so-called tangent half angle substitution. By suitable application of this substitution, every sc-polynomial can be converted into a rational function $F(x)$ in a new variable $x = tan(\frac{\theta}{2})$, with denominator $(1+x^2)^m$. If this rational function is decomposable as $F = G(H)$ such that G is a polynomial and H is a rational function with denominator of the form $(1+x^2)^r$ then, the equation $f(\theta) \equiv 0$ can be numerically and symbolically solved more efficiently (see also Gathen&Weiss (1993)). To solve this problem, first of all, we use the procedure **interm** and then the procedure **equals** that checks when two fields are equal and finally the procedure **leftcomponent**.

We consider the sc-polynomial $f(\theta) \equiv 0$ as in Kovacs& Hommel(1992),

$$f(\theta) \equiv 9 + 20c + 10^2 + 10s + 16cs - 2s^2$$

By applying the half angle subtitution to $f(\theta)$, we have the rational function $F(x)$:

$$F = \frac{-10\,x^2 + 39 - x^4 + 52\,x - 12\,x^3}{(1+x^2)^2}$$

Using the FRAC procedure **interm** on the rational function F:

interm$(F) =$

$$[\frac{x(2\,x - 1)}{1 + x^2}]$$

time 2.66 words 612373.

There exits only an intermediate subfield generate by the rational function $h = \dfrac{x(2\,x - 1)}{1 + x^2}$; now applying the FRAC procedure **leftcomponent** to this h we get:

lefcomponent$(h, F, x) =$

$$39 - 52\,x + 16\,x^2$$

time 0.65 words 1254.

Recoverting the rational function h to a sc-polynomial by applying the inverse of the half angle subtitution, we get:

$$f(\theta) \equiv g(h(\theta))$$

where $g = 39 - 52\,x + 16\,x^2$ and $h = 1 - c + \frac{s}{2}$. This last equation is simpler.

1.5 Integrating

Assume we want to integrate an indefinite integral of the form (see Zippel (1991), (1992)):

$$\int F(x)H(x)^{\frac{1}{s}}dx,$$

where $F(x)$ and $H(x)$ are rational functions. If the rational function $H(X)$ is a suitable leftcomponent of $F(x)$, i.e. if there exists $G(X)$ such that $F = G(H)$, then

$$\int F(x)H(x)^{\frac{1}{s}}dx = \int G(y^s)dy.$$

If such a $G(y)$ exists satisfying some added easy condition, then the above integral can be reduced to the integral of a rational function, which is simpler. We can do that using the procedure **leftcomponent**.

1.6 Manipulating Parametric Curves

A parametrization $(G(t), H(t))$ of an algebraic plane curve $F(x, y) = 0$ is called simpler or faithful if every point (x, y) of the curve (except a finite number of them) corresponds to a unique value of the parameter t, i.e. given a point (x_0, y_0) there exists a unique t_0 such that

$$x_0 = G(t_0), \quad y_0 = H(t_0)$$

Not every algebraic curve has a parametrization, but if it has one then there exists a faithful one. The natural question arising fromm the above definition is: given a parametrization of a curve, how to check if it is simpler or not, and in the negative case to determine a simpler one. This is a very important topic in computer aided design (CAD); see for example Sederberg (1986), Farin (1988) and Alonso (1994). We use the procedures **netto** or **sederberg** for solving that problem.

A parametrization $(G(t), H(t))$ is quasi-polynomial if the union field $\mathbb{K}(G, H)$ contains a non-constant polynomial (the procedure **quasipolynomial** checks that). In this particular case, we can use the faster procedure **maxcompol** to test simplicity.

Finally, these procedures can be understood as tests of birationallity, i.e. checking if

$$\mathbb{K}(F_1(t), ..., F_m(t)) = \mathbb{K}(t).$$

See Ollivier (1989), Gutierrez&Recio (1992–I).

In order to illustrate the use of the FRAC package for solving the above problem, we consider the parameterization of the unit circle $x^2 + y^2 - 1$:

$$F_1(t) = \frac{(2 + 2t)t^2}{1 + 2t + t^2 + t^4}, F_2(t) = \frac{+1 + 2t + t^2 - t^4}{1 + 2t + t^2 + t^4}$$

First of all, we check if the given parameterization is quasipolynomial, via:

quasipolynomial(F_1, F_2)

> false

time 2.18 words 49298.

Then, we must apply the procedures **netto** (see the last section):

netto$(F_1, F_2) =$

$$\frac{t^2}{(1 + t)}$$

time 3.25 words 89268.

So, the parameterization is no simpler and the new parameter is $s = \frac{t^2}{(1 + t)}$.

1.6 Near-separated polynomials and other applications

Decomposition of a polynomial $F(x)$ is strongly related to the separated bivariate polynomial $F(x) - F(y)$ (see Fried and MacRae (1969)). This fact together with fast polynomial decomposition algorithms is used by Gathen (1990) in order to find efficiently separated polynomial factors of the bivariate polynomial $F(x) - F(y)$. In the same way, a decomposition of a rational function $F(x) = \frac{r(x)}{s(x)}$ such that $gcd(r(x), s(x)) = 1$ is strongly related to the associated near-separated bivariate

polynomial $r(x)s(y) - r(y)s(x)$ (see Alonso (1994), Gutierrez and Recio (1992–II). Finding near-separated polynomial factors of the bivariate polynomial $r(x)s(y) - r(y)s(x)$ is equivalent to decomposing the associated rational function $F(x) = \frac{r(x)}{s(x)}$. This fact is the basic idea of some of the procedures that we are presenting.

Other applications of the functional decomposition problem are: the n-partition problem (see Lenstra&Lenstra &Lovasz (1982)), the problem of characterizing the class of automorphisms of $\mathbb{K}[x_1, ..., x_m]$ and computing their inverses (see Dickerson (1989), Gutierrez (1991)) and the problem of finding public key ciphers in algebraic cryptography (see Cade (1985)).

Finally, assumimg that we want to know if the bivariate polynomial:

$$f = 3x^2y - 2x^2 - x + y - 3y^2x + 2y^2$$

over the rational number field is a near-separated one, we will use the procedure **nearseparated**, given an upper bound of the degree:

nearseparated$(f, 3, x, y) =$
$$\frac{2x^2 + x}{1 + 3x^2}$$

time 0.21 words 2222.

The polynomial is near-separated and the associated rational function is the output of the procedure.

2 Description of implemented procedures

In this section we succinctly describe most of the procedures contained in the FRAC package, giving the arguments, calling sequence and a brief comment about the procedures. The package also contains a help : Help(FRAC). The binary code FRAC.m is about 150K.

PROCEDURE: leftcomponent

Arguments: Two univariate rational functions $F_1(x)$, $F_2(x)$ and the variable x

Calling sequence:

$leftcomponent(F_1(x), F_2(x), x)$

Synopsis:

• The procedure **leftcomponent** computes, solving a linear system of equations, a rational function $G(x)$ such that $G(F_1(x))$ equals $F_2(x)$. If such a function $G(x)$ does not exists, the procedure returns an error message.

PROCEDURE: nearseparated

Arguments: A bivariate polynomial $f(x, y)$, a positive integer s and the variables x, y.

Calling sequence:

$nearseparated(f(x, y), s, x, y)$

Synopsis:

• The procedure **nearseparated** decides if the polynomial $f(x, t)$ is a near–separated polynomial. In the affirmative case it returns the associated rational function $\frac{r(x)}{s(x)}$ and otherwise returns 0. The argument s must be an upper bound of the degree of a posible associated rational function.

PROCEDURE: decompol

Arguments: A polynomial $f(x)$

Calling sequence: $decompol(f(x))$

Synopsis:

• The procedure **decompol** computes a complete decomposition of $f(x)$ following the algorithm of Gutierrez et al.(1989). This procedure decomposes polynomials of degree 120 on a Macintosh Powerbook 180 in 50 seconds.

PROCEDURE: decomposition

Arguments: A rational function $F(x)$

Calling sequence: $decomposition(F(x))$

Synopsis:

• The procedure **decomposition** computes a complete decomposition of $F(x)$. During such computation the factorization of a bivariate polynomial is required, plus the procedures **leftcomponent** and **nearseparated** described above. If the input is a polynomial, the **decomposition** program uses instead the procedure **decompol**. A typical performance of **decomposition** computes a complete decomposition of a rational function of degree 40 in 145s.

PROCEDURE: maxcompol

Arguments: List of polynomials

Calling sequence:

$maxcompol([pol.1, \ldots, pol.n])$

Synopsis:

• The procedure **maxcompol** computes the greatest right common polynomial component of a list of polynomials, using techniques of polynomial decomposition.

PROCEDURE: netto

Arguments: List of rational functions

Calling sequence:

$netto([rat.func.1, \ldots, rat.func.n])$

Synopsis:

• The procedure **netto** computes the greatest right common component of a list of rational functions following the proof of Luroth's theorem by Netto (Schinzel (1982), Alonso (1994)). This procedure requires computing the G.C.D. of bivariate polynomials.

111

PROCEDURE: sederberg

Arguments: List of rational functions

Calling sequence:

$sederberg([rat.func.1,\ldots,rat.func.n])$

Synopsis:

• The procedure **sederberg** computes the greatest right common component of a list of rational functions, this time using theoretical results appearing in Sederberg (1986). This procedure uses only G.C.D. computations of univariate polynomials. It's faster, but random.

PROCEDURE: interpol

Arguments: A polynomial $f(x)$

Calling sequence: $interpol(f(x))$

Synopsis:

• The procedure **interpol** computes all intermediate subfields between $\mathbb{K}(f(x))$ and $\mathbb{K}(x)$. This procedure uses some subroutines of the procedure **decompol**.

PROCEDURE: interm

Arguments: A rational function $F(x)$

Calling sequence: $interm(F(x))$

Synopsis:

• The procedure **interm** computes all intermediate subfields between $\mathbb{K}(F(x))$ and $\mathbb{K}(x)$. Uses **leftcomponent** and **nearseparated** described above. If the input is a polynomial, then the **interm** uses instead the procedure **interpol**.

PROCEDURE: intermorden

Arguments: A rational function $F(x)$

Calling sequence: $intermorden(F(x))$

Synopsis:

• The procedure **intermorden** computes the ordered intermediate subfields according to the inclusion relation. This procedure returns a list of lists, with a sublist for each intermediate field between $\mathbb{K}(F(x))$ and $\mathbb{K}(x)$. In each sublist the first element is the generator of this intermediate subfield, and the other elements are the generators of the intermediate fields that are contained in the first one. No intermediate fields are between the first and the others. So, we can describe the lattice graph and we also can determine all non-equivalent complete decompositions of a rational function.

PROCEDURE: interm2

Arguments: Two rational functions $F_1(x)$ and $F_2(x)$

Calling sequence: $interm(F_1(x), F_2(x))$

Synopsis:

• The procedure **interm2** returns all ordered intermediate subfields between $\mathbb{K}(F_1(x))$ and $\mathbb{K}(F_2(x))$. Note that the degree of $F_1(x)$ must be greater than or equal to the degree of $F_2(x)$.

PROCEDURE: intersectionpol

Arguments: Two polynomials $f_1(x)$ and $f_2(x)$

Calling sequence:

$intersectionpol(f_1(x), f_2(x))$

Synopsis:

• The procedure **intersectionpol** computes, if it exists, the polynomial generator of the intersection field $\mathbb{K}(f_1(x)) \cap \mathbb{K}(f_2(x))$

PROCEDURE: intersection

Arguments: Two rational functions $F_1(x)$
,
$F_2(x)$ and a positive integer s

Calling sequence: $intersection(F_1(x), F_2(x))$

Synopsis:

- The procedure **intersection** computes, if it exists, the rational function $G(x)$ of degree least than the positive integer s and generator of the intersection field $\mathbb{K}(F_1(x)) \cap \mathbb{K}(F_2(x))$. If such a function $G(x)$ does not exits, the procedure returns the message "may be the ground field".

PROCEDURE: SSfaithful

Arguments: A list of multivariate rational functions and the list of the variables, $x_1, \ldots, x_n$, involved.

Calling sequence:

$SSfaithful([rat.func.1, \ldots, rat.func.n], [x_1, \ldots, x_n])$

Synopsis:

- The procedure **SSfaithful** decides if a parametrization of a curve given by the rational functions of the input is faithful or not, following results of Shannon and Sweedler (1988). This procedure uses computation of Gröbner basis.

PROCEDURE: Ollifaithful

Arguments: Same of the precedent function

Calling sequence:

$Ollifaithful([rat.func.1, \ldots, rat.func.n], [x_1, \ldots, x_n])$

Synopsis:

- The procedure **Ollifaithful** computes the same things **SSfaithful**, but now using results of Ollivier appearing in Ollivier (1989). It also computes some Gröbner basis.

PROCEDURE: TRfaithful

Arguments: Two rational functions $F_1(x)$ and $F_2(x)$

Calling sequence: $TRfaithful(F_1(x), F_2(x))$

Synopsis:

- The procedure **TRfaithful** decides if a parameterization of a plane curve given by $F_1(x)$ and $F_2(x)$ is faithful or not. This procedure uses a Taylor resultant computation of two bivariate polynomials. (see Abhyankar and Bajaj (1989), Abhyankar (1990), Alonso (1994)).

PROCEDURE: quasipolynomial

Arguments: A list of univariate rational functions

Calling sequence:

$quasipolynomial([rat.func.1, \ldots, rat.func.n])$

Synopsis:

- The procedure **quasipolynomial** decides if the
parametrization of a curve given by the list of the rational functions is quasipolynomial or not.

PROCEDURE: equals

Arguments: Two rational functions $F_1(x)$ and $F_2(x)$

Calling sequence: $equals(F_1(x), F_2(x))$

Synopsis:

- This useful procedure decides, by solving a system of linear equations, when two rational functions generate the same field. It returns "false" if $\mathbb{K}(F_1(x))$ is different from $\mathbb{K}(F_2(x))$; and otherwise returns a linear fraction $U(x)$ such that $U(F_1(x)) = F_2(x)$.

PROCEDURE: Implicit

Arguments: A list of multivariate rational functions and the list of the variables, $t_1, \ldots, t_m$, involved.

Calling sequence:

Implicit([rat.func.1,$\ldots$,rat.func.n],[$t_1, \ldots, t_m$])

Synopsis:

• This procedure computes the implicitization ideal of the rational funcions following the algorithm of Alonso& Gutierrez&Recio (1994). This procedure uses computations of some quotient ideals.

Acknowledgment

The authors are indebted to Prof. Carlos Ruiz de Velasco for his suuggestions regarding the implementation on Maple.

This work is partially supported by CICyT PB 92/0498 /C02/01 and Esprit Bra-POSSO. 6846.

References.

Abhyankar, S., Bajaj, C.:*Computations with Algebraic Curves*. ISSAC–89. L.N.C.S. No. 358, pp.274–284, Springer–Verlag, 1989.

Abhyankar, S.:*Algebraic Geometry for scientists and engineers*. Math. Surveys and Monographs N.35. American Math. Society 1990.

Alonso, C.,Gutierrez, J., Recio, T.:*An Implicitization Algorithm with fewer variables* . To appear in Computer Aided Geometric Design, 1994.

Alonso, C.: *Desarrollo Análisis e implementación de algoritmos para la manipulación de variedades paramétricas*. Ph. dissertation, Dep. Math. and Computing, Universidad de Cantabria, Mayo 1994.

Cade, J. J.:*A new public–key cipher which allows signatures*. Proc. 2nd SIAM Conf. on Appl. Linear Algebra, Raleigh NC, 1985.

Dickerson,M.: *Functional Decomposition of Polynomials* . Tech. Rep. 89-1023, Dep. of Computer Science, Cornell University, Ithaca NY (1989).

Farin, G.:*Curves and Surfaces for Computer Aided Geometric Design*. Academic Press, Boston 1988.

Fried, M., MacRae, R.:*On curves with separated variables*. Math. Ann., 180, pp. 220–226, 1969.

Gathen, J. von zur.: *Functional decomposition of polynomials: the tame case* . J. of Symbolic Computation 9, pp. 281-299 (1990).

Gathen, J.&Weiss, J.:*Homogeneus bivariate decompositions*. Preprint, Dep. of Computer Science, University of Toronto, 1993.

Gutierrez,J.: *A polynomial decomposition algorithm over factorial domains*. Compt. Rendues Math. Acad. Science Canada, Vol. xIII-2, pp. 437-452 (1991).

Gutierrez,J.&Recio,T.&Ruiz de Velasco.: *A polynomial decomposition algorithm of almost quadratic complexity*. Proc. AAECC-6/88. L. N. Computer Science 357, pp. 471-476 (1989).

Gutierrez,J.&Recio, T.: *Rational function decomposition and Groebner Bases in the parameterization of plane curves*. Proc. of LATIN'92. L. N. Computer Science 583, pp. 231-245 (1992–I).

Gutierrez,J.&Recio,T.: *A Practical Implementation of two rational function decomposition Algorithms* . Proc. of ISSAC'92. ACM (1992–II).

Helmke, U. *The variety of subfields of* $\mathbb{K}(x)$. Comm. in Algebra, 18(11) pp. 3775-3789, 1990.

Kovacs, P.&Hommel, G.:*Simplification of Symbolic Inverse Kinematic Transformations through Functional Decomposition*. Adv. in Robotics. Ferrara Sept. 1992.

Kozen,D.&Landau,S.: *Polynomial decomposition algorithms*. J. of Symbolic Computation 7, pp. 445-456 (1989).

Lenstra, A. K., Lenstra, H. W., Lovasz, L.:*Factoring Polynomials with Rational Coefficients*. Math. Ann. 261, pp.515–534, 1982.

Ollivier,F. *Inversibility of rational mappings and structural identifiability in Automatics*. Proc. IS-SAC'89, pp. 43–53, ACM; 1989.

Ritt,F.: *Prime and Composite polynomials*. Trans. Amer. Math. Society 23, pp. 51-66 (1922).

Schinzel, A.:*Selected topics on polynomials*. Ann Arbor, University of Michigan press, 1982.

Sederberg, T. W.:*Improperly parametrized rational cur- ves*. Computer Aided Geometric Design, 3, pp. 67–75, 1986.

Shannon, D., Sweedler, M.:*Using Gröbner bases to determine algebra membership, split surjective algebra homomorphisms, determine birational equivalence*. J. Symbolic Computation, 6, pp. 267–273; 1988.

Zippel,R.:*Rational Function Decomposition* . Proc. of ISSAC-91. ACM press, 1991. Technical report, Cornell University, 1992.

About the Authors

Cesar Alonso is currently an Assistant Professor of Computer Science at the Universidad de Oviedo. He received his Ph.D in Mathematics from Universidad de Cantabria in the area of Algebraic Symbolic Computation. His primary interest are in the area of Algebraic Algorithms. He can be reached by electronic mail sent to:

calonso@aicvax.aic.uniovi.es

Jaime Gutierrez is an Associate Professor of Applied Mathematics at the Universidad de Cantabria. After receiving a Ph.D in theoretical and applied mathematics from the Universidad de Cantabria, he has continued to pursue research interest in pure mathematics, symbolic computation, solid modeling and applications to robotic. His e-mail address is:

gutierrez@ccucvx.unican.es

Tomas Recio is a full Professor of Mathematics at the Universidad de Cantabria. He received his Ph.D. in Mathematics from Universidad Complutense de Madrid in the area of Real Algebraic Geometry. He has published extensively in the fields of real algebraic geometry, computational geometry, symbolic computation, robotics and complexity theory. He can be reached by electronic mail sent to:

recio@ccucvx.unican.es

GROUP RINGS AND HOPF-GALOIS THEORY IN MAPLE

Timothy Kohl

Department of Mathematics and Statistics, SUNY at Albany, Albany, NY

Abstract

As the title implies, the subject is known as Hopf-Galois theory which attempts to expand the notions of classical Galois theory to more general settings. While the particulars of the subject are not the main focus of this discussion, the principal objects under study, namely group rings and certain subrings of these group rings are of interest. The computational demands involved in describing them and performing computations in them, are what motivated the development, in Maple, of a collection of tools to accomplish this task. Before going further, some background would be appropriate and will help frame the discussion to follow.

I. GROUP RINGS

Definition : Given a ring A with identity and a finite group G the *group ring* AG is defined as the set of all elements of the form:

$$\sum_{i=1}^{n} a_i g_i$$

where $G = \{ g_1, g_2, \ldots, g_n \}$ and $a_i \in A$. Addition is defined by the rule:

$$\sum_{i=1}^{n} a_i g_i + \sum_{i=1}^{n} b_i g_i = \sum_{i=1}^{n} (a_i + b_i) g_i$$

(i.e. The group elements act like a vector space basis or more precisely as a basis for a free A-algebra of rank n, since A need not be a field.)

Multiplication of elements in AG is similar to polynomial multiplication (as will be important in Maple) but not quite. Specifically,

$$\left(\sum_{i=1}^{n} a_i g_i \right) \left(\sum_{j=1}^{n} b_j g_j \right) = \sum_{i=1}^{n} \sum_{j=1}^{n} a_i b_j g_i g_j$$
$$= \sum_{k=1}^{n} c_k g_k$$

for some $c_1, c_2, \ldots, c_n \in A$. Since the g_i's are elements of the group G then $g_i g_j \in G$ and $a_i b_j \in A$ so the product of two elements in AG is again an element of AG. One defines the additive identity element of AG to be $\sum_i 0_A g_i$ and the multiplicative identity to be $1_A e$ where 0_A and 1_A are respectively the additive and multiplicative identities of A and e is the identity element of G. It is easy (although tedious) to verify that AG satisfies all the conditions, such as the associative law $\alpha (\beta \gamma) = (\alpha \beta) \gamma$, in the definition of a ring and we shall assume this since it is well known.

Now the particular group rings under investigation are those of the form $\mathbb{Q}(\zeta_n) C_n$ where n is an integer greater than 1 and:

- $\mathbb{Q}(\zeta_n)$ - extension field of the rationals $\mathbb{Q}$, generated by all n-th roots of unity e.g. $\zeta_n = e^{2\pi i / n}$.

- C_n - the cyclic group of order n, $C_n = \langle \sigma \rangle = \{1, \sigma, \sigma^2, \ldots, \sigma^{n-1}\}$ with $\sigma^i \sigma^j = \sigma^{[i+j]}$ where [i+j] denotes the class of i+j modulo n.

Representing this group ring and determining how to do calculations within it were accomplished by using the following three observations:

(1) By default, Maple works with expressions in as many indeterminates as desired and treats them as polynomials over the rationals.

(2) Any finite algebraic extension of the rational numbers $K/\mathbb{Q}$ can be viewed as the quotient ring $\mathbb{Q}[z]/(f(z))$ where $f(z)$ is an irreducible polynomial in $\mathbb{Q}[z]$. e.g. $\mathbb{Q}(\sqrt{3}) \cong \mathbb{Q}[z]/(z^2 - 3)$

(3) If C_n denotes the cyclic group of order n then we can represent the group ring KC_n as the quotient ring $K[x]/(x^n - 1)$. Since in this quotient ring, $x^n = 1$, every element is a polynomial expression of degree $< n$, with coefficients in K, and multiplication of two such expressions is carried out in the usual way except that the relation $x^n = 1$ is imposed to give us another 'polynomial' of degree $< n$. That is, we identify x with σ and reduce all exponents modulo n.

Of course the first two observations are what lie at the heart of *RootOf()* and other related commands, but it is the third observation which yields the following representation of $\mathbb{Q}(\zeta_n)C_n$

$$\mathbb{Q}(\zeta_n)C_n \cong \mathbb{Q}[z, x]/(\Phi_n(z), x^n - 1)$$

where $\Phi_n(z)$ is the n-th cyclotomic polynomial, the minimal polynomial of ζ_n.

Since

$$\sum_{j=0}^{n-1}\left(\sum_{i=0}^{\phi(n)-1} a_{ij}\zeta_n^i\right)\sigma^j = \sum_{j=0}^{n-1}\sum_{i=0}^{\phi(n)-1} a_{ij}\zeta_n^i\sigma^j$$

where $a_{ij} \in \mathbb{Q}$, then $\{\zeta_n^i\sigma^j \,|\, i = 0, \ldots, \phi(n) - 1, j = 0, \ldots, n - 1\}$ is a $\mathbb{Q}$-basis for $\mathbb{Q}(\zeta_n)C_n$. Hence, elements of $\mathbb{Q}(\zeta_n)C_n$ can be represented in Maple as follows:

$$(*) \qquad \sum_{j=0}^{n-1}\sum_{i=0}^{\phi(n)-1} a_{ij}z^i x^j \qquad a_{ij} \in \mathbb{Q}$$

where addition is just ordinary polynomial addition. Multiplication of two elements of the form (*) is carried out like polynomial multiplication except that we make the formal substitutions, $\Phi_n(z) = 0$ (since we identify z with ζ_n) and $x^n - 1 = 0$, so that the result is another expression of the form (*).

With this in mind all that was needed was to code the routines for performing the 'arithmetic' in $\mathbb{Q}(\zeta_n)C_n$. Although the subsequent package eventually grew in size, the code for these operations resides in these routines which are given here:

```
with(numtheory);

simp := proc(expr,n)
local e,ee,temp,ne,c,i,j,degx,degz:
e := d + dd + expand(expr):
ne := nops(e):
ee := 0:
for i from 1 to ne do
temp := op(i,e):
degx := degree(temp,x):
degz := degree(temp,z):
temp := subs(x∧degx = x∧(degx mod n),temp):
temp := subs(z∧degz = z∧(degz mod n),temp):
ee := ee + temp:
od:
ee := ee - d - dd:
e := collect(ee,x):
for j from 0 to n-1 do
c[j] := coeff(e,x,j):
c[j] := cyclosimp(c[j],n):
od:
e :=c[0]:
for j from 1 to n-1 do
e := e + c[j]*(x∧ j):
od:
e := sort(e,x);
RETURN(e):
end:

cyclosimp := proc(expr,n)
local d,j,e,cyp:
d := phi(n):
cyp := z∧d-cyclotomic(n,z):
e := expr:
for j from d to n-1 do
e := subs(z∧j=z∧(j-d)*cyp,e):
od:
e := collect(e,z):
RETURN(e):
end:

mu := proc(expr1,expr2,n)
RETURN(simp(expr1*expr2,n));
end;

ad := proc(expr1,expr2,n)
RETURN(simp(expr1+expr2,n));
end;
```

The main procedure is *simp()* which reduces a given expression to one of the correct form to represent an element of the group ring. The only parameters that are passed are an expression to be simplified and the particular choice of 'n'. The procedure *cyclosimp()* is used so that all powers of z (think ζ_n) that appear in an expression are of exponent less that $\phi(n)$. For example, $\phi(9) = 6$ and $\Phi_9(z) = z^6 + z^3 + 1$ hence $\zeta_9^6 + \zeta_9^3 + 1 = 0$ so $\zeta_9^6 = -\zeta_9^3 - 1$ and thus ζ_9^7, for instance, may be written as $\zeta_9\zeta_9^6 = \zeta_9\left(-\zeta_9^3 - 1\right) = -\zeta_9^4 - \zeta_9$

An initial example of a calculation that can be done is to determine a set of orthogonal idempotents in $\mathbb{Q}(\zeta_n)C_n$, that is n elements $e_1, \ldots, e_n$ such that $e_i^2 = e_i$ and $e_i e_j = 0$ for $i \neq j$. For example, if n=3 we may define $e_1 = \frac{1}{3}(\sigma^2 + \sigma + 1)$, $e_2 = \frac{1}{3}(\zeta^2\sigma^2 + \zeta\sigma + 1)$, $e_3 = \frac{1}{3}(\zeta\sigma^2 + \zeta^2\sigma + 1)$ and readily verify that these satisfy the above properties. Keep in mind that ζ is a primitive cube root of unity and as such $1+\zeta+\zeta^2 = 0$ so that $\zeta^2 = -\zeta - 1$.

>e1 := 1/3*(x∧2+x+1);

$$e1 := \frac{1}{3}x^2 + \frac{1}{3}x + \frac{1}{3}$$

>e2 := 1/3*(z∧2*x∧2+z*x+1);

$$e2 := \frac{1}{3}z^2x^2 + \frac{1}{3}zx + \frac{1}{3}$$

>e3 := 1/3*(z*x∧2+z∧2*x+1);

$$e3 := \frac{1}{3}zx^2 + \frac{1}{3}z^2x + \frac{1}{3}$$

>mu(e1,e1,3);

$$\frac{1}{3}x^2 + \frac{1}{3}x + \frac{1}{3}$$

>mu(e2,e2,3);

$$(-\frac{1}{3}z - \frac{1}{3})x^2 + \frac{1}{3}zx + \frac{1}{3}$$

>mu(e3,e3,3);

$$\frac{1}{3}zx^2 + (-\frac{1}{3}z - \frac{1}{3})x + \frac{1}{3}$$

>mu(e1,e2,3);mu(e1,e3,3);mu(e2,e3,3);

$$0$$

$$0$$

$$0$$

Calculations such as these are interesting, but it was a more complex problem which spurred the development of a larger package which included these routines.

II. HOPF ALGEBRAS

In this section we shall describe how Maple was used to compute bases for certain Hopf algebras over $\mathbb{Q}$ which are given as subrings of group rings of the form $\mathbb{Q}(\zeta_{p^k})C_{p^k}$ where p is an odd prime. These Hopf algebras act analgously to a Galois group on the field extension, $\mathbb{Q}(a^{1/p^k})/\mathbb{Q}$ where $a \in \mathbb{Q}$ but $a \neq b^p$ for any $b \in \mathbb{Q}$. This extension is not Galois in the usual sense, but is an example of what is called a Hopf-Galois extension. (See the references for further information).

Without going into excessive detail, the problem is this. Given the group ring $\mathbb{Q}(\zeta_{p^k})C_{p^k}$, there is a group $\Delta_{p^k} = \langle\delta\rangle$ cyclic of order $\phi(p^k) = p^k - p^{k-1}$ which acts on $\mathbb{Q}(\zeta_{p^k})C_{p^k}$ as follows:

$$\delta(a\zeta_{p^k}^i\sigma^j) = a(\zeta_{p^k}^{i\pi}\sigma^{j\pi}) \text{ for } a \in \mathbb{Q}$$

where π is the least primitive root of p^k which in Maple is given by the *primroot()* function in the *numtheory* package. This in mind, we can ask for a determination of $\left(\mathbb{Q}(\zeta_{p^k})C_{p^k}\right)^{\Delta_{p^k}}$, that is the subring of elements fixed by all of Δ_{p^k}. Specifically, $\left(\mathbb{Q}(\zeta_{p^k})C_{p^k}\right)^{\Delta_{p^k}} = \{\alpha \in \mathbb{Q}(\zeta_{p^k})C_{p^k} | \delta^r\alpha = \alpha \; \forall\delta^r \in \Delta_{p^k}\}$ which is a Hopf algerbra over $\mathbb{Q}$ of rank p^k (as a vector space) and shall be denoted by $_{p^k}H$. The problem is how to compute a basis for $_{p^k}H$ over $\mathbb{Q}$. Now the initial approach was to take each element of the set of $\phi(p^k)p^k$ basis elements $\{\zeta_{p^k}^i\sigma^j | i = 0, \ldots, \phi(p^k) - 1, j = 0, \ldots, p^k - 1\}$ and compute the 'trace' of it under the action of Δ_{p^k}. That is, given $\zeta_{p^k}^i\sigma^j$ define:

$$(*) \qquad tr(\zeta_{p^k}^i\sigma^j) = \sum_{r=0}^{\phi(p^k)-1} {}^{\delta^r}\left(\zeta_{p^k}^i\sigma^j\right)$$

$$= \sum_{r=0}^{\phi(p^k)-1} \zeta_{p^k}^{i\pi^r}\sigma^{j\pi^r}$$

It is easy to see that $tr(\zeta_{p^k}^i\sigma^j)$ is fixed by every element of Δ_{p^k} and some basic theory tells us that this set spans $_{p^k}H$. However, this set is not linearly independant (over $\mathbb{Q}$) so we must throw out enough elements to yield a basis. It was easy to define, in Maple, the action of Δ_{p^k} on the basis elements by defining how each power of the generator δ acts on a basis element and sum these together as in (*) to define the trace function. The two procedures are *delta(expr,n,r)* which returns $^{\delta^r}(expr)$ for $\delta^r \in \Delta_n$, and *tr(expr,n)* which returns $tr(expr)$.

Note the following sample calculations where we take the image (under the trace map) of a generic basis element of $\mathbb{Q}(\zeta_9)C_9$ and demonstrate that this image is invariant under the action of Δ_9.

>simp(tr(z*x,9),9);

$$\left(-z^5 - z^2\right) x^8 + \left(-z^4 - z\right) x^7 + z^5 x^5 + z^4 x^4$$
$$+ z^2 x^2 + zx$$

>simp(delta(",9,1),9);

$$\left(-z^5 - z^2\right) x^8 + \left(-z^4 - z\right) x^7 + z^5 x^5 + z^4 x^4$$
$$+ z^2 x^2 + zx$$

So we combine the above observations to get a procedure *genspan(n)* , where n = p^k, of course, which will generate the aforementioned spanning set and another procedure *linearize()* to reduce this spanning set to a basis. For example, in the following sample calculations we generate a spanning set s_3 (resp. s_9) and pass it to the *linearize* routine which outputs a basis b_3 (resp. b_9) for $_3H$ (resp. $_9H$). The use of *nops()* is to compare the cardinality of each spanning set as compared with that of the basis derived from it.

>s3:=genspan(3);

$$s3 := \{1, x^2+x, (-z-1)x^2+zx, zx^2+(-z-1)x\}$$

>nops(s3);

$$4$$

>b3 := linearize(s3,3);

$$b3 := \{1, x^2 + x, (-z-1)x^2 + zx\}$$

>nops(b3);

$$3$$

>s9:=genspan(9);

$s9 := \{1, z^4 x^8 + \left(-z^5 - z^2\right) x^7 + \left(-z^4 - z\right) x^5$
$\quad + z^2 x^4 + zx^2 + z^5 x, \left(-z^5 - z^2\right) x^8$
$\quad + \left(-z^4 - z\right) x^7 + z^5 x^5 + z^4 x^4 + z^2 x^2 + zx,$
$\quad z^3 x^8 + \left(-z^3 - 1\right) x^7 + z^3 x^5$
$\quad + \left(-z^3 - 1\right) x^4 + z^3 x^2 + \left(-z^3 - 1\right) x,$
$\quad \left(-z^4 - z\right) x^8 + z^5 x^7 + zx^5 + \left(-z^5 - z^2\right) x^4$
$\quad + z^4 x^2 + z^2 x, z^5 x^8 + zx^7 + z^2 x^5$
$\quad + \left(-z^4 - z\right) x^4 + \left(-z^5 - z^2\right) x^2 + z^4 x,$
$\quad \left(-z^3 - 1\right) x^6 + z^3 x^3, z^3 x^6 + \left(-z^3 - 1\right) x^3,$
$\quad zx^8 + z^2 x^7 + z^4 x^5 + z^5 x^4 + \left(-z^4 - z\right) x^2$
$\quad + \left(-z^5 - z^2\right) x, z^2 x^8 + z^4 x^7$
$\quad + \left(-z^5 - z^2\right) x^5 + zx^4 + z^5 x^2$
$\quad + \left(-z^4 - z\right) x, x^6 + x^3,$
$\quad x^8 + x^7 + x^5 + x^4 + x^2 + x,$
$\quad \left(-z^3 - 1\right) x^8 + z^3 x^7 + \left(-z^3 - 1\right) x^5$
$\quad + z^3 x^4 + \left(-z^3 - 1\right) x^2 + z^3 x\}$

>nops(s9);

$$13$$

>b9 := linearize(s9,9);

$b9 := \{1, z^4 x^8 + \left(-z^5 - z^2\right) x^7 + \left(-z^4 - z\right) x^5$
$\quad + z^2 x^4 + zx^2 + z^5 x, \left(-z^5 - z^2\right) x^8$
$\quad + \left(-z^4 - z\right) x^7 + z^5 x^5 + z^4 x^4 + z^2 x^2$
$\quad + zx, z^3 x^8 + \left(-z^3 - 1\right) x^7$
$\quad + z^3 x^5 + \left(-z^3 - 1\right) x^4 + z^3 x^2$
$\quad + \left(-z^3 - 1\right) x, \left(-z^4 - z\right) x^8 + z^5 x^7$
$\quad + zx^5 + \left(-z^5 - z^2\right) x^4 + z^4 x^2$
$\quad + z^2 x, z^5 x^8 + zx^7 + z^2 x^5 + \left(-z^4 - z\right) x^4$
$\quad + \left(-z^5 - z^2\right) x^2 + z^4 x, \left(-z^3 - 1\right) x^6 + z^3 x^3,$
$\quad z^3 x^6 + \left(-z^3 - 1\right) x^3,$
$\quad x^8 + x^7 + x^5 + x^4 + x^2 + x\}$

>nops(b9);

$$9$$

Although not immediately obvious, distinct $\zeta_{p^k}^i \sigma^j$'s can have identical traces and since the output of *genspan()* is a set, in the theoretical sense and as the Maple data type, duplicate traces do not appear in the spanning set. Furthermore, some additional conditioning is performed so that, among other things, the constant 1 always appears in the basis and if $tr\left(\zeta_n^i \sigma^j\right)$ is any other rational number, it is not added to the spanning set. Hence the spanning set in general has fewer than $\phi(p^k)p^k$ elements yet more than p^k elements, hence the need for the *linearize()* routine. Consider the following example where we generate a basis for $_{27}H$. Again note the size of the spanning set as compared to the basis.

>s27 := genspan(27):
>nops(s27);

$$40$$

>b27 := linearize(s27,27);

$$b27 := \{1, z^6 x^{24} + z^{12} x^{21} + \left(-z^{15} - z^6\right)x^{15}$$
$$+ z^3 x^{12} + z^{15} x^6 + \left(-z^{12} - z^3\right)x^3, \ldots$$

$$\vdots$$

$$z^{13} x^{26} + \left(-z^{17} - z^8\right)x^{25} + \left(-z^{16} - z^7\right)x^{23}$$
$$+ z^{11} x^{22} + z^{10} x^{20} + \left(-z^{14} - z^5\right)x^{19}$$
$$+ \left(-z^{13} - z^4\right)x^{17} + z^8 x^{16}$$
$$+ z^7 x^{14} + \left(-z^{11} - z^2\right)x^{13}$$
$$+ \left(-z^{10} - z\right)x^{11} + z^5 x^{10} + z^4 x^8 + z^{17} x^7$$
$$+ z^{16} x^5 + z^2 x^4 + zx^2 + z^{14} x\}$$

>nops(b27);

$$27$$

By hand, even computing b9 was rather difficult (let alone b27), so this program had already proved its value. Now when computing the basis for $_{27}H$, the excess elements that this algorithm generated caused not only *genspan()* to take a long time to work but *linearize()* as well. With this method, computing a basis for $_{81}H$ was prohibitively slow, but with the results this initial setup yielded, patterns began to emerge which indicated that a much more efficient algorithm was possible. Using a variation of a result that appears in [1] together with a number of unpublished results of the author, a much improved algorithm was implemented which yielded a very rapid calculation of the bases for not

only $_{27}H$ but also $_{81}H$, with none of the overhead induced by looking at spanning sets. The culmination of this is the procedure *fastbasis(b,p,k)* which recursively generates a basis for $_{p^{k+1}}H$ given a basis b of $_{p^k}H$. If we make the convention $_{p^0}H = {}_1H = \mathbb{Q}$ and let $b_1 = \{1\}$ then applying *fastbasis()* to b_1 yields a basis b_p of $_pH$ and applying *fastbasis()* to b_p yields a basis b_{p^2} of $_{p^2}H$ and so on.

We illustrate this process with p=3 as before:

>b1 := {1}

$$b1 := \{1\}$$

>b3 := fastbasis(b1,3,0);

$$b3 := \{1, \left(-z - 1\right)x^2 + zx, zx^2 + \left(-z - 1\right)x\}$$

>b9 := fastbasis(b3,3,1);

$$b9 := \{1, z^4 x^8 + \left(-z^5 - z^2\right)x^7 + \left(-z^4 - z\right)x^5$$
$$+ z^2 x^4 + zx^2 + z^5 x, \ldots$$

$$\vdots$$

$$x^8 + x^7 + x^5 + x^4 + x^2 + x\}$$

>b27 := fastbasis(b9,3,2);

$$b27 := \{1, z^6 x^{24} + z^{12} x^{21} + \left(-z^{15} - z^6\right)x^{15}$$
$$+ z^3 x^{12} + z^{15} x^6 + \left(-z^{12} - z^3\right)x^3, \ldots$$

$$\vdots$$

$$z^{13} x^{26} + \left(-z^{17} - z^8\right)x^{25} + \left(-z^{16} - z^7\right)x^{23}$$
$$+ z^{11} x^{22} + z^{10} x^{20} + \left(-z^{14} - z^5\right)x^{19}$$
$$+ \left(-z^{13} - z^4\right)x^{17} + z^8 x^{16} + z^7 x^{14}$$
$$+ \left(-z^{11} - z^2\right)x^{13} + \left(-z^{10} - z\right)x^{11}$$
$$+ z^5 x^{10} + z^4 x^8 + z^{17} x^7 + z^{16} x^5$$
$$+ z^2 x^4 + zx^2 + z^{14} x\}$$

>b81 := fastbasis(b27,3,3):

We need not display all of b81, but we can show some typical basis elements and some calculations to demonstrate that these are indeed elements of $_{81}H$. Specifically, $\alpha \in {}_{p^k}H$ means $\delta^r(\alpha) = \alpha$ (i.e. $\delta^r(\alpha) - \alpha = 0$) for all $\delta^r \in \Delta_{p^k}$, and since Δ_{p^k} is cyclic, it suffices to check this for r=1.

```
>b81[13];
```

$$\left(-z^{39} - z^{12}\right) x^{78} + z^{51}x^{75} + z^{21}x^{69} + z^6x^{66}$$
$$+ \left(-z^{30} - z^3\right) x^{60} + z^{42}x^{57} + z^{12}x^{51}$$
$$+ \left(-z^{51} - z^{24}\right) x^{48} + z^{48}x^{42}$$
$$+ z^{33}x^{39} + z^3x^{33} + \left(-z^{42} - z^{15}\right) x^{30}$$
$$+ z^{39}x^{24} + z^{24}x^{21} + \left(-z^{48} - z^{21}\right) x^{15}$$
$$+ \left(-z^{33} - z^6\right) x^{12} + z^{30}x^6 + z^{15}x^3$$

```
>simp(delta(b81[13],81,1)-b81[13],81);
```

$$0$$

```
>b81[59];
```

$$z^{43}x^{80} + z^5x^{79} + z^{10}x^{77} + z^{53}x^{76}$$
$$+ \left(-z^{31} - z^4\right) x^{74} + z^{20}x^{73} + z^{25}x^{71}$$
$$+ \left(-z^{41} - z^{14}\right) x^{70} + \left(-z^{46} - z^{19}\right) x^{68}$$
$$+ z^{35}x^{67} + z^{40}x^{65} + z^2x^{64} + z^7x^{62}$$
$$+ z^{50}x^{61} + \left(-z^{28} - z\right) x^{59} + z^{17}x^{58}$$
$$+ z^{22}x^{56} + \left(-z^{38} - z^{11}\right) x^{55}$$
$$+ \left(-z^{43} - z^{16}\right) x^{53} + z^{32}x^{52} + z^{37}x^{50}$$
$$+ \left(-z^{53} - z^{26}\right) x^{49} + z^4x^{47} + z^{47}x^{46}$$
$$+ z^{52}x^{44} + z^{14}x^{43} + z^{19}x^{41}$$
$$+ \left(-z^{35} - z^8\right) x^{40} + \left(-z^{40} - z^{13}\right) x^{38}$$
$$+ z^{29}x^{37} + z^{34}x^{35} + \left(-z^{50} - z^{23}\right) x^{34}$$
$$+ zx^{32} + z^{44}x^{31} + z^{49}x^{29} + z^{11}x^{28}$$
$$+ z^{16}x^{26} + \left(-z^{32} - z^5\right) x^{25}$$
$$+ \left(-z^{37} - z^{10}\right) x^{23} + z^{26}x^{22} + z^{31}x^{20}$$
$$+ \left(-z^{47} - z^{20}\right) x^{19} + \left(-z^{52} - z^{25}\right) x^{17}$$
$$+ z^{41}x^{16} + z^6x^{14} + z^8x^{13} + z^{13}x^{11}$$
$$+ \left(-z^{29} - z^2\right) x^{10} + \left(-z^{34} - z^7\right) x^8$$
$$+ z^{23}x^7 + z^{28}x^5 + \left(-z^{44} - z^{17}\right) x^4$$
$$+ \left(-z^{49} - z^{22}\right) x^2 + z^{38}x$$

```
>simp(delta(b81[59],81,1)-b81[59],81);
```

$$0$$

Conclusion

With Maple and a polynomial-based description
of these group rings, we have been able to construct

a framework in which the complex expressions that
arise are no longer difficult to manipulate. In doing so, we have freed ourselves to go after the main
problem and deal with it within this same environment without having to worry about tedious intermediate calculations. Furthermore, as mentioned
earlier, preliminary calculations led to preliminary
results which in turn led to not only increased insight into the problem itself, but an improved computational scheme. It is in this respect that Maple
stands out, not just as a computational tool but
as a conduit of empirical information from which
observant users may derive deeper understanding
of their subject.

REFERENCES

[1] Lindsay N. Childs, *On the Hopf Galois Theory for Separable Field Extensions*, Communications in Algebra
17(4) (1989), 809–825.

[2] C. Greither and B. Pareigis, *Hopf Galois Theory for
Separable Field Extensions*, Journal of Algebra **106**
(1987), 239–258.

[3] Thomas W. Hungerford, *Algebra*, Springer-Verlag,
Berlin, 1974.

The author Timothy Kohl is currently working
on his Ph.D in mathematics at the State University
of New York at Albany. He can be reached at the
following address:

Timothy Kohl

Department of Mathematics and Statistics

State University of New York at Albany

1400 Washington Avenue

Earth Science 110

Albany, New York 12222

RATIONAL GENERATING FUNCTION APPLICATIONS IN MAPLE

Robert A. Ravenscroft, Jr.
Department of Computer Science and Statistics, University of Rhode Island,
Kingston, RI

1.   Introduction

Generating function techniques have been used throughout the literature to evaluate summations and recurrences.   Rational generating functions are a particularly effective computation tool, as they encode all *linear recurrence sequences*. These are sequences defined by homogeneous linear recurrences with constant coefficients.   There are known generating function techniques for evaluating summations and convolutions that involve linear recurrence sequences. Unfortunately, these methods do not lend themselves well to hand calculation.

The Maple **genfunc** package provides tools for manipulating rational generating functions and the sequences that they encode. This paper illustrates how Maple is used for quick and easy evaluation of sums and convolutions of linear recurrence sequences. Section 2 looks at encoding and decoding generating functions. Summations and convolutions are considered in Section 3. Techniques for hybrid terms and linear indexing are presented in Section 4. *Hybrid terms* are terms defined by the term-wise product of two or more sequences. *Linear indexing* defines a new sequence by using a linear polynomial to index a known linear recurrence sequence. To illustrate these applications of Maple, we consider several summations and convolutions, including some well known problems involving the Fibonacci numbers.

2.   Basic Maneuvers

To use generating functions, we must first be able to map between closed forms and generating functions and between recurrences and generating functions. The **genfunc** package functions **rgf_encode**, **rgf_expand**, and **rgf_sequence** and the Maple library function **rsolve** perform these tasks.

To encode a generating function we use the **rgf_encode** function. As with most of the functions in the **genfunc** package, **rgf_encode** requires that the index variable of the sequence and the generating function variable be passed as arguments. In the examples in this paper, we use **n** as the index variable and **z** as the generating function variable.

Consider the sequence $\langle n^2 + 2n + 3 \rangle$. We employ **rgf_encode** to find its generating function $G(z)$.

```
> rgf_encode(n^2+2*n+3, n, z);
```

$$\frac{3}{1-z} + 2\frac{z}{(1-z)^2} + z\left(\frac{1}{(1-z)^2} + 2\frac{z}{(1-z)^3}\right)$$

```
> Gz := rgf_norm(", z);
```

$$Gz := -\frac{-3 + 3z - 2z^2}{1 - 3z + 3z^2 - z^3}$$

The function **rgf_norm** puts a rational generating function into a "normal" form with respect to the generating function variable **z**. The function **rgf_expand** reverses the process to find the closed form of a sequence defined by a rational generating function. Consider the expansion of $G(z)$.

```
> rgf_expand(Gz, z, n);

2 (n + 1) (1/2 n + 1) - n + 1

> expand(");

  2
 n  + 2 n + 3
```

To determine if an expression is a valid closed form for a linear recurrence sequence, the type **rgf_seq** is defined by the **genfunc** package. The index variable must be passed as an argument to **rgf_seq** in the type expression. Consider the sequences $\langle n^2 + 2^n \rangle$ and $\langle 1/n \rangle$.

```
> type(n^2+2^n, rgf_seq(n));

true

> type(1/n, rgf_seq(n));

false
```

The sequence $\langle n^2 + 2^n \rangle$ is a linear recurrence sequence. The sequence $\langle 1/n \rangle$ has generating function $-\ln(1 - z)$, thus it is not a linear recurrence sequence.

Since all linear recurrence sequences are defined by homogeneous linear recurrences with constant coefficients, we often need to map a recurrence to a generating function. Maple's library function **rsolve** provides a 'genfunc' option that allows us to find a generating function from a recurrence. Consider the sequence $\langle F_n \rangle$ of Fibonacci numbers which is defined by the recurrence $F_n = F_{n-1} + F_{n-2}$, with boundary conditions $F_1 = F_2 = 1$. We can use **rsolve** to find the generating function $F(z)$ of the Fibonacci numbers.

```
> Fn := F(n) = F(n-1) + F(n-2);

Fn := F(n) = F(n - 1) + F(n - 2)

> Fz := rsolve({Fn, F(1)=1, F(2)=1}, F(n),
    'genfunc'(z));
```

```
                 z
Fz := - ------------
               2
        - 1 + z + z
```

To find the recurrence and boundary conditions for the Fibonacci numbers from their generating function we use the **rgf_sequence** function. This function answers a number of queries about the sequence encoded by a rational generating function. We use the 'recur' and 'boundary' queries here.

```
> rgf_sequence('recur', Fz, z, F, n);

F(n) = F(n - 1) + F(n - 2)

> rgf_sequence('boundary', Fz, z, F, n);

F(1) = 1, F(2) = 1
```

With the few tools considered here, we are ready to evaluate summations and convolutions of linear recurrence sequences. A well known result states that the generating function of the sequence $\langle s_n \rangle$ defined by $s_n = \sum_{k=0}^{n} a_k$, is $S(z) = A(z)/(1-z)$, where $A(z)$ is the generating function of $\langle a_k \rangle$. To evaluate this summation, we find the generating function of the solution, $S(z) = A(z)/(1 - z)$, and then expand $S(z)$ to find s_n.

Consider the summation $s_n = \sum_{k=0}^{n} 2^k$. To compute s_n, we use **rgf_encode** to find the generating function of $\langle 2^n \rangle$, divide this generating function by $1 - z$, and expand the result using **rgf_expand**.

```
> Az := rgf_encode(2^n, n, z);

           1
Az := -------
        1 - 2 z

> rgf_expand(Az/(1-z), z, n);

   n
 2 2  - 1
```

A similar technique applies to convolutions of linear recurrence sequences. The convolution is defined by $s_n = \sum_{k=0}^{n} a_k b_{n-k}$. Summation is the special case where $b_k = 1$. The sequence $\langle s_n \rangle$ is encoded by the generating function $S(z) = A(z)B(z)$, where $A(z)$ and $B(z)$ are the generating functions of $\langle a_k \rangle$ and $\langle b_k \rangle$. Expanding $S(z)$ gives s_n.

Consider the convolution $s_n = \sum_{k=0}^{n} k 2^{n-k}$. To evaluate s_n, we use **rgf_encode** to find the generating functions of $\langle n \rangle$ and $\langle 2^n \rangle$, multiply them to find the generating function of s_n, and expand the resulting generating function with **rgf_expand**.

```
> Az := rgf_encode(n, n, z);

            z
Az :=  --------
              2
         (1 - z)

> Bz := rgf_encode(2^n, n, z);

           1
Bz :=  -------
        1 - 2 z

> rgf_expand(Az*Bz, z, n);

                    n
 - n - 2 + 2 2
```

3. Summations and Convolutions Involving Symbolic Function Names

Often, a summation involves a symbolic function name instead of a closed form expression. For example, the problem might be to evaluate the summation $s_n = \sum_{k=1}^{n} F_k$, where $\langle F_k \rangle$ is the sequence of Fibonacci numbers. If a recurrence or generating function is given for the named function, we can use the functions in Section 2 to find a generating function for the solution and then expand it. However, this may not be desirable or feasible. The solution might be more understandable if it can be expressed in terms of the named function. For example, the sum of the first n Fibonacci numbers is known to be $F_{n+2} - 1$. The corresponding closed form is $(((1 + \sqrt{5})/2)^{n+2} - ((1 - \sqrt{5})/2)^{n+2})/\sqrt{5} - 1$. It is not unreasonable to regard $F_{n+2} - 1$ as a more informative solution. In other cases, a named function used in a problem may have a generating function that cannot be expanded by **rgf_expand**. Thus, we cannot find a closed form solution for the problem.

The sum $s_n = \sum_{k=0}^{n} F_k$ is the indefinite summation of the sequence $\langle F_k \rangle$. Evaluating the indefinite summation $s_n = \sum_{k=0}^{n} a_k$ of a linear recurrence sequence $\langle a_n \rangle$ is equivalent to solving the linear recurrence $s_n = s_{n-1} + a_n$, for boundary condition $s_0 = a_0$. For the sum of the Fibonacci numbers, we noted that the indefinite sum can be expressed using the symbolic name F_n. In general, the indefinite sum of any linear recurrence sequence $\langle a_n \rangle$ can be expressed in finite terms using the symbolic function name a_n [2, 3].

We use the function **rgf_relate** to evaluate indefinite summations that involve symbolic names of linear recurrence sequences. This function is a general purpose routine that relates two sequences that have common factors in the denominator of their generating functions. Let the sequence $\langle a_n \rangle$ be encoded by the generating function $A(z)$. Let $B(z)$ be a generating function whose denominator has factors in common with the denominator of $A(z)$. The function **rgf_relate** expresses the sequence encoded by $B(z)$ in terms of the symbolic sequence name a_n. The first four arguments of **rgf_relate** describe the generating function $A(z)$ and the sequence $\langle a_n \rangle$. The last two arguments describe the generating function $B(z)$ of the sequence that is to be expressed in terms of a_n.

Consider $s_n = \sum_{k=1}^{n} F_k$, the summation of the first n Fibonacci numbers. We found the generating function $F(z)$ of $\langle F_n \rangle$ in Section 2. To find an expression for the indefinite summation s_n, we divide $F(z)$ by $1 - z$ and relate s_n to the symbolic name F_n.

```
> rgf_relate(Fz, z, F(n), n, Fz/(1-z), z);

2 F(n) + F(n - 1) - 1
```

The expression for s_n can be left in this form, or the Fibonacci recurrence can be applied twice to express this in the simpler form $F_{n+2} - 1$. The function **rgf_simp** does this simplification for us. The function has 5 arguments and an optional sixth argument. It simplifies an expression with respect to a symbolic sequence name. The first argument is the expression to be simplified. The second and third arguments define the generating function of the sequence being simplified. The fourth and fifth arguments describe the sequence name. If the sixth argument is omitted, the expression is simplified relative to the nth term of the sequence. If the sixth argument is present, it provides the target index, and the sequence is simplified relative to that index.

```
> rgf_simp(Sn, Fz, z, F(n), n, n+2);
```

F(n + 2) - 1

This technique works for all linear recurrence sequences. We examine one other example here. Consider the sum $s_n = \sum_{k=0}^n t_k$, where the sequence $\langle t_k \rangle$ is defined by the recurrence $t_n = t_{n-1} + 2t_{n-2} + 3t_{n-3}$ with boundary conditions $t_0 = 0$, $t_1 = 1$, and $t_2 = 2$. Following the steps above, we can express s_n in finite terms using the symbolic name t_n.

```
> Tn := T(n)=T(n-1)+2*T(n-2)+3*T(n-3);

Tn := T(n) = T(n - 1) + 2 T(n - 2)

              + 3 T(n - 3)

> Tz := rsolve({Tn, T(i=0..2)=i}, T(n),
             'genfunc'(z));

                     2
             z + z
Tz := - ----------------------
              2       3
        - 1 + z + 2 z  + 3 z

> rgf_relate(Tz, z, T(n), n, Tz/(1-z), z);
```

6/5 T(n) + T(n - 1) + 3/5 T(n - 2) - 2/5

Convolutions involving symbolic names of linear recurrence sequences can also be expressed in finite terms using those symbolic function names. In some cases, **rgf_relate** can be applied immediately to evaluate the convolution. Consider the convolution of the Fibonacci sequence with itself, $s_n = \sum_{k=1}^n F_k F_{n-k}$. The generating function of s_n is $F^2(z) = z^2/(1 - z - z^2)^2$. We can relate s_n to F_n.

```
> rgf_relate(Fz, z, F(n), n, Fz^2, z);
```

(1/5 n - 1/5) F(n) + 2/5 F(n - 1) n

For other convolutions, this approach does not give satisfactory results. Consider the convolution $s_n = \sum_{k=0}^n t_k F_{n-k}$, where $\langle t_k \rangle$ is encoded by the generating function $T(z) = 1/(1 - 2z + 3z^2)$. The generating function of s_n is $T(z)F(z) = z/(1 - z - z^2)/(1 - 2z + 3z^2)$. We can relate s_n to either the symbolic name F_n or the name t_n. However, the most desirable result relates s_n to both function names, which **rgf_relate** cannot do. To find this

result, we first expand $S(z)$ using partial fractions with respect to z.

```
> Tz := 1/(1-2*z+3*z^2);

                1
Tz := ---------------
              2
        1 - 2 z + 3 z

> Sz := convert(Fz*Tz, 'parfrac', z);

                    5 + 4 z
Sz := - 1/11 ------------
                      2
              - 1 + z + z

                 - 5 + 12 z
      + 1/11 --------------
                      2
              1 - 2 z + 3 z
```

The left hand term can be related to F_n and the right hand term can be related to t_n.

```
> rgf_relate(Fz, z, F(n), n, op(1, Sz), z);
```

9/11 F(n) + 5/11 F(n - 1)

```
> rgf_relate(Tz, z, T(n), n, op(2, Sz), z);

                 12
- 5/11 T(n) + ---- T(n - 1)
                 11
```

Adding the two expressions gives s_n.

In general, all convolutions involving symbolic names of linear recurrence sequences can be expressed in finite terms using the symbolic names of the sequences involved in the convolution [2, 3]. We can evaluate such convolutions by applying partial fractions and then relating each term of the partial fraction expansion to the appropriate sequence.

4. Hybrid Terms and Linear Indexing

Hybrid terms are sequences defined by the termwise product of two or more parent sequences. For example, $F_n 2^n$ is a hybrid term derived from the sequences $\langle F_n \rangle$ and $\langle 2^n \rangle$. If $\langle a_n \rangle$ and $\langle b_n \rangle$ are linear recurrence sequences, then the hybrid sequence $\langle a_n b_n \rangle$ is a linear recurrence sequence and has a rational generating function [1, 2, 3].

The function `rgf_hybrid` computes the generating function of a hybrid sequence. The first argument to `rgf_hybrid` is the generating function variable. The remaining arguments are the generating functions of the hybrid term's parent sequences. Once the generating function of a hybrid term is found, we can employ the techniques of Section 3 to evaluate summations and convolutions involving that hybrid term.

Consider the summation $s_n = \sum_{k=1}^{n} F_k^2$. The sequence $\langle F_n^2 \rangle$ is a hybrid sequence defined by the term-wise product of two Fibonacci sequences. We employ `rgf_hybrid` to find its generating function.

```
> F2z := rgf_hybrid(z, Fz, Fz);
```

$$
F2z := -\frac{-z + z^2}{1 - 2z - 2z^2 + z^3}
$$

Dividing this generating function by $1 - z$ gives the generating function for s_n. The value of s_n can then be related to the summand F_n^2.

```
> rgf_relate(F2z, z, F(n)^2, n,
    F2z/(1-z), z);
```

$$
3/2\,F(n)^2 + 1/2\,F(n-1)^2 - 1/2\,F(n-2)^2
$$

This result expresses s_n in terms of the squares of the Fibonacci numbers. We can leave the result as is, or "simplify" it to involve just F_n and F_{n-1}.

```
> rgf_simp(``, Fz, z, F(n), n);
```

$$
F(n)^2 + F(n)\,F(n-1)
$$

A further "simplification" can be obtained if we simplify with respect to F_{n+1}.

```
> rgf_simp(``, Fz, z, F(n), n, n+1);
```

$$
F(n)\,F(n+1)
$$

We note that while these simplifications work out well for summations involving Fibonacci numbers, results for other sequences are not so compact. In general, if a sequence is defined by an order m recurrence, simplified expressions involving

that sequence will require at most m consecutive terms from the symbolic sequence name [2].

The techniques for hybrid sums work for all hybrid sequences defined by the term-wise product of two or more linear recurrence sequences. We examine one other example here. Consider the hybrid sum $s_n = \sum_{k=0}^{n} a_k b_k$, where the sequence $\langle a_k \rangle$ is encoded by the generating function $A(z) = 1/(1 - z + 4z^2)$ and the sequence $\langle b_k \rangle$ is encoded by the generating function $B(z) = 1/(1 - 5z - 2z^2)$. Following the steps above, we can express s_n in finite terms using the symbolic function names a_n and b_n.

```
> Az := 1/(1-z+4*z^2);
```

$$
Az := \frac{1}{1 - z + 4z^2}
$$

```
> Bz := 1/(1-5*z-2*z^2);
```

$$
Bz := \frac{1}{1 - 5z - 2z^2}
$$

```
> ABz := rgf_hybrid(z, Az, Bz);
```

$$
ABz := \frac{1 + 8z^2}{1 - 5z + 114z^2 + 40z^3 + 64z^4}
$$

```
> Sn := rgf_relate(ABz, z, AB(n), n,
                ABz/(1-z), z);
```

$$
Sn := \frac{213}{214}AB(n) + \frac{52}{107}AB(n-2)
$$
$$
+ \frac{109}{107}AB(n-1) + \frac{32}{107}AB(n-3) + 9/214
$$

```
> Sn := subs({seq(AB(n-i)=A(n-i)*B(n-i),
            i=0..3)}, Sn);
```

$$
Sn := \frac{213}{214}A(n)\,B(n) + \frac{52}{107}A(n-2)\,B(n-2)
$$

126

```
            109
        +   --- A(n - 1) B(n - 1)
            107

             32
        +   --- A(n - 3) B(n - 3) + 9/214
            107
```

We can leave the expression in this form, or "simplify" it relative to a_n and b_n.

```
> Sn := rgf_simp(Sn, Az, z, A(n), n);


Sn := 9/214


   /213        13
 + |--- B(n) - --- B(n - 2)
   \214        107

                                  \
              - 2/107 B(n - 3)| A(n)
                                  /

   /109
 + |--- B(n - 1) - 6/107 B(n - 3)
   \107

                      13          \
                  +   --- B(n - 2)| A(n - 1)
                      107         /
```

```
> Sn := rgf_simp(Sn, Bz, z, B(n), n);

              /205        14        \
Sn := 9/214 + |--- A(n) + --- A(n - 1)| B(n)
              \214        107       /

       / 19        36        \
     + |--- A(n) + --- A(n - 1)| B(n - 1)
       \107       107       /
```

```
> expand("");

        205            19
9/214 + --- A(n) B(n) + --- A(n) B(n - 1)
        214            107


     14                      36
 +   --- A(n - 1) B(n - 1) + --- A(n - 1) B(n)
     107                     107
```

Linear indexing uses a linear polynomial to index a linear recurrence sequence. The sequence $\langle F_{2n+1} \rangle$ uses the linear polynomial $2n + 1$ to index the Fibonacci numbers. If $\langle a_n \rangle$ is a linear

recurrence sequence, then the sequence $\langle a_{\alpha n+\beta} \rangle$, for integer constants $\alpha > 0$ and β, is a linear recurrence sequence and has a rational generating function [2, 3].

We need one other piece of information before we can compute the generating function of a sequence defined by linear indexing. If a sequence $\langle a_n \rangle$ is defined by an order m recurrence, then the sequence $\langle a_{\alpha n+\beta} \rangle$, for integer $\alpha > 0$ and β, is defined by an order m recurrence [3]. Knowing this, we can use the function rgf_findrecur to compute the order m recurrence defining $\langle a_{\alpha n+\beta} \rangle$.

The function rgf_findrecur computes the order m recurrence that defines a sequence from $2m$ consecutive values of the sequence. Once the recurrence is found, rsolve is used to find the generating function for the sequence. The techniques of Section 3 can then be used to evaluate summations and convolutions that involve the linearly indexed sequence.

Consider the summation $s_n = \sum_{k=0}^{n} t_k$, where $t_k = F_{2k+1}$. The sequence $\langle F_k \rangle$ is defined by an order 2 recurrence, therefore $\langle t_k \rangle$ is defined by an order 2 recurrence. The first four terms of the sequence $\langle t_k \rangle$ are $\langle 1, 2, 5, 13 \rangle$. The function rgf_findrecur computes its recurrence and rsolve computes its generating function $T(z)$.

```
> rgf_findrecur(2, [1, 2, 5, 13], T, n);

T(n) = 3 T(n - 1) - T(n - 2)

> rsolve({", T(0)=1, T(1)=3}, T(n),
    'genfunc'(z));


               1
Tz := -------------
           2
          z  - 3 z + 1
```

The generating function of s_n is $T(z)/(1 - z)$. The value of s_n can be related to F_{2n+1} and the resulting expression simplified relative to F_{2n}.

```
> rgf_relate(Tz, z, F(2*n+1), n,
    Tz/(1-z), z);

2 F(2 n + 1) - F(2 n - 1) - 1

> rgf_simp('', Fz, z, F(n), n, 2*n);

2 F(2 n) + F(2 n - 1) - 1
```

Not surprisingly, this can be further simplified to yield a more compact result.

```
> rgf_simp(", Fz, z, F(n), n, 2*n+2);
```

$$F(2n + 2) - 1$$

5.   Conclusion

This paper considered the use of rational generating functions in Maple to find finite expressions for summations and convolutions of linear recurrence sequences. Techniques were presented that handled hybrid terms and linear indexing. The techniques presented in this paper are general enough to solve quickly and conveniently any summation or convolution involving linear recurrence sequences.

The author would like to thank Edmund Lamagna, John Savage, and Keith Geddes for their useful feedback on this work.

REFERENCES

[1] Robert A. Ravenscroft, Jr. and Edmund A. Lamagna, "Symbolic Summation with Generating Functions," *Proceedings of the ACM-SIGSAM 1989 International Symposium on Symbolic and Algebraic Computation*, ACM Press, 1989.

[2] Robert A. Ravenscroft, Jr., "Generating Function Algorithms for Symbolic Computation," Ph. D. Thesis, Department of Computer Science, Brown University, 1991.

[3] Robert A. Ravenscroft, Jr., "Symbolic Computation with Generating Functions," submitted to *Journal of Symbolic Computation* special issue *Symbolic Computation in Combinatorics* Δ_1, the Proceedings of the ACSyAM Workshop September 21-24, 1993, Mathematical Sciences Institute, Cornell University.

Robert A. Ravenscroft, Jr. received his Ph.D. in computer science from Brown University. His research interests include the analysis of algorithms and symbolic computation. In 1991–92, he was a visiting post-doctoral fellow with the Symbolic Computation Group at the University of Waterloo. While there, he implemented the **genfunc** package in Maple and redesigned and reimplemented much of the **rsolve** library function. He was most recently employed as a lecturer in computer science at the University of Rhode Island.

Robert Ravenscroft
Department of Computer Science and Statistics
Tyler Hall, University of Rhode Island
Kingston, RI 02881, USA

IIIB. MAPLE IN SCIENCE

SYMBOLIC-NUMERIC COMPUTATIONS FOR PROBLEM-SOLVING IN PHYSICAL CHEMISTRY AND BIOCHEMISTRY

J. Grotendorst[1], J. Dornseiffer[2] and S. M. Schoberth[3]
Central Institute for Applied Mathematics[1], Institute for Applied Physical Chemistry[2], Institute for Biotechnology[3], Research Center Jülich, Jülich, Germany

1 Introduction

This article presents two illustrative examples, from Physical Chemistry and Biochemistry, showing how the use of modern computing tools such as Maple changes the approach to solving mathematical problems in science. The concept of scientific computation has evolved from a field encompassing primarily numerical methods to a much broader field that includes algebraic and analytical methods, numerical methods, and graphics. The combination of all these computing techniques facilitates efficient and accurate problem-solving. It allows the development of software systems for the automatic solution of problems in ways that are not possible with conventional computing systems. Figure 1 displays typical steps in a computerized solution process.

First, we study the thermodynamic behavior of steam reforming reactions. The product composition at equilibrium is calculated by solving a system of nonlinear equations which itself is derived by reformulating the equilibrium condition for each separate reaction. The Newton-Raphson method and symbolic computation techniques are used to solve this system of nonlinear equations.

In our second example we demonstrate how the McConnell equations in Biochemistry, a linear inhomogeneous system of differential equations with constant coefficients, can be solved elegantly by using symbolic linear algebra alone. This method utilizes the special structure of the ODE system and therefore is faster and more direct than simply using a general ODE solver. It is also an independent method compared to the solution techniques found in the literature [1-3]. In addition, methods of mixed symbolic-numeric type for the determination of the formal parameters involved in the analytical solutions are described.

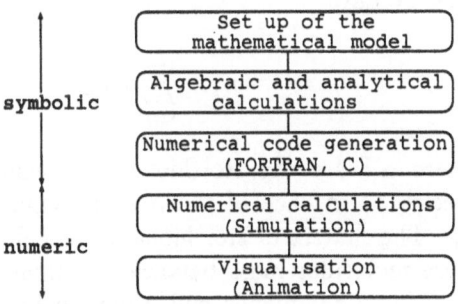

Figure 1: Computing steps in a scientific problem-solving process.

2 Physical Chemistry

The catalytic reaction between steam and hydrocarbon into mixtures of hydrogen, carbon monoxide, carbon dioxide and methane forms the basic feedstock (synthesis gas) to produce ammonia, methanol and other chemicals. To find the most economic reaction conditions, especially carbon boundaries, it is necessary to study theoretically the reaction behavior with respect to the operating parameters. In the present thermodynamic study we examine the temperature and pressure dependence of chemical compositions at equilibrium. We consider the following simple system of steam reforming reactions to demonstrate the mathematical model and the computation techniques involved:

1. The methane-steam equality

$$CH_4 + H_2O \rightleftharpoons CO + 3H_2$$

2. The water-gas shift

$$CO + H_2O \rightleftharpoons CO_2 + H_2$$

3. The carbon-methanation equality

$$CH_4 \rightleftharpoons C + 2H_2$$

To set up a practical computation model the equilibrium condition for each separate reaction is formulated in terms of the reaction extents ξ_i, $i = 1..3$. If $ne_j, j = 1..5$, is the number of moles of chemical species S_j present at equilibrium and n_j is the initial number of moles of that species, then we have

$$ne_j = n_j + \sum_{i=1}^{3} \alpha_{ji}\xi_i,$$

where α_{ji} are the stoichiometric coefficients of species S_j in each of the reaction equations (i) [4]. The matrix of stoichiometric coefficients for these molecular mass balance equations is given by (Maple commands will occur on lines beginning with " > "):

```
> with(linalg):
  alpha:=matrix([[-1,0,-1],[-1,-1,0],
                 [1,-1,0],[3,1,2],
                 [0,1,0]]);
```

$$\alpha := \begin{bmatrix} -1 & 0 & -1 \\ -1 & -1 & 0 \\ 1 & -1 & 0 \\ 3 & 1 & 2 \\ 0 & 1 & 0 \end{bmatrix}$$

Therefore, the composition of the participating chemical components at equilibrium is evaluated by the following matrix algebra:

```
> xi:=vector(3):
  print(xi);
  n :=vector(5):
  print(n);
  ne:=add(n,multiply(alpha,xi)):
  print(ne);
```

$$[\xi_1, \xi_2, \xi_3]$$

$$[n_1, \dot{n}_2, n_3, n_4, n_5]$$

$$[n_1 - \xi_1 - \xi_3, n_2 - \xi_1 - \xi_2, n_3 + \xi_1 - \xi_2,$$

$$n_4 + 3\xi_1 + \xi_2 + 2\xi_3, n_5 + \xi_2]$$

The equilibrium constants of the reactions in terms of the partial pressures

```
> p:=vector([P[CH4],P[H2O],P[CO],P[H2],
             P[CO2]]);
```

$$p := [P_{CH4}, P_{H2O}, P_{CO}, P_{H2}, P_{CO2}]$$

are calculated by

```
> eqn[i]:=K[i]=
  Product(p[j]^alpha[j,i],j=1..5);
  print(");
```

$$K_i = \prod_{j=1}^{5} p_j^{\alpha_{ji}}$$

```
> for i from 1 to 3 do
     print(value(eqn['i']))
  od:
```

$$K_1 = \frac{P_{CO}P_{H2}^3}{P_{CH4}P_{H2O}}$$

$$K_2 = \frac{P_{H2}P_{CO2}}{P_{H2O}P_{CO}}$$

$$K_3 = \frac{P_{H2}^2}{P_{CH4}}$$

Now, if we insert the relation $p_j = x_j P$, where p_j denotes the partial pressure, P the total pressure and x_j the mole fraction of chemical component S_j, then we obtain

```
> for j from 1 to 5 do
     p[j]:=x[j]*P
  od:
  j:='j':
  for i from 1 to 3 do
     print(value(eqn['i']))
  od:
```

$$K_1 = \frac{P^2 x_3 x_4^3}{x_1 x_2}$$

$$K_2 = \frac{x_4 x_5}{x_2 x_3}$$

$$K_3 = \frac{P x_4^2}{x_1}$$

The individual mole fractions x_j are calculated by dividing ne_j by the total number of moles. Thus, the sum over the component index should yield unity.

```
> nsum := sum(ne[j],j=1..5):
  x := map(y -> y/nsum,ne);
```

$$x := [\frac{n_1 - \xi_1 - \xi_3}{\%1}, \frac{n_2 - \xi_1 - \xi_2}{\%1}, \frac{n_3 + \xi_1 - \xi_2}{\%1},$$

$$\frac{n_4 + 3\xi_1 + \xi_2 + 2\xi_3}{\%1}, \frac{n_5 + \xi_2}{\%1}]$$

$$\%1 = n_1 + 2\xi_1 + \xi_3 + n_2 + n_3 + n_4 + n_5$$

```
> Sum(x[j],j=1..5)=
  simplify(sum(x[j],j=1..5));
```

$$\sum_{j=1}^{5} x_j = 1$$

The equilibrium mixture defined by $\xi_i, i = 1..3$, must simultaneously satisfy the equilibrium condition for each reaction, i.e. since $x_i = x_i(\xi_1, \xi_2, \xi_3)$ we have to solve the following equations for the reaction extents simultaneously:

```
> for i from 1 to 3 do
    f[i] := ln(value(rhs(eqn['i'])))-
            ln(lhs(eqn['i']))=0:
    print(");
  od:
```

$$ln\left(\frac{P^2(n_3 + \xi_1 - \xi_2)(n_4 + 3\xi_1 + \xi_2 + 2\xi_3)^3}{(n_1 - \xi_1 - \xi_3)}\right) -$$

$$ln(n_1 + 2\xi_1 + \xi_3 + n_2 + n_3 + n_4 + n_5)^2(n_2 - \xi_1 - \xi_2)) -$$

$$ln(K_1) = 0$$

$$ln\left(\frac{(n_4 + 3\xi_1 + \xi_2 + 2\xi_3)(n_5 + \xi_2)}{(n_2 - \xi_1 - \xi_2)(n_3 + \xi_1 - \xi_2)}\right) - ln(K_2) = 0$$

$$ln\left(\frac{P(n_4 + 3\xi_1 + \xi_2 + 2\xi_3)^2}{(n_1 - \xi_1 - \xi_3)}\right) -$$

$$ln(n_1 + 2\xi_1 + \xi_3 + n_2 + n_3 + n_4 + n_5) - ln(K_3) = 0$$

Before solving this system of nonlinear equations we determine the temperature dependence of the constants $ln(K_i), i = 1..3$. To obtain an expression for the reaction enthalpy we use Kirchhoff's law with a polynomial ansatz for the description of the molar heat capacity. We have

```
> diff(H(T),T)=
  Sum(A[j]*(s*T)^j,j=0..6);
```

$$\frac{d}{dT}H(T) = \sum_{j=0}^{6} A_j (sT)^j$$

```
> expand(dsolve(value("),H(T)));
  assign("):
```

$$H(T) = A_0 T + \frac{A_1 s T^2}{2} + \frac{A_2 s^2 T^3}{3} + \frac{A_3 s^3 T^4}{4} +$$

$$\frac{A_4 s^4 T^5}{5} + \frac{A_5 s^5 T^6}{6} + \frac{A_6 s^6 T^7}{7} + \_C1$$

Here, s $(= 10^{-3})$ denotes a scaling factor of the temperature T as used in thermochemical tables. Inserting the expression for $H(T)$ into the equation of van't Hoff and then integrating yields

```
> diff(lnK(T),T)='H(T)'/(R*T^2);
```

$$\frac{d}{dT}lnK(T) = \frac{H(T)}{RT^2}$$

```
> expand(dsolve(", lnK(T)));
  assign("):
```

$$lnK(T) = \frac{A_6 s^6 T^6}{42 R} + \frac{A_5 s^5 T^5}{30 R} + \frac{A_4 s^4 T^4}{20 R} + \frac{A_3 s^3 T^3}{12 R} +$$

$$\frac{A_2 s^2 T^2}{6 R} + \frac{A_1 s T}{2 R} + \frac{A_0 \ln(T)}{R} - \frac{\_C1}{RT} + \_C2$$

The coefficients A_j are determined by the corresponding coefficients of the pure substances in each reaction. These coefficients can be obtained from thermochemical data prepared for computer calculations [5]. The following coefficient matrix is used:

$$\begin{bmatrix} a0_{CH_4} & a0_{H2O} & a0_{CO} & a0_{H2} & a0_{CO2} & a0_C \\ a1_{CH_4} & a1_{H2O} & a1_{CO} & a1_{H2} & a1_{CO2} & a1_C \\ a2_{CH_4} & a2_{H2O} & a2_{CO} & a2_{H2} & a2_{CO2} & a2_C \\ a3_{CH_4} & a3_{H2O} & a3_{CO} & a3_{H2} & a3_{CO2} & a3_C \\ a4_{CH_4} & a4_{H2O} & a4_{CO} & a4_{H2} & a4_{CO2} & a4_C \\ a5_{CH_4} & a5_{H2O} & a5_{CO} & a5_{H2} & a5_{CO2} & a5_C \\ a6_{CH_4} & a6_{H2O} & a6_{CO} & a6_{H2} & a6_{CO2} & a6_C \end{bmatrix}$$

Now, reading in the coefficients

```
> readlib(readdata):
  U:=readdata('prothero.dat',6):
  U:=convert(U,matrix):
```

and then including carbon (graphite) into the stoichiometric coefficient matrix α

```
> beta:=stack(alpha,[0,0,1]);
```

$$\beta = \begin{bmatrix} -1 & 0 & -1 \\ -1 & -1 & 0 \\ 1 & -1 & 0 \\ 3 & 1 & 2 \\ 0 & 1 & 0 \\ 0 & 0 & 1 \end{bmatrix}$$

allows calculation of the coefficients $A_j, j = 0..6$, for the three reactions by the following matrix multiplication:

```
> V:=multiply(U,beta): print(V);
```

$$\begin{bmatrix} 10.45411100 & -5.293134000 & 2.846880000 \\ 20.38716900 & 33.69417100 & 36.52153400 \\ -85.87703600 & -57.81158600 & -98.41227000 \\ 100.2483700 & 51.32875000 & 103.6469300 \\ -57.17781800 & -25.50851700 & -55.79341600 \\ 16.39330720 & 6.730507200 & 15.30281000 \\ -1.890068240 & -0.7339833400 & -1.698663200 \end{bmatrix}$$

The integration constants _C1 and _C2 can be calculated by using special values of $H(T)$ and $lnK(T)$ at standard conditions $(T = 298.15°K)$ for the three equations [6]. Inserting the value of the gas constant R and taking into account the conversion factor F between the unit cal (used in Prothero's thermochemical table [5]) and the SI unit Joule we eventually arrive at:

```
> HO[1]:=206185: HO[2]:=-41165:
  HO[3]:= 74873:
  lnK0[1]:=-57.3621: lnK0[2]:=11.546:
  lnK0[3]:=-20.47458665:
  s:=10^(-3): R:=8.3143: F:=4.1868:
  for i from 1 to 3 do
    for j from 0 to 6 do
      A[j]:=F*V[j+1,i];
    od;
    solve({HO[i]=subs(T=298.15,
         H(T))},_C1);assign(");
    solve({lnK0[i]=subs(T=298.15,
         lnK(T))},_C2);assign(");
```

```
    ln(K[i])=eval(lnK(T));
    print(");assign(");
    _C1:='_C1':_C2:='_C2':
  od:
```

$$ln(K_1) =$$
$$-2.266129470 \; 10^{-20}T^6 + 2.751704836 \; 10^{-16}T^5 -$$
$$1.439640670 \; 10^{-12}T^4 + 4.206807101 \; 10^{-9}T^3 -$$
$$7.207461326 \; 10^{-6}T^2 + 5.133144050 \; 10^{-3}T +$$
$$5.264336374 \; ln(T) - 23067.92495\frac{1}{T} - 10.97646894$$

$$ln(K_2) =$$
$$-8.800218119 \; 10^{-21}T^6 + 1.129751855 \; 10^{-16}T^5 -$$
$$6.422612785 \; 10^{-12}T^4 + 2.153951730 \; 10^{-9}T^3 -$$
$$4.851992918 \; 10^{-6}T^2 + 8.483621895 \; 10^{-3}T -$$
$$2.665443083 \; ln(T) + 4698.749275\frac{1}{T} + 8.82258362$$

$$ln(K_3) =$$
$$-2.036641141 \; 10^{-20}T^6 + 2.568659013 \; 10^{-16}T^5 -$$
$$1.404783771 \; 10^{-12}T^4 + 4.349423747 \; 10^{-9}T^3 -$$
$$8.259514570 \; 10^{-6}T^2 + 9.195504045 \; 10^{-3}T +$$
$$1.433592386 \; ln(T) - 8107.550186\frac{1}{T} - 3.56194983$$

This completes the thermodynamic model. Next, we consider a feed mixture of $n_1 = 1$ moles of methane and $n_2 = 1$ moles of steam under a pressure of $P = 5$ bar and solve the equilibrium equations for the reaction extents $\xi_i, i = 1..3$. In Maple we applied the procedure fsolve which permits the numerical solution of equations. Assuming a step size of 10 degrees the program fsolve had to solve 46 times a system of three nonlinear equations within an accuracy of 6 digits over the interval $T = [750..1200]$. On a RS/6000-32H workstation this computing task requires about 80 seconds. The temperature dependence of the reaction extents and the resulting mole fractions for steam, hydrogen, carbon monoxide, carbon dioxide, and methane are displayed in Figures 2 and 3. When the reaction extent ξ_3 is zero or negative, the condition for no carbon formation (see Fig. 2), then a two dimensional calculation model has to be applied. Setting $\xi_3 = 0$ we obtain the two roots of $\xi_3(T)$ as follows:

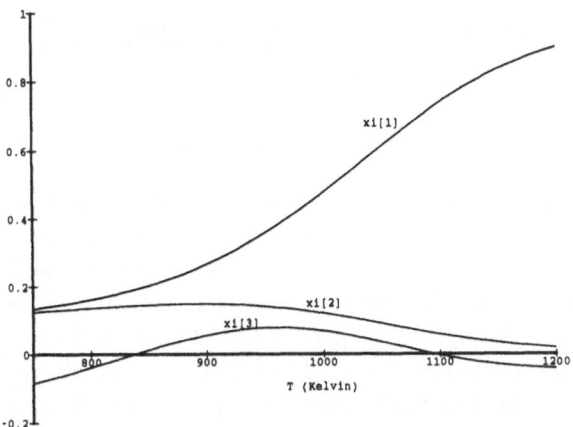

Figure 2: Temperature dependence of the reaction extents $\xi_i, i = 1..3$, under the conditions $P = 5\ bar$ and a steam to carbon atom ratio $H_2O : C = 1 : 1$.

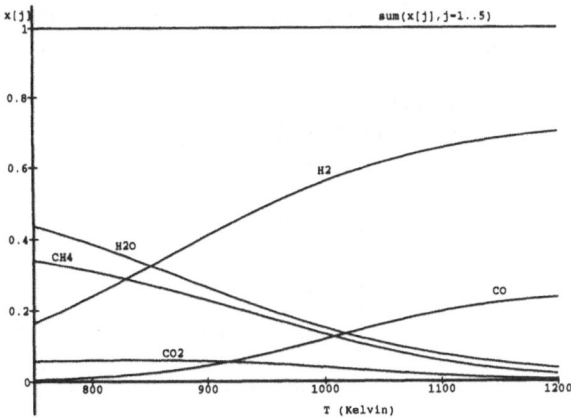

Figure 3: Temperature dependence of the mole fractions $x_j, j = 1..5$. Same conditions as in Figure 2.

```
> P:=5: n[1]:=1: n[2]:=1:
  n[3]:=0: n[4]:=0: n[5]:=0:
  Digits:=6;
  f1:=lhs(f[1]): f2:=lhs(f[2]):
  f3:=lhs(f[3]): xi[3]  := 0:
  fsolve({f1,f2,f3}, {xi[1],xi[2],T},
       {xi[1]=0.6..0.8,xi[2]=.0..0.1,
       T=1070..1100}):
  op(");

  fsolve({f1,f2,f3}, {xi[1],xi[2],T},
       {xi[1]=0.15..0.25,
       xi[2]=.05..0.2, T=820..860}):
  op(");
```

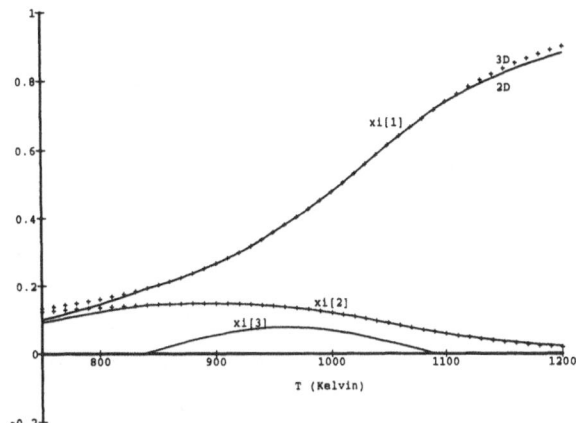

Figure 4: Temperature dependence of the reaction extents $\xi_i, i = 1..3$. Same conditions as in Figure 2. In regions with no carbon formation a two dimensional calculation model is applied.

$$\xi_1 = .729953, \xi_2 = .0616126, T = 1095.76$$

$$\xi_1 = .190220, \xi_2 = .143352, T = 837.633$$

Figure 4 shows the effect on the reaction extents ξ_1 and ξ_2 when a two dimensional calculation is done in regions with no carbon formation.

For the fast numerical solution of the nonlinear equations we used the Newton-Raphson algorithm and the programming language FORTRAN. The Jacobian matrix needed for this algorithm is calculated in exact form by Maple. The system of linear equations encountered in the Newton iteration is solved numerically by applying the LU decomposition routine *dgetrs* and the general linear equation solver *dgetrf* from LAPACK [7]. A preprocessor for FORTRAN code generation and optimization was implemented in Macrofort [8], a macro language for FORTRAN code generation in Maple (see Appendix). This preprocessor provides the capability to construct complete and ready-to-compile FORTRAN code for a given problem size, i.e. larger systems of reaction equations can be studied in the same way as demonstrated here. The generated numerical programs were used to investigate the temperature and pressure dependence of the reaction extents and the resulting mole fractions (see Fig. 5 and 6).

$$\frac{d}{dt}MA(t) = -k_{A1}\,MA(t) + k_{-1}\,MB(t) + d_A$$

$$\frac{d}{dt}MB(t) = k_1\,MA(t) - k_{B1}\,MB(t) + d_B$$

The functions $MA(t)$ and $MB(t)$ describe the time dependence of the nuclear spin magnetization of a single nuclear species which is transferred back and forth between chemically distinct sites ($A \rightleftharpoons B$) by kinetic processes. The magnetic resonance signals measured in NMR experiments are quantified as peak areas. The parameters k_1 and k_{-1} characterize the first-order rate constants for the forward and reverse reactions. The meaning of the other parameters will become clear in the sequel. From linear algebra we know that ODE systems with constant coefficients can be solved with the help of matrix exponentials [12]. To do it in Maple we first have to define the appropriate matrices.

```
> with(linalg):
  C:=matrix([[-k[A1], k[-1]],
          [ k[1], -k[B1]]]);
  D:=vector([d[A], d[B]]);
  Y0:=vector([MA0, MB0]);
```

$$C := \begin{bmatrix} -k_{A1} & k_{-1} \\ k_1 & -k_{B1} \end{bmatrix}$$

$$D := [d_A, d_B]$$

$$Y0 := [MA0, MB0]$$

Applying the Maple function for the evaluation of matrix exponentials leads to the following solution of the inhomogeneous ODE system:

```
> F:=multiply(exponential(C, t), Y0):
  f:=s -> multiply(exponential(C,
               t-s), D):
  P:=vector([int(f(s)[1], s=0..t),
          int(f(s)[2], s=0..t)]):
  S1:= add(F,P):
```

It should be noted that the solution of the analogous one-dimensional ODE is found in a formally identical way. The correctness of the one dimensionl solution is easily verified:

```
> exp(c*t)*y0+int(exp(c*(t-s))*d,
  s=0..t):
  evalb(expand(diff(",t)=c*"+d));
```

Figure 5: Temperature and pressure dependence of the reaction extent ξ_3. The steam to carbon atom ratio is $H_2O : C = 1 : 1$.

Figure 6: Temperature and pressure dependence of the mole fractions $x_j, j = 1..5$. Same conditions as in Figure 5.

3   Biochemistry

In recent years, in vivo nuclear magnetic resonance (NMR) spectroscopy has allowed measuring rate constants of transport and diffusion across living cell membranes [9-10]. The theoretical basis for the analysis of magnetization-transfer experiments is a system of differential equations first introduced by McConnell [11]. These equations, here formulated in a slightly different way as in Ref. [3], are given by

```
> diff(MA(t),t)=-k[A1]*MA(t) +
               k[-1]*MB(t) + d[A];
  diff(MB(t),t)= k[1]*MA(t) -
               k[B1]*MB(t) + d[B];
```

true

If matrix C is invertible then the solution of the inhomogeneous ODE system can be computed with matrix algebra alone. We have

```
> CD:=multiply(inverse(C), D):
  S2:=add(multiply(exponential(C,t),
     add(Y0,CD)), -CD):
```

Again, this direct solution can be motivated by the one dimensional case:

```
> dsolve({diff(y(t),t)=
        c*y(t)+d, y(0)=y0}, y(t));
```

$$y(t) = -\frac{d}{c} + e^{ct}\left(\frac{d}{c} + y0\right)$$

Comparing the solutions S1 and S2 yields

```
> map(simplify, add(S1,-S2));
```

$$[0, 0]$$

Next, we introduce physical boundary conditions to obtain a special regular solution. Let *MeA* and *MeB* denote the unperturbed equilibrium magnetizations of A and B respectively, i.e. the limits of *MA(t)* and *MB(t)* for t → ∞. These limits are given by the components of the constant vector -CD in solution S2

```
> MeA := -CD[1]; MeB := -CD[2];     #(*)
```

$$MeA := \frac{d_A k_{B1} + d_B k_{-1}}{k_{A1} k_{B1} - k_{-1} k_1}$$

$$MeB := \frac{k_1 d_A + d_B k_{A1}}{k_{A1} k_{B1} - k_{-1} k_1}$$

provided that both eigenvalues of matrix C

```
> eigenvals(C);
```

$$-\frac{k_{A1}}{2} - \frac{k_{B1}}{2} + \frac{\sqrt{k_{A1}^2 - 2 k_{A1} k_{B1} + k_{B1}^2 + 4 k_{-1} k_1}}{2}$$

$$-\frac{k_{A1}}{2} - \frac{k_{B1}}{2} - \frac{\sqrt{k_{A1}^2 - 2 k_{A1} k_{B1} + k_{B1}^2 + 4 k_{-1} k_1}}{2}$$

are negative. For realistic parameter values this condition holds. We have $k_1 \geq 0, k_{-1} \geq 0, k_{A1} = 1/T_1^A + k_1 \geq k_1$, and $k_{B1} = 1/T_1^B + k_{-1} \geq k_{-1}$. Here, $1/T_1^A$ and $1/T_1^B$ denote the relaxation rates of the spins in the two environments A and B. Under condition (*) the solution of the original inhomogeneous ODE system can be simplified to a sum where one term is the solution of a homogeneous ODE system and the other term, usually

the particular solution, is constant. Then, linear algebra computations in Maple lead to the following compact analytical solution (see Ref. [13] for details):

$$MA(t) = \left(\frac{\left(\sqrt{\%1} - k_{B1} + k_{A1}\right) MA10}{2\sqrt{\%1}} - \frac{k_{-1} MB10}{\sqrt{\%1}}\right) e^{-\frac{(k_{A1}+k_{B1}+\sqrt{\%1})t}{2}} + \left(\frac{\left(-k_{A1} + k_{B1} + \sqrt{\%1}\right) MA10}{2\sqrt{\%1}} + \frac{k_{-1} MB10}{\sqrt{\%1}}\right) e^{\frac{(-k_{A1}-k_{B1}+\sqrt{\%1})t}{2}} + MeA$$

$$MB(t) = \left(\frac{\left(-k_{A1} + k_{B1} + \sqrt{\%1}\right) MB10}{2\sqrt{\%1}} - \frac{k_1 MA10}{\sqrt{\%1}}\right) e^{-\frac{(k_{A1}+k_{B1}+\sqrt{\%1})t}{2}} + \left(\frac{\left(\sqrt{\%1} - k_{B1} + k_{A1}\right) MB10}{2\sqrt{\%1}} + \frac{k_1 MA10}{\sqrt{\%1}} + \right) e^{\frac{(-k_{A1}-k_{B1}+\sqrt{\%1})t}{2}} + MeB$$

where

$$\%1 = k_{A1}^2 - 2 k_{A1} k_{B1} + k_{B1}^2 + 4 k_{-1} k_1,$$

and

$$MA10 = MA0 - MeA,$$

$$MB10 = MB0 - MeB.$$

Solving the ODE system via the Maple function dsolve needs much more time compared to the more direct method using the matrix exponentials. On a RS/6000 workstation we measured a factor greater than 60 between these two methods. This is explained by the more general algorithms used in dsolve which work for nonlinear equations and may not be efficient for homogeneous linear systems.

The analytical solutions for *MA(t)* and *MB(t)* depend on the eight parameters *MA0*, *MB0*, *MeA*, *MeB*, T_1^A, T_1^B, k_1, and k_{-1}. Here, *MA0* and *MB0* denote the initial values of *MA(t)* and *MB(t)* at time 0. Usually, the parameters involved in *MA(t)* and $MB(t)$ are determined by a nonlinear least-squares analysis, i.e. by fitting the model parameters to experimental data for $MA(t)$ and $MB(t)$ obtained at different values of time [1-3]. For the numerical parameter fitting we used the ACM program NL2SOL [14] which is based on the Levenberg-Marquardt algorithm and which needs an analytic Jacobian matrix as input. The symbolic evaluation of the Jacobian matrix is done by the Maple procedure *jacobian*. Again, for the translation into optimized FORTRAN code and the generation of a driver program for the fitting routine we implemented a preprocessor in Macrofort [8]. This program generator is capable of constructing complete and ready-to-compile FORTRAN code for a given set of functions and parameters.

Now, for plotting the functions $MA(t)$ and $MB(t)$ we substitute special values for the formal parameters. These values were computed by fitting data from inversion transfer experiments to investigate transport processes in a special biological system [15] (see Fig.7).

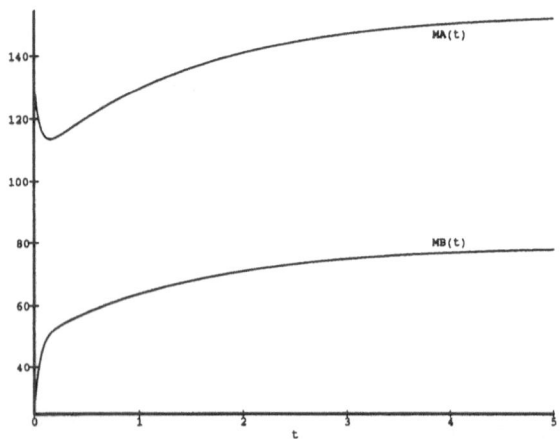

Figure 7: Plot of $MA(t)$ and $MB(t)$ for the special parameter values $k_1 = 6.6, k_{-1} = 10.5, MeA = 153.5, MeB = 78.8, MA0 = 130.4, MB0 = 27.08, T_1^A = 1.4$, and $T_1^B = 1.8$.

4 Conclusion

The combination of symbolic and numeric computation techniques leads to new approaches in areas of applied mathematics and science. It is shown, in examples from Physical Chemistry and Biochemistry, how the use of a modern computer algebra system such as Maple enables an automatic and computerized solution of problems in ways that are not possible with conventional computing systems.

In our first example we study the thermodynamic behavior of steam reforming reactions and determine the temperature and pressure dependence of the chemical compositions at equilibrium. Using Maple's symbolic computation capabilities a system of nonlinear equations for the reactions extents is derived. For the fast numerical solution of the nonlinear equations we applied the Newton-Raphson algorithm and the programming language FORTRAN. A FORTRAN preprocessor in Maple calculates the Jacobian matrix needed for this algorithm and generates a complete and ready-to-compile FORTRAN program.

In the second example we demonstrate how the McConnell equations in Biochemistry, a linear inhomogeneous system of differential equations with constant coefficients, can be solved analytically in a direct way by using Maple's symbolic linear algebra routines. In addition, the automatic generation of numerical code for the determination of the formal parameters involved in the solutions is described.

References

[1] J. J. Led and H. Gesmar, J. Magn. Res. 49, 444-464, (1982).

[2] P. W. Kuchel and B. E. Chapman, J. Theor. Biol. 105, 569-589, (1983).

[3] G. Robinson, B. E. Chapman, and P. W. Kuchel, Eur. J. Biochem. 143, 643-649, (1984).

[4] A. Ovenston and J.R. Walls, Chem. Engineering Sci. 35, 627-633, (1980).

[5] A. Prothero, Comb. Flame 13, (1969).

[6] I. Barin "Thermochemical Data of Pure Substances", VCH Verlagsgesellschaft, Weinheim (1993).

[7] J. Dongarra et al., "The LAPACK user's guide", SIAM publications, Philadelphia, (1992).

[8] C. Gomez, "Macrofort: a FORTRAN code generator in Maple", Maple Share Library, Version 5.0.

[9] K. Kirk, NMR Biomed. 3, 1-16, (1990).

[10] P. W. Kuchel, NMR Biomed. 3, 102-119, (1990).

[11] H. M. McConnell, J. Chem. Phys. 28, 430-431, (1958).

[12] P. Hartman, "Ordinary Differential Equations", John Wiley & Sons, Inc., (1964).

[13] J. Grotendorst, P. Jansen, and S. M. Schoberth, The Maple Technical Newsletter, 1, 56-62, (1994).

[14] J. E. Dennis, D. M. Gay, and R. E. Welsch, ACM Trans. Math. Software, Vol. 7, No. 3, (1981).

[15] S. M. Schoberth, A. A. de Graaf, B. E. Chapman, P. W. Kuchel, and H. Sahm, XV International Conference on Magnetic Resonance in Biological Systems, Jerusalem, Israel, August 16-21, (1993).

5 Appendix

It follows a listing of the FORTRAN code generator gen_fort in Macrofort and an example how it was used within Maple:

```
> with(share):
  readlib(macrofor):
  read(gen_fort):
  f1:=map(y->lhs(y),f):
  gen_fort(f1,xi,3,x,5,'extents.f'):
  # Maple program gen_fort in Macrofort
  gen_fort:=proc(ff,xx,d,x1,dd,pg_name)
  local f,fo,ii,jac,jj,j,k,linit,
        lwhile,m,pg,xi,zj,y,ipiv,n,
        nwhile1,llwhile,forloop,fw,eps,
        maxiter,start,step,nsteps,P,T,
        znorm;
  # calculate the jacobian matrix
  jac:=linalg[jacobian]
        ([seq(ff[ii],ii=1..d)],
         [seq(xx[ii],ii=1..d)]);
  # pg: MAPLE list describing main pgm
  pg:=[[declaref,
        'implicit double precision',
        ['(a-h,o-z)']],
       [declaref,'dimension',
       [f[d],zj[d,d],xi[d],y[dd]]],
       [declaref,dimension,[ipiv[d],
       n[dd]]],
       [writem,output,
        [''P,vector n:''],[]],
       [readm,input,
        ['d22.15','.dd.'i6'],
        [P,seq(n[ii],ii=1..dd)]],
       [writem,output,
        [''eps,maxiter:''],[]],
       [readm,input,['d22.15',i6],
        [eps,maxiter]],
       [writem,output,
        [''start,step,nsteps:''],[]],
       [readm,input,['2e14.7,i6'],
        [start,step,nsteps]]];
  # while instruction
  linit:=[[writem,output,
        [''Inital vector xi:''],[]],
        [readm,input,[''.d.'e14.7'],
        [seq(xi[ii],ii=1..d)]]];
  for ii from 1 to d do
     pushe([equalf,f[ii],ff[ii]],
           'linit')
  od;
  lwhile:=[[matrixm,zj,jac],
        [callf,dgetrf,[d,d,zj,d,
         ipiv,info]],
        [callf,dgetrs,[''N'',d,1,zj,
         d,ipiv,f,d,info]],
        [dom,k,1,d,[equalf,xi[k],
         -f[k]+xi[k]]]];
  for ii from 1 to d do
     pushe([equalf,f[ii],ff[ii]],
           'lwhile')
  od;
  llwhile:=[whilem,znorm(f,d) >= eps,
        linit, lwhile,maxiter];
  # translate the mole fractions
  fw:=[];
  for ii from 1 to dd do
     pushe ([equalf,y[ii],x1[ii]],
           'fw')
  od;
  fw:=[op(fw),[writem,4,['9(1x,e14.7),
     1x,i6'],[T,y,xi,nwhile1]]];
  forloop:=[dom,j,1,nsteps,[[equalf,T,
     start+(j-1)*step]],llwhile,fw]];
  pg:=[op(pg),[openm,4,'result.dat',
     UNKNOWN,forloop]];
  pg:=[programm,newton,pg];
  # fo: quadratic norm of f
  fo:=[[declaref,
        'implicit double precision',
        ['(a-h,o-z)']],
       [declaref,'dimension',[f[m]]],
       [equalf,znorm,0.],
       [dom,j,1,d,[equalf,znorm,
        znorm+f[j]**2]],
       [equalf,znorm,sqrt(znorm)]];
  fo:=[functionm,'double precision',
        znorm,[f,m],fo];
  # FORTRAN code generation
  writeto(pg_name);
  init_genfor();
  precision := double;
  optimized := true;
  interface(quiet=true);
  genfor(pg);
  genfor(fo);
  interface(quiet=false);
  writeto(terminal);
  end:
```

Biographies

Johannes Grotendorst obtained his doctoral degree in theoretical chemistry at the University of Regensburg and his background also includes studies in mathematics and physics at the University of Bielefeld. Since 1988, he has been at the Institute for Applied Mathematics at the Jülich Research Center (KFA). His current research interests include integration of symbolic and numeric computation techniques for mathematical problem-solving in applied science, extrapolation methods in theory and practice and parallel computing in computational chemistry.

Jürgen Dornseiffer obtained his diploma in chemistry at the Technical University of Aachen (RWTH) and is currently a Ph.D. student in the Institut for Applied Physical Chemistry at the Jülich Research Center. His research interests include the conversion of hydrocarbon waste materials by steam reforming.

Siegfried Schoberth obtained his doctoral degree in microbiology at the University of Göttingen. He is currently a research scientist in the Institute for Biotechnology at the Jülich Research Center. His research interests include "in vivo" NMR spectroscopy and general microbiology.

The authors may be reached at:

Research Center Jülich
D-52425 Jülich
Fed. Rep. Germany

j.grotendorst@kfa-juelich.de (J. Grotendorst)
j.dornseiffer@kfa-juelich.de (J. Dornseiffer)
s.schoberth@kfa-juelich.de (S.M. Schoberth)

MAPLE V ANALYTICALLY CALCULATING PLANE STRAIN FE-MATRICES WITH ROTATIONAL DEGREES OF FREEDOM

G. Zirwas
Lehrstuhl für Baumechanik, München, Germany

Finite Element Method (FEM) in Structural Mechanics

Structural mechanics in the civil or mechanical engineering domain typically poses a variety of systems of partial differential equations on arbitray shaped domains with mutiple boundary conditions, constraints and couplings. Due to their flexibility in solving these problems, FE Methods have become very popular methods of approximation.

Triangular Finite Elements

There are many strategies to calculate the stiffness matrices of finite elements [1][2][3]. A rough workhorse for plane strain or plane stress in two dimensional continuum mechanics is the CST formulation with three nodes and six degrees of freedom.

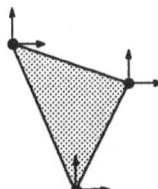

It shows, due to a Constant STrain and STress distribution in the element, only poor numerical behaviour. Furthermore no rotational degrees of freedom are handled, which leads to spurious problems when combined in three dimensions with elements which have, like beams, plates and shells, more than displacements as degrees of freedom at the nodes[4][5]. To overcome the situation it is not necessary to give up the simple triangular geometry, which has advantages for example in meshing and mesh refinements.

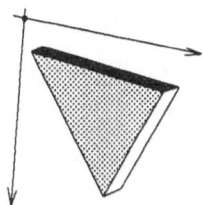

A complete step forward is made when the unknowns at the nodes are extended from the displacements u^i to the gradient $u^i\big|_j$ of the displacements. This means that all rotations at the nodes are known, but furtheremore the complete state of strain is involved and thereby the state of stress too. No jumps in stress or strain between the elements as in common formulations arise at the nodes because all state variables of interest are involved in the degrees of freedom. With MapleV it has become possible to provide a triangular element (TRT18 Traingular element with RoTations and 18 degrees of freedoms at three nodes) with a complete formulation of u^i and $u^i\big|_j$. Starting with the formalism of continuum mechanics, we change over to a symbolic solution in MapleV which is slightly different from typical FEM algorithms which do numerical integrations. At the end MapleV automatically codes the body of a TRT18 subroutine in C or FORTRAN for FEM-programs.

Continuum Mechanics of Plane Strain/Stress

The static and dynamic [6] situation of plane strain and plane stress are described by the well known coupled system of partial differential equations of Lamé or Navier.

$$(\lambda+\mu)u_i\big|_j^{\;i}+\mu u_j\big|_i^{\;i}=0.$$

The gradient

$$J^{ij}=u^i\big|^j$$

of the fields of displacements u^i is built up by an antisymmetric part of rotations

$$\omega_{ij} = \frac{1}{2}(u_i\big|_j - u_j\big|_i)$$

and a symmetric part of strain

$$\varepsilon_{ij} = \frac{1}{2}(u_i\big|_j + u_j\big|_i)$$

$$J^{ij} = \omega^{ij} + \varepsilon^{ij}.$$

The two independent constants of Lamé determine, in linear elasticity, the strain related stresses

$$\sigma_{ij} = \lambda\,\varepsilon_1^l\,g_{ij} + 2\mu\,\varepsilon_{ij}.$$

The scalar product of strain and stress

$U_{in} = \frac{1}{2}\sigma_i^j\,\varepsilon_j^i$ provides the free inner energy as

Lagrangian density for an approximation or a

strong solution. The inner energy [6] $U_{in} = \frac{1}{2}\sigma_i^j\,\varepsilon_j^i$

in technical (physical) strain and stress is

$$U_{in} = \frac{1}{2}\left[\sigma_x\,\varepsilon_x + \sigma_y\,\varepsilon_y + \tau_{xy}\,\gamma_{xy}\right].$$

To get the energy $U_{Element}$ stored in the finite area of an element needed for the approximation with the FE-Method, U_{in} has to be integrated either numerically or analytically (symbolically)

$$U_{Element} = \int\limits_{(V)} U_{in}\; dV.$$

In the case of flat coordinates, like Cartesian coordinates, where the base vectors do not change,

$$\bar{g}_n = \bar{e}_n = \mathrm{const}$$

the covariant and contravariant derivatives

$$(\;)\big|^n\;;\;(\;)\big|_n$$

simplify significantly because the Christoffel symbols vanish

$$\bar{g}_{n,\,k} = \Gamma_n{}^m{}_k\,\bar{g}_m \equiv 0.$$

Not only do triangles easily mesh arbitrary geometries with arbitrary refinements, triangular element formulations have the main advantage, in contrast to other isoparametric deformed ones, of vanishing covariant and contravariant derivatives.

Calculating with MapleV

We start with initializations and prepare some variables .
All parts of the entire MapleV input file are boxed.

```
restart; with(plots): with(linalg):
versn:=`### TRT_MSWS94 03.11.1994 12:00 GZ ###`;

FAZ     := 9;                   ### number of shape functions
DOF     := 2*FAZ;               ### number of Degrees Of Freedom
x       := vector(2);           ### global cartesian coordinates 2D.
s       := vector(2);           ### unit coordinates 2D.
ri      := matrix(3,2);         ### 3 coordinates of the element nodes.
g       := vector(6);           ### straight lines <=> planes.
fm      := vector(FAZ);         ### shape functions.
xtri    := vector(2);           ### element geometry; global coordinates.
Can     := vector(DOF);         ### coefficients for the shapefunctions.
GEO     := matrix(2,2);         ### base vectors.
u       := vector(2);           ### global field of displacement.
Jus     := matrix(2,2);         ### diff(ui,sj) gradients local derivation
Jux     := matrix(2,2);         ### diff(ui,xj) gradients global derivation
d       := vector(DOF);         ### degrees of freedom at nodes.
K1      := matrix(DOF,DOF);     ### one part of the Element Matrix
DC      := matrix(DOF,DOF);     ### d-Can-trans-matrix.
Txst    := matrix(DOF,DOF):     ### trans-matrix for Jacobian degrees.
```

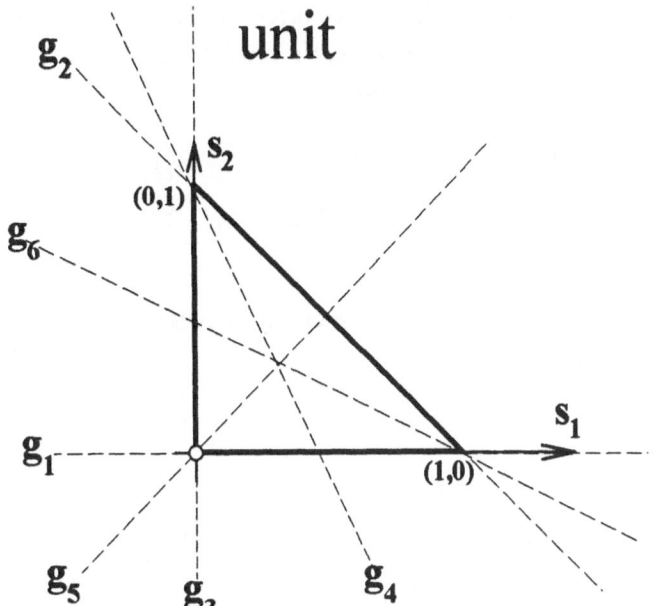 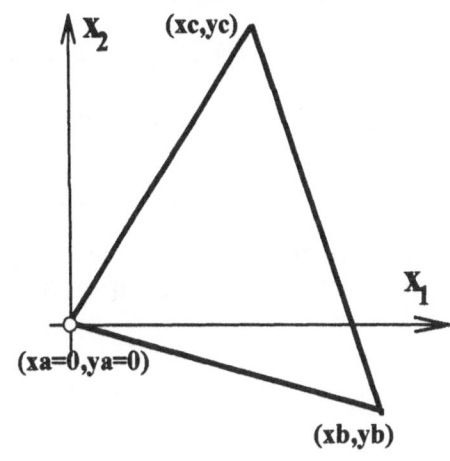

Projection from unit to global coordinates

Three lines at the boundary and three going through the midpoint are used

```
g[1]:=s[2];
g[2]:=1-s[1]-s[2];
g[3]:=s[1];
g[4]:=1-2*s[1]-s[2];
g[5]:=s[1]-s[2];
g[6]:=1-s[1]-2*s[2];
```

to build up as equations of planes the shape-functions fm[1] to fm[9].

```
fm[1]:=g[2] ;
fm[2]:=g[3] ;
fm[3]:=g[1] ;
fm[4]:=g[2]*g[3] ;
fm[5]:=g[2]*g[3]*g[4] ;
fm[6]:=g[1]*g[3] ;
fm[7]:=g[1]*g[3]*g[5] ;
fm[8]:=g[1]*g[2] ;
fm[9]:=g[1]*g[2]*g[6] ;
```

Nine shape functions for both of the displacement fields $u_{x_1}(s_1, s_2)$ and $u_{x_2}(s_1, s_2)$ in global directions x_1 and x_2 leads to 18 degrees of freedom. To evaluate the projection of a triangle from unit coordinates to global coordinates only the linear shape functions are needed. To decrease the number

of unknowns a shifted element with xa=0 and ya=0 is used. The three nodes are:

```
ri[1, 1]:=0;
ri[1, 2]:=0;
ri[2, 1]:=xb;
ri[2, 2]:=yb;
ri[3, 1]:=xc;
ri[3, 2]:=yc;
```

The global geometry depends on the three nodes and the unit coordinates. It is a projection from the s_1, s_2 to the $x^{i_2}(s_1, s_2) = x_{i_2}(s_1, s_2)$

$$x_{triangl}^{i_2}(s_1, s_2) = r_{corner}^{i_2 \, i_1} \cdot fm_{i_1}(s_1, s_2).$$

```
xtri[1]:=0; xtri[2]:=0;
for i1 to 3 do      ### 3 nodes.
  for i2 to 2 do    ### 2 coordinates.
   xtri[i2]:=xtri[i2]+ri[i1,i2]*fm[i1];
  od;
od:
```

A well known equation of tensor analysis states how to derive the base vectors. Because the curvature is zero, they are constant and the situation becomes simple.

$$\bar{g}^{i_2} = \frac{\partial\left(x^{i_1}_{\text{triangl}}(s_1, s_2)\, \vec{e}_{i_1}\right)}{\partial s_{i_2}} = \frac{\partial\left(x^{i_1}_{\text{triangl}}(s_1, s_2)\right)}{\partial s_{i_2}}\vec{e}_{i_1}$$

All is stored in the matrix GEO for later usage.

```
for i1 to 2 do      ### flat base vectors g_i2..
  for i2 to 2 do
   GEO[i1,i2]:=diff(xtri[i1],s[i2]);
  od;
od:
```

The determinant dF measures the changed size of differential areas $dx_1 dx_2 = dF \cdot ds_1 ds_2$ for later

integration. We seek the inverse relation from $x^{i_2}(s_1, s_2)$ to $s^{i_2}(x_1, x_2)$ for the chain rule in partial derivations that is

$$\begin{bmatrix} dx_1 \\ dx_2 \end{bmatrix} = [GEO]\begin{bmatrix} ds_1 \\ ds_2 \end{bmatrix} = \begin{bmatrix} \dfrac{\partial x_1}{\partial s_1} = xb & \dfrac{\partial x_1}{\partial s_2} = xc \\[2mm] \dfrac{\partial x_2}{\partial s_1} = yb & \dfrac{\partial x_2}{\partial s_2} = yc \end{bmatrix}\begin{bmatrix} ds_1 \\ ds_2 \end{bmatrix}$$

and vice versa for the total derivatives written in a matrix notation:

$$\begin{bmatrix} ds_1 \\ ds_2 \end{bmatrix} = \underbrace{\frac{1}{[GEO]}}_{[IG]}\begin{bmatrix} dx_1 \\ dx_2 \end{bmatrix} = \begin{bmatrix} \dfrac{\partial s_1}{\partial x_1} = \dfrac{yc}{xb\,yc - xc\,yb} & \dfrac{\partial s_1}{\partial x_2} = -\dfrac{xc}{xb\,yc - xc\,yb} \\[3mm] \dfrac{\partial s_2}{\partial x_1} = -\dfrac{yb}{xb\,yc - xc\,yb} & \dfrac{\partial s_2}{\partial x_2} = \dfrac{xb}{xb\,yc - xc\,yb} \end{bmatrix}\begin{bmatrix} dx_1 \\ dx_2 \end{bmatrix}.$$

```
IG:=inverse(GEO);
dF:=det(GEO);
```

The global fields of displacements $u_{x_1}(s_1, s_2)$ and $u_{x_2}(s_1, s_2)$ are superposed by the shape functions $fm_{i_1}(s_1, s_2)$ with respect to 18 generalized coefficients or degrees of freedom Can.

```
u[1]:=0: u[2]:=0:
for i1 to FAZ do
 u[1]:=u[1]+Can[2*i1-1]*fm[i1];
 u[2]:=u[2]+Can[2*i1  ]*fm[i1];
od:
```

Two material constants of Lamé adjust the energy density in linear elasticity. It is a linear combination $U_{in} = c_1\, U1 + c_2\, U2$ of two parts. The combination depends on whether you have plane strain or plane stress. The Jacobian Jux

$$Jux_{i_1 i_2} = \frac{\partial u_{x_{i_1}}(x_1, x_2)}{\partial x_{i_2}}$$

has to be used. U1 and U2 are positiv definite in the coefficients of the Jacobian Jux.

```
### c1=mu.
U1:=Jux[1,1]^2+Jux[2,2]^2+
```

```
     1/2*(Jux[1,2]+Jux[2,1])^2;   ### c1.
### c2=lam for Plane Strain or
### c2=2*lam*mu/(lam+2*mu) for Plane Stress
U2:=1/2*(Jux[1,1]+Jux[2,2])^2;    ### c2
```

The chain rule expands in the prepared derivatives with respect to the local coordinates to Jux:

$$Jux_{i_1 i_2} = \frac{\partial u_{x_{i_1}}(s_1, s_2)}{\partial x_{i_2}} = \underbrace{\frac{\partial u_{x_{i_1}}(s_1, s_2)}{\partial s_{i_3}}}_{Jus_{i_1 i_3}}\frac{\partial s_{i_3}(x_1, x_2)}{\partial x_{i_2}}$$

```
for i1 to 2 do
 for i2 to 2 do
   Jus[i1,i2]:=diff(u[i1],s[i2]);
 od;
od;

for i1 to 2 do
 for i2 to 2 do
  Jux[i1,i2]:=0;
   for i3 to 2 do
   Jux[i1,i2]:=Jux[i1,i2]+Jus[i1,i3]*IG[i3,i2];
            ### chain rule for partial derivation
  od;
 od;
od;
```

The integration with respect to unit coordinates s_1 and s_2 has to be taken into account by scaled differential areas. The term U1n that remains from the U1 portion of the energy is ready for integration (numerical or symbolic(analytical)).

```
U1n:=U1*dF:
```

FE-Method extremises, as an approximation, the inner energy

$$U_{element} = \frac{1}{2} K_{i_1 i_2} d^{i_1} d^{i_2}$$

which depends on the stiffness matrix and the displacements d, which are up to now the generalized ones, Can. Therefore the stiffness coefficients are the second derivatives of the stored energy

$$K_{i_1 i_2} = \frac{\partial^2 (U_{element})}{\partial d^{i_1} \, \partial d^{i_2}}$$

and here

$$K1_{i_1 i_2} = \frac{\partial^2 (U_{element})}{\partial Can^{i_1} \, \partial Can^{i_2}}$$

In the nonlinear situation this is the tangent stiffness matrix. The double integration over the unit triangular area is split up for every coefficient in the FE-Matrix, in order to manage the integration workload. What we get in this step is a stiffness matrix K1 for generalized displacements Can.

```
for i1 to DOF do
 for i2 to DOF do
  K1I[i1,i2]:=diff(U1n,Can[i1],Can[i2]);
                 ### stiffness-Integrand.
  K1[i1,i2]:=int(int(K1I[i1,i2],s[1]=0..(1-
s[2])),s[2]=0..1);
  od;
od:
```

We start to substitute Can by enumerated nodal displacements, which are the coefficients of the displacement vector and the Jacobian at the nodes. The coefficients in the Jacobian can be realized as degrees of rotations and stresses. To manage an inversion of this 18 by 18 matrix the mixed Jacobian Jus at the nodes has to be used.

```
d[ 1]:=subs(s[1]=0,s[2]=0,u[1]):    ### 1. node
d[ 2]:=subs(s[1]=0,s[2]=0,u[2]):
d[ 3]:=subs(s[1]=0,s[2]=0,Jus[1,1]):
d[ 4]:=subs(s[1]=0,s[2]=0,Jus[1,2]):
d[ 5]:=subs(s[1]=0,s[2]=0,Jus[2,1]):
```

```
d[ 6]:=subs(s[1]=0,s[2]=0,Jus[2,2]):
d[ 7]:=subs(s[1]=1,s[2]=0,u[1]):    ### 2. node
d[ 8]:=subs(s[1]=1,s[2]=0,u[2]):
d[ 9]:=subs(s[1]=1,s[2]=0,Jus[1,1]):
d[10]:=subs(s[1]=1,s[2]=0,Jus[1,2]):
d[11]:=subs(s[1]=1,s[2]=0,Jus[2,1]):
d[12]:=subs(s[1]=1,s[2]=0,Jus[2,2]):
d[13]:=subs(s[1]=0,s[2]=1,u[1]):    ### 3. node
d[14]:=subs(s[1]=0,s[2]=1,u[2]):
d[15]:=subs(s[1]=0,s[2]=1,Jus[1,1]):
d[16]:=subs(s[1]=0,s[2]=1,Jus[1,2]):
d[17]:=subs(s[1]=0,s[2]=1,Jus[2,1]):
d[18]:=subs(s[1]=0,s[2]=1,Jus[2,2]):
```

The dependency DC[i1,i2] of nodal degrees of freedom d[i1] on the Can[i2] are inverted to the dependency CD[i1,i2] of the Can[i1] on the nodal degrees of freedom d[i2]. The linear transformation is applied.

```
for i1 to DOF do
 for i2 to DOF do
  DC[i1,i2]:=diff(d[i1],Can[i2]);
  od;
od:
CD:=inverse(DC):
CDt:=transpose(CD):
KG:=multiply(CDt,multiply(K1,CD)):
```

The Jacobians at the nodes are, for KG, still in a mixed formualtion. In a last transformation Txst the displacements u are left invariant but the Jacobian Jus is expressed by the Jacobian Jux. The transformation appears as a block matrix for all three nodes

$$Jus_{i_1 i_2} = \frac{\partial u_{x_{i_1}}}{\partial s_{i_2}} = \underbrace{\frac{\partial u_{x_{i_1}}}{\partial x_{i_3}}}_{Jux_{i_1 i_3}} \underbrace{\frac{\partial x_{i_3}}{\partial s_{i_2}}}_{GEO_{i_3 i_2}}$$

or in matrix notation for every node

$$
\begin{bmatrix} u_1 \\ u_2 \\ u_{1,s_1} \\ u_{1,s_2} \\ u_{2,s_1} \\ u_{2,s_2} \end{bmatrix} = \begin{bmatrix} 1 & & & & & \\ & 1 & & & & \\ & & \dfrac{\partial x_1}{\partial s_1} & \dfrac{\partial x_2}{\partial s_1} & & \\ & & \dfrac{\partial x_1}{\partial s_2} & \dfrac{\partial x_2}{\partial s_2} & & \\ & & & & \dfrac{\partial x_1}{\partial s_1} & \dfrac{\partial x_2}{\partial s_1} \\ & & & & \dfrac{\partial x_1}{\partial s_2} & \dfrac{\partial x_2}{\partial s_2} \end{bmatrix} \begin{bmatrix} u_1 \\ u_2 \\ u_{1,x_1} \\ u_{1,x_2} \\ u_{2,x_1} \\ u_{2,x_2} \end{bmatrix}
$$

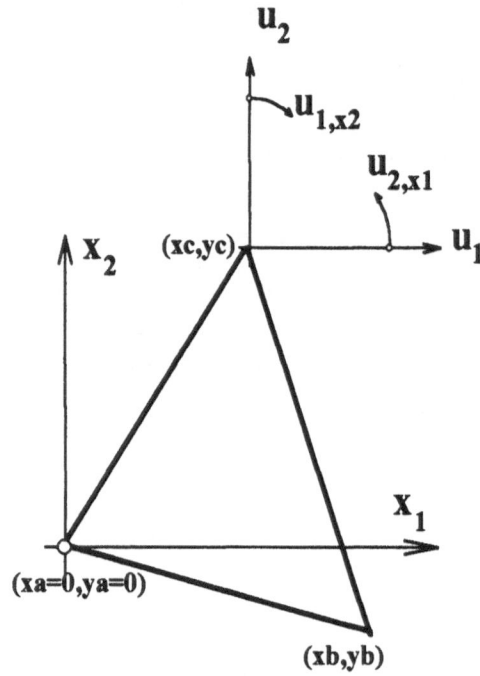

The degrees of freedom are at every node

$$
\begin{bmatrix} u_1 \\ u_2 \\ \varepsilon_x = u_{1,x_1} \\ \varphi_1 = u_{1,x_2} \\ \varphi_2 = u_{2,x_1} \\ \varepsilon_y = u_{2,x_2} \end{bmatrix}
$$

for TRT18. Condesing u_{1,x_1} and u_{2,x_2} the element TRT12 with 12 degrees of freedom can be derived. Here the drilling degrees of freedom are the side rotations. The degrees of freedom

$$
\begin{bmatrix} u_1 \\ u_2 \\ \varphi_1 = u_{1,x_2} \\ \varphi_2 = u_{2,x_1} \end{bmatrix}
$$

are shown at node c:

```
for i1 to DOF do
 for i2 to DOF do
  Txst[i1,i2]:=0;
 od:
 Txst[i1,i1]:=1;
od:
for i1 to 2 do
 for i2 to 2 do
  Txst[ 2+i1, 2+i2]:=GEO[i2,i1];
  Txst[ 4+i1, 4+i2]:=GEO[i2,i1];
  Txst[ 8+i1, 8+i2]:=GEO[i2,i1];
  Txst[10+i1,10+i2]:=GEO[i2,i1];
  Txst[14+i1,14+i2]:=GEO[i2,i1];
  Txst[16+i1,16+i2]:=GEO[i2,i1];
 od;
od:
Txstt:=transpose(Txst):
```

Applying this transformation to KG, the entire to global coordinates related stiffness-matrix TRT is achieved.

```
TRT:=multiply(Txstt,multiply(KG,Txst)):
```

We can look at the entire analytically calculated stiffness coefficients, which are functions of the geometry of the element, normalized to the Lamé constant μ and the thickness of the element

$$
TRT_{ij} = TRT_{ij}(xb, yb, xc, yc).
$$

A few results:

$$TRT_{[2,8]} = \frac{98x_c{}^2 - 98x_c x_b + 49y_c{}^2 - 49y_c y_b + 2x_b{}^2 + y_b{}^2}{90(x_c y_b - y_c x_b)}$$

$$TRT_{[8,10]}(yb = 0) = \frac{33x_b y_c{}^2 - 31x_c y_c{}^2}{360 x_b y_c}$$

Automated code generation is, in MapleV, a sophisticated way to plug the element into FEM-programs.

```
readlib(C):
# C(TRT);
# fortran(TRT);
```

The complete matrix or single coefficients can be coded in C

```
     TRT[18][8] = -(-46.0*yb*xc*xc+60.0*yb*xc*xb-63.0*yb*yc*yc+34.0*yc
*yb*yb+4.0*yb*xb*xb+2.0*yb*yb*yb-80.0*yc*xc*xb+8.0*yc*xb*xb-4.0*yc*xc
*xc-2.0*yc*yc*yc)/(xb*yc-xc*yb)/360;
                          . . . . . . .
                          . . . . . . .
```

or FORTRAN.

Concluding Remarks

Without influencing the basic solution process a **varying family** of element formulations are derived. **Condensing** degrees of freedoms at the nodes, the number of unknowns can be lowered from 18 (TRT18 complete gradient) to 12 (TRT12 only rotational degrees of freedom) and to 6 (CST no rotational degrees of freedom).
The MapleV solution can easily be applied to a pyramid shaped brick element (PRT48), with 4 nodes and 48 degrees of freedom. Downward we can have a standard beam element in 3D.
All these elements fit together, because the fields of displacements at their boundaries are **natural splines** of degree three.

Application of computational resources during the production of FE-codes leads to faster response at the application afterwards. **No numerical** integration has to be passed for **every** single element.

Both

analytical symbolic integration with MapleV

and

automated and optimized C or FORTRAN code of the matrices

leads to a rational and fast application of computational resources.

References

[1] K.J. Bathe. Finite Element Procedures in Engineering Analysis. Prentice-Hall, Englewood Cliffs, N.J., 1982.
[2] K.J. Bathe and E. Dvorkin. A formulation of general shell elements - the use of mixed interpolation of tensioral components. Int. J. Numer. Methods in Eng., 22, 697-722, 1986.
[3] B. Szabo and I. Babuska. Finite element analysis, John Wiley Sohns, 1991.

[4] T.J.R. Hughes and F. Brezzi, On the drilling degrees of freedom, Comput. Meths. Appl. Mech. Engrg., 72, 105-121, 1989.

[5] M. Iura and S. N. Atluri, Geometrically nonlinear membrane elements with drilling degrees of freedom, Proc. of the Int. Conf. on Comp. Eng. Sci. (ICES 1991), 393-398, Atlanta, 1991.

[6] Flügge W., Tensor analysis and continuum mechanics, Springer, N.Y. 1986.

[7] G. Zirwas, Impedance Matrix of Viscoelastic (layered) Halfspace, Proc. of the Int. Conf. on Comp. Eng. Sci. (ICES 1991), 1293-1296, Atlanta, 1991

The author Gerhard Zirwas studied Civil Engineering at the Technical University of Munich in Germany. He is presently an Assistant at this University. He recently works on continuum mechanics of large scale structures and soil structure interaction [7].

G. Zirwas
Lehrstuhl für Baumechanik
Technische Universität München
Arcisstr. 21
D-80333 München FRG

email: t5121ak@sunmail.lrz-muenchen.de
Tel: +49 89 2105 8341

CHEMICAL ENGINEERING WITH MAPLE

Ross Taylor[1] and Katherine Atherley[2]

Department of Chemical Engineering[1], Clarkson University, Potsdam, NY;

Waterloo Maple Software[2], Waterloo, Ontario, Canada

Introduction

Chemical engineering students are required (by accreditation agencies) to make appropriate use of computers throughout their program. Appropriate use is defined as including most of the following: (1) programming in a high level language; (2) use of software packages for analysis and design; (3) use of appropriate utilities; (4) simulation of engineering problems. Maple (Char, 1991) is a powerful and flexible computing tool that has the potential of becoming the software of choice for much scientific and engineering work, perhaps replacing, at least in part, other computer based methods such as traditional programming languages and special purpose analysis and design programs. The fact that Maple is designed for symbolic manipulation should not be taken to imply that it is unsuitable for the numerical calculations that dominate engineering computing today. Maple's symbolic mathematical abilities combined with numerical capabilities and sophisticated graphics allow new approaches to the teaching of traditional materials. In this paper we focus on a few ways in which Maple can be used in selected courses in the chemical engineering curriculum.

Computer Programming

Undergraduate engineering students are required to do some programming in a high-level language. Often, the programming course is the first one to expose students to elements of engineering problem solving. Most students learn Fortran, but C or Basic are included in some curricula. With more and more students learning to use Maple in their calculus classes (and fewer and fewer of them using any kind of traditional programming language after they graduate), it makes sense to consider adopting Maple as the programming language for use in their engineering courses (see the box *Maple at Clarkson*). Maple is built around a programming language that is custom designed for symbolic mathematical calculations and manipulations. Unlike Fortran, the Maple language supports standard mathematical structures as data-types, and can work with them in sensible mathematical ways. The language is more natural for mathematical work, and produces (much) shorter, easier to understand programs than Fortran.

Material and Energy Balances

Maple allows you to solve elementary material balance problems in a systematic way that makes it almost impossible to get the problem formulation wrong. Just a few lines of Maple code can set up the material balances and mole fraction summation equations for any process unit regardless of the number of components and input and output streams. Systems with chemical reactions also can be modeled. Maple's ability to handle symbolic indices makes it possible to identify components with a number, name, chemical formula, or any other convenient label. The number of unknown variables and independent equations can be quickly counted and, hence, the number of degrees of freedom determined. Specification equations can be added to the model equations and Maple asked to solve the entire set of equations in one go. Problems often can be solved symbolically in terms of an unspecified parameter (reactor conversion, say). This is useful if it is desirable to evaluate the solution at several parameter values. It is unnecessary even to choose a basis as the actual specification of interest can be included among the set of equations.

Maple at Clarkson

In common with many other schools, undergraduate engineering students at Clarkson University are required to take a programming course. Until this year, the language taught in this course was Fortran (although some students received instruction in Basic). No longer will Clarkson students take courses in Fortran (unless it be by choice); the introductory computing course has been completely revised and Clarkson students will now be programming using the computer algebra system Maple.

The decision to abandon Fortran in favor of Maple was not reached lightly or without considerable debate within the engineering school at Clarkson. Factors in favor of Maple included the fact that the students were receiving some instruction in the use of Maple in their calculus classes and that the site license allows Maple to be installed on nearly all machines owned by Clarkson. All students (regardless of discipline) currently receive a PC when they enroll as freshmen. Clarkson's PC program has been in place for over a decade and this year the freshmen received their computers with Maple already installed on the hard drives of their PC.

The fact that Maple can exchange files across platforms is a great advantage for a computationally diverse campus like Clarkson; in addition to the huge number of PCs (> 3000) there are nearly two hundred IBM RS/6000 workstations and more than a few Suns and DECs.

Thermodynamics

The calculation of the critical constants for a cubic equation of state is a classic problem in thermodynamics, one that is covered in most thermodynamics textbooks (see, e.g . Walas, 1985; Sandler, 1989). Usually, textbook examples include finding the critical constants for the simplest cubic equation of state, that of Van der Waals, or those of the Redlich-Kwong family. With Maple, however, it is possible to obtain explicit expressions for the critical constants for a generic cubic equation of state (Taylor and Monagan, 1993). Constants for particular equations of state can be obtained as special cases of the general result.

Maple is a useful tool for the visualization of thermodynamic functions. Figure 1 shows the roots of the compressibility polynomial for the Soave-Redlich-Kwong equation as a function of reduced temperature at a reduced pressure of 0.75. The well-known fact that a cubic equation can have complex roots at certain parameter values is well-illustrated here. This figure was obtained with about 30 lines of Maple code which included the calculation and ordering of the roots themselves.

Figure 2 shows the Gibbs energy surface for a ternary liquid mixture computed with the NRTL model (see WalasFig, 1985). 3D diagrams in Maple can be rotated and viewed from other angles; in this case it is possible to show graphically the conditions under which liquid-liquid phase splitting can occur.

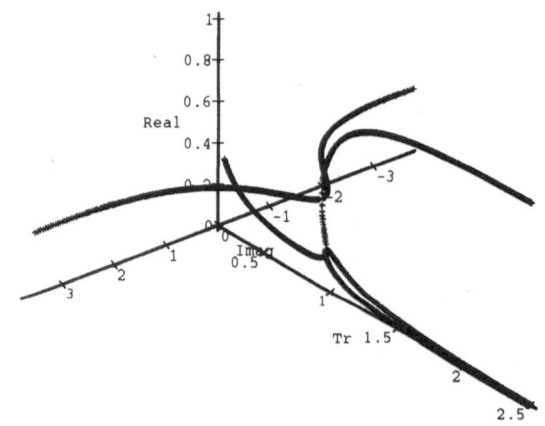

Figure 1: Compressibility as a function of reduced temperature (Tr) at a reduced pressure of 0.75 computed using the Soave-Redlich-Kwong cubic equation of state. The vertical axis is the real part of the compressibility, the horizontal axis at the back of the figure is the imaginary part of the compressibility. All axes are dimensionless. The region of three real roots is shown between a region posessing a real liquid-like root and complex vapor-like roots and another region with a real vapor-like root and two complex liquid-like roots.

150

what equations, and what numerical methods should be used to solve (each subset of) the equations have been thoroughly explored. Figure 3 shows an incidence matrix (sparsity pattern) for a column distilling a nonideal binary mixture in 10 stages. The equations and variables are grouped by stage leading to the familiar block tridiagonal pattern shown in the figure; however, just two lines of Maple is all it takes to reorder the equations so that they are grouped by type rather than by stage. It is pertinent to point out that although with Maple it is simple to explore different computational strategies, Maple also makes topics like this largely irrelevant (even if they are interesting).

Figure 2: Gibbs energy surface for acetone(1), benzene(2), carbon tetrachloride(3). x1 and x2 are the mole fractions of components 1 and 2. The vertical axis is the dimensionless Gibbs energy.

A Maple procedure to obtain *expressions* for the activity coefficients from any model of the excess Gibbs energy and for a specified number of components can be written in about 10 lines of Maple code. Students can investigate different models of the Gibbs energy function without running the risk of getting the derivation incorrect.

Simple phase equilibrium calculations and the creation of phase diagrams for binary systems using Maple have been discussed by Taylor (1994).

Figure 3: Incidence matrix for a small distillation problem (2 components, 10 stages). The pattern is "upside-down" because Maple plots cannot (yet) go from high to low on any axis.

Separation Processes

Multicomponent distillation simulations require the numerical solution of a large set of equations: Material balances, energy balances, and equilibrium (thermodynamics) equations. These equations are sparse, nonlinear, and can easily number in the hundreds and sometimes in the thousands. The literature on distillation contains scores of papers discussing methods of solving these equations (see Seader, 1985). Issues such as what form the equations should take, what variables should be used, in what way should the equations and variables be ordered, what variables should be computed from

Four pages of Maple code is all that is required to solve most multicomponent distillation problems (including the *derivation* of *all* of the equations). This compares to the many hundreds (or even thousands) of lines that would be needed to solve the same problem using Fortran. Interlinked columns and nonstandard specifications also are simple to deal with (and only slightly more difficult to solve). Figure 4 shows the McCabe-Thiele diagram plotted from the results obtained by numerically solving the equations that gave us Figure 3.

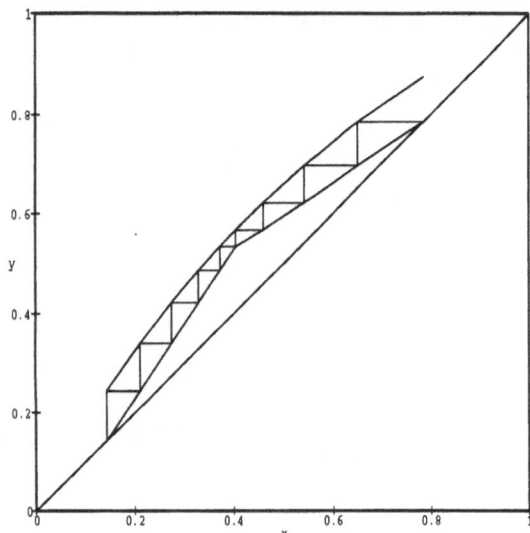

Figure 4: McCabe-Thiele diagram constructed from the solution obtained with Maple to a binary distillation problem. The y-axis is the mole fraction of component 1 in the vapor, the x-axis is the mole fraction of component 2 in the liquid. Both axes are dimensionless. The upper curve is the equilibrium line, the staircase represents the equilibrium stages, 9 of them here (the column has a total condenser).

Chemical Reaction Engineering

Chemical reaction engineering problems often require the solution of systems of coupled differential equations. Textbook problems sometimes are specially simplified so that the equations can be solved analytically. While such solutions can also be obtained with Maple, it is no longer necessary to simplify problems in this way. Software packages that possess numerical methods for solving ODEs can be used to solve more realistic problems. This point of view has already been expressed by Fogler (1992) who uses Mathematica and Polymath for solving reaction engineering problems. The advantages of a CAS over a purely numerical method of solution include the fact that the reaction system can be analyzed symbolically and the material and energy balance equations also may be derived.

Figure 5 is the well-known illustration of multiple steady states in a nonisothermal continuous stirred tank reactor (CSTR). This plot was created using the parameters given by Shacham et al. (1994) who also considered the CSTR dynamics and noted that

the stability of each steady state could be determined by computing the eigenvalues of the state matrix using Polymath. All of these things are possible with Maple; however, Maple is capable of evaluating the eigenvalues of the state matrix symbolically as well!

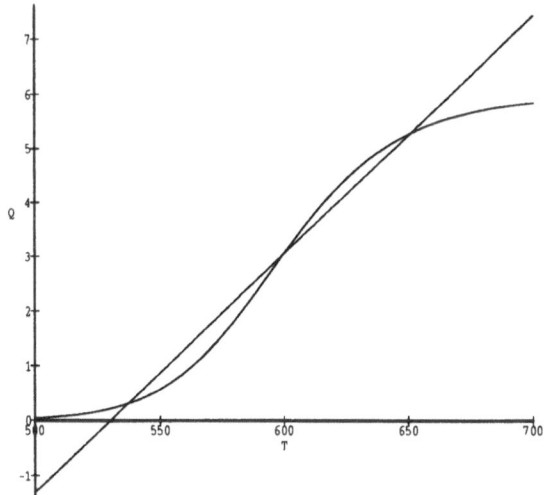

Figure 5: Multiple steady states in a stirred tank reactor. The vertical axis is the heat duty (Btu/hr x 10^{-5}), the x-axis is the reactor temperature (K). The straight line represents the heat lost due to cooling the reactor. The S-curve is the heat generated by the reaction. Each intersection of these lines represents a solution of the steady-state material and energy balances. The middle solution is unstable.

Pocess Design

We do not think it requires too great a leap of imagination to expect students to set-up and solve entire process flowsheeting problems using Maple. The techniques that are useful for material and energy balances around simple units are readily applied to process flowsheets with any number of units and their interconnections. Flowsheets containing recycle streams are easy to handle.

While Maple will not (nor should it) replace specialized programs designed for large-scale plant simulation, Maple is a useful tool for teaching students how flowsheeting simulations work. The Maple programming language encourages problems to be formulated in a way that is reminiscent of equation oriented flowsheeting (Westerberg et al., 1979) but it is not too difficult to instruct Maple to solve flowsheet problems using tearing or simultaneous modular strategies. Maple's open interface and powerful language makes it possible for engineers to create their own unit models in the Maple programming language; others would then be able to use them as plug-in modules in their own problems if these modules were to be made available.

More Maple

There are, of course, a number of ways in which Maple needs to be improved.

Maple needs to gain the ability to read and write binary direct access files (as is possible in Fortran and C). This will make it possible to access databanks of physical property data for direct use in engineering calculations.

Maple needs improved symbolic capabilities. There are many engineering formulae where it is necessary to differentiate arbitrary sums and products (i.e. a sum or product of indexed variables where the index range is non-numeric: i.e. $i = 1..c$). The fact that Maple cannot do this is a very serious impediment to using Maple for certain important problems (e.g. the derivation of thermodynamic properties of mixtures). We also need to be able to (elegantly) exclude selected elements from sums and products of indexed variables.

A great many problems in chemical engineering require finding numerical solutions to large (or small) systems of (sparse) nonlinear equations. This is not one of Maple's strengths. The floating point solver built into Maple lacks some of the features that one would like to see. In particular, it is not possible to provide the initial estimates or to control the iteration history. On the other hand, it is possible to program Newton's method in Maple so that all the user must provide is a set of equations, a list of unknown variables and a starting point; Maple can compute the Jacobian symbolically thereby removing one of the major chores that must be faced when using the method as part of a Fortran program. In fact, Maple is used for precisely this purpose within some companies that write software for process engineering simulations. Differential arc-length homotopy continuation methods, recommended by Seader (1985) for solving difficult nonlinear problems, also may be easily (and elegantly) programmed in the Maple programming language.

Unfortunately, Maple currently is many times slower than compiled Fortran when carrying out large scale numerical computations. Maple needs better (i.e. faster) routines for purely numerical computing. A fast sparse linear equation solver, for example, would go a long way to making large scale flowsheeting problems and multicomponent distillation problems a practical proposition. For now, you can set up the problem with Maple and then translate the results into Fortran (or C) so that you can compile the application yourself.

Many models in chemical engineering consist of large sets of (stiff) ordinary differential equations, (ODEs) mixed systems of differential and algebraic equations (DAEs) or partial differntial equations (PDEs). Fast numerical methods for stiff ODEs, DAE systems, and for solving PDEs by, for example, the method of lines would be very welcome.

Maple's graphics capabilities, although quite good, could be improved by adding more basic plot types that are encountered often in (chemical) engineering (triangular diagrams and their three-dimensional counterparts are examples that come to mind).

Conclusion

In this article we have highlighted only a few ways in which Maple can be used in chemical engineering education. Additional applications of Maple are listed in Table 1. Some of these worksheets are included in the Maple share library (and which is provided as part of the new Release 3 of Maple V), others are available from the first author. We have also identified a few areas where Maple needs improved capabilities.

The fact is that Maple can have a significant impact in almost all areas of chemical engineering education. However, there are some problems associated with using Maple in existing courses.

Using Maple to derive expressions that are standard fare in current engineering textbooks will rapidly demonstrate that it is not always easy to get from Maple an answer that you recognize. It may be quite trivial for Maple to solve your problem correctly but it can require considerable skill in expression manipulation in order to get a familiar result. While this may not be important in solving original problems, it can make matching the results

Table 1: Maple Worksheets in Chemical Engineering

1. Material balance calculations on a variety of chemical process units

2. Thermodynamics
   (a) Critical constants for cubic equations of state (Taylor and Monagan, 1993)
   (b) Phase equilibrium calculations and phase diagrams for ideal systems (Taylor, 1994)
   (c) Activity coefficients in binary and multicomponent systems
   (d) Gibbs free energy surfaces
   (e) Phase equilibrium calculations for nonideal systems
   (f) Flash calculations for ideal systems
   (g) Advanced flash calculations
   (h) Thermodynamic property relations and the Maxwell equations

3. Reaction Engineering
   (a) Material balances in tubular reactors
   (b) Isothermal tubular reactor (multiple reactions, numerical integration)
   (c) Nonisothermal tubular reactor
   (d) Multiple steady states in a CSTR
   (e) CSTR dynamics
   (f) Fitting reaction rate coefficients to rate data.

4. Equilibrium Stage Separations
   (a) Constant molar overflow in distillation
   (b) Multicomponent distillation - stage-to-stage calculations
   (c) Multicomponent distillation - simultaneous solution
   (d) McCabe-Thiele diagrams

5. Numerical Methods
   (a) Newton's method for systems of equations
   (b) Homotopy-continuation for systems of equations

in established textbooks a frustrating experience. Perhaps we will have to get used to new ways of looking at old results. The problem of simplifying the chore of obtaining recognizable results remains as a challenge for the computer algebra community.

The exercises and examples in many standard textbooks (see, for example, Felder and Rousseau, 1986; Reklaitis, 1983) were designed to be solved by hand. Many (if not most) of these problems are far too simple if Maple is on hand to assist with the problem solving; when you have solved one, you have solved them all.

This brings us to some important questions: Do we want students to use Maple for solving engineering problems? Can the use of a CAS prevent students from mastering essential skills that are better assimilated when solving problems by hand? Computer algebra is finding increasing use in the teaching of calculus at many schools. It is impossible to turn back the clock and abandon the use of computer algebra in mathematics courses thereby making its use in engineering a nonissue. It will not be possible to prevent students from using tools they have learned once it has become clear that they are useful. It will be up to us as educators to find the proper time and place in our courses to introduce students to engineering problem solving with Maple.

References

Char, B.W., K.O. Geddes, G.H. Gonnet, B.L. Leong, M.B. Monagan, and S.M. Watt, Maple V Language Reference Manual, Springer-Verlag (1991).

Felder, R.M. and R. W. Rousseau, Elementary principles of Chemical Processes, 2nd Ed., Wiley, New York (1986).

Fogler, H.S., Elements of Chemical Reaction Engineering, 2nd Edition, Prentice-Hall, Englewood Ciffs, NJ (1993)

Reklaitis, G.V., Introduction to Material and Energy Balances, McGraw-Hill, New York, 1983.

Sandler, S.I. Chemical and Engineering Thermodynamics, Wiley, 2nd Edition. New York (1989).

Schacham, M., N. Brauner, and M.B. Cutlip, "Exothermic CSTRs. Just how stable are those steady states," Chem. Eng. Ed., pp. 30-35, Winter (1994).

Seader. J.D. "The BC and AD of Equilibrium Stage Operations," Chem. Eng. Ed., XIX(2), 88 (1985).

Taylor, R. and Monagan, M.B., "Thermodynamics with Maple. I - Equations of State," Maple Tech, Vol. 10, (1993).

Taylor, R., "Thermodynamics with Maple. II - Phase Equilibria in Binary Systems," Maple Tech, Vol. 11, (1994).

Walas, S.M. Phase Equilibria in Chemical Engineering, Butterworths, Stoneham, MA (1985)

Westerberg, A.W., H.P. Hutchison, R.L. Motard, and P. Winter, Process Flowsheeting, Cambridge University Press, Cambridge (1979).

The Authors

Ross Taylor is a professor of Chemical Engineering at Clarkson University. His interests are in the areas of mass transfer and separation processes and he is the coauthor (with Professor R. Krishna of the University of Amsterdam) of Multicomponent Mass Transfer (Wiley, 1993) and (with Harry Kooijman) of *ChemSep*, a software package for separation process simulation used in universities in several countries. He can be contacted by email at taylor@sun.soe.clarkson.edu.

Katherine Atherley is Head of Documentation at Waterloo Maple Software (WMS). She has a degree in pure mathematics from the University of Waterloo, where she first encountered Maple. She worked with the Maple research group before joining WMS. Since then she has been involved with many aspects of Maple, from programming and writing documentation to providing user support, and teaching Maple courses.

The authors are writing a book entitled "Chemical Engineering with Maple".

IVA. MAPLE IN EDUCATION

MAPLE LABS AND PROGRAMS FOR CALCULUS

William W. Farr and Michael VanGeel
Department of Mathematical Sciences, Worcester Polytechnic Institute,
Worcester, MA

Abstract

This paper describes work done over the past three years at Worcester Polytechnic Institute aimed at integrating Maple computer labs and projects into freshman calculus. We give an overview of the course structure and then provide examples drawn from labs we have used. We also describe the Maple CalcP package, a set of about thirty Maple procedures written at WPI for use in the course.

Introduction

For the past three years, we have been working on introducing Maple into the calculus course taken by nearly all entering freshmen at WPI. Most of our students are engineering majors of one sort or another, so they need to be able to use calculus to solve problems. Our main goals in introducing Maple into the course were to expose the students to a CAS, which we see becoming a standard tool for any technical profession, and, more importantly, to focus the students' attention away from the details of manipulation and onto the fundamentals of using calculus to solve problems.

In the course of developing labs and projects for use in the course, we have written about thirty different Maple procedures for use by the students. They range from visualization tools for topics like Newton's method and solids of revolution to routines for computing curvature and normal and tangential acceleration for parametric curves in two and three dimensions. This paper gives a brief summary of all the routines in the CalcP package and provides examples of how some them have been used in the course. Our main motivation in writing the procedures was to keep the focus of the students on calculus and to keep the Maple component of the course from becoming too burdensome. On the other hand, we do encourage the students to learn Maple by using it to solve problems and students with programming backgrounds are encouraged to learn a little about programming. In fact, the second author of this paper was a student in the Maple calculus course who went on to help with the programming in his second year at WPI.

The routines we have written have been collected together in the form of the Maple package CalcP, including help pages for each procedure in the package. (For calculus with projects - see the course description below.) They have been tested with the X11/UNIX and PC-Windows versions of Maple V, Release 2 and Release 3. They are available by anonymous ftp from the host wpi.wpi.edu in the files Calc/CalcP.R2 and Calc/CalcP.R3. If there is sufficient interest, the package will be submitted to the Maple share library. Maple users are encouraged to obtain and use this package. Report any problems of bugs to the first author, preferably via e-mail at the address bfarr@wpi.edu.

Course Format

Calculus classes at WPI normally meet four times a week. The regular academic year consists of four seven-week terms, each roughly equivalent to a quarter in terms of contact hours. For the Maple calculus sections, one of the four weekly class meetings takes place in our computer lab, equipped with 22 DECstation UNIX workstations.

When students arrive at the lab, they are given a two-to-four page handout that contains background material on the lab topic, examples, and a set of exercises for them to complete. The students work in pairs on the lab. During the lab period, the instructor and student helpers circulate around the room, providing help and answering questions. Students generally do not complete the lab in one hour. On the contrary, our intention is to provide a mix of routine computations and challenging exercises in the labs. The student teams have about a week to finish the exercises and write them up in the form of a joint lab report.

To encourage reflection, investigation, and understanding on the part of the students, we require that the lab reports consist of three parts. In the first part, the students are to describe in their own words the main ideas of the lab. The second part consists of their answers to the exercises, including explanations of what they did and why. The last part is a Maple worksheet containing the details of their work on the lab. We note that requiring this format makes the labs much easier to grade; we don't have to dig through the worksheet to figure out what they did.

Students in the Maple course also spend about three weeks per term working in teams of three to four students on a project. These are generally more involved applications of calculus, requiring modeling and analysis on the part of the students. Recent projects have ranged from design of a roller coaster loop to analysis of crank shaft diameter - connecting rod length tradeoffs in engine design. Each group produces a joint report, with the teams having about two weeks to produce a draft report, which is criticized and returned. The teams then have about a week to produce a final draft upon which their grade is based.

Students in the Maple course also have traditional homework and exams, counting for approximately 30% of their final grades. The five labs and one project per term count for the remainder.

Maple programs

This section provides a brief description of the each of the Maple procedures defined in the CalcP package. Examples are provided in the next few sections.

- The CalcP package contains three procedures that are useful in the visualization and manipulation of polyline approximations to a function.

  ArcInt displays the integral to calculate the *actual* arc length of a given function. It can be used to find a numerical value as well.

  PolyLength gives the numerical value of the arc length of a polyline approximation to a function.

  PolyLine displays the graph of a polyline approximation to a given function.

- Sequences defined by functional recursion are handled by the following four procedures.

  PlotSeq plots the points of a recursive sequence.

  Recur returns only the nth value of a recursive sequence.

  Sequence displays the first n values of a recursive sequence.

  ShowSeq displays a graph of the values of a recursive sequence. It differs from **PlotSeq** in that it graphs the sequence generator and the line $y = x$. It then connects the points of the sequence to these two curves by line segments. The resulting plot is a graphical interpretation of the convergence of a recursive sequence.

- Plotting of parametric curves in two and three dimensions is performed by the following four procedures.

  ParamPlot animates the drawing of a two dimensional parametric plot.

  ParamPlot3D animates the drawing of a three dimensional parametric plot.

  SurfLoop plots a two dimensional parametric curve on a three dimensional surface.

  Loopimate animates the drawing of a parametric curve on a three dimensional surface. Used in the study of surfaces and their parameterization.

- Solids of revolution can be explored using the following `CalcP` procedures.

  LeftDisk graphically displays the solid of revolution for a function approximated by disks or washers.

  LeftInt returns the volume of the solid of revolution approximated by n disks or washers.

  RevInt displays the integral to calculate the volume of the solid of revolution of a given function. The numerical value can also be computed.

  revolve plots the solid of revolution formed by rotating a given function about a specified axis parallel to the x-axis.

- Several procedures for studying parametric curves can be found in the `CalcP` package. In the following descriptions, "vector function" refers to a function of a single real variable, with values in $\mathbf{R}^2$ or $\mathbf{R}^3$.

  Curvature computes the curvature of a two or three dimensional vector function.

  NormAccel returns the magnitude of the normal acceleration along a vector function.

  normalvect returns the unit normal vector to a given two or three dimensional vector function.

  Speed returns the scalar speed of a particle moving along a two or three dimensional vector function.

  TanAccel returns the magnitude of the tangential acceleration along a vector function.

  tanvect returns the unit tangent vector to a vector function.

  unitvect creates a unit vector out of a vector or two endpoint of a vectors.

  VDiff returns a vector which is the derivative of a vector function.

  VMag calculates the magnitude of a vector.

  VPlot plots the graph of a two or three dimensional vector function.

- The following are related to linear and higher order approximations to functions, including applications like Newton's method for solving single equations.

  Newton computes the Newton series approximation for a root of a function.

  NewtonPlot plots the function and the Newton series approximation to the root of the function.

  secantline returns an expression for a line that intersects a given curve in exactly two places.

  tangentline returns the expression for a line that is tangent to a given curve.

  TanPlane returns the expression for a plane that is tangent to a given three dimensional surface.

  Taylor returns the expression for a truncated Taylor series approximation to a function. Unlike the standard Maple procedure **taylor**, Taylor is a polynomial that can be manipulated in the usual manner.

  TayPlot shows the effectiveness of varying Taylor polynomial approximations. It plots the function itself along with specified Taylor polynomial approximations.

Examples

In this section, we provide examples of labs and Maple programs.

Definition of the derivative

In this lab, students use the `secantline` procedure from the `CalcP` package to investigate and compare the algebraic definition of the derivative as the limit of the difference quotient and the geometric definition as the limiting slope of secant lines. As the examples given below show, the `secantline` procedure takes three arguments: an expression or procedure f, a base point a, and an increment h. The result of the procedure is the Maple expression for the straight line between the points $(a, f(a)$ and $(a+h, f(a+h))$. As also shown below, the result can be displayed, plotted, and even animated. Figure 1 below shows the result of the `plot` command in the example.

```
> f := proc(x)  x^3+2*x+1 end;
f := proc(x) x^3+2*x+1 end
> secantline(f,x=0,1.0);

   3.000000000 x + 1
> secantline(f,x=0,0.5);
```

```
   2.250000000 x + 1
> plot({f(x),secantline(f,x=0,1),
   secantline(f,x=0,0.5)},x=0..1);
> with(plots):
> animate({f(x),secantline(f,x=1,1-t)},
x=0.5..2.5,t=0..0.99);
```

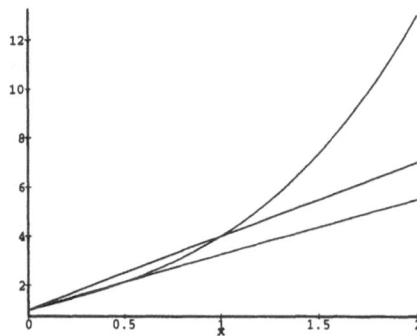

Figure 1: Example plot generated by the **secantline** command.

Solids of revolution

This topic appears about halfway through integral calculus. The Maple programs help the students by displaying the solid of revolution (**revolve**), the integral for the volume of the solid (**RevInt**), and by plotting and computing disk approximations to the solid (**LeftDisk** and **LeftInt**). Examples are shown in the Maple session and figures below.

```
> f := proc(x)  x^2+1 end;
f := proc(x) x^2+1 end
> revolve(f(x),x=-2..2);
> RevInt(f(x),x=-2..2);
     2
    /
   |        2    2
   |   Pi (x  + 1)  dx
   |
   /
   -2
> evalf(RevInt(f(x),x=-2..2));
   86.28907824
> LeftDisk(f(x),x=-2..2,5);
> LeftDisk(f(x),x=-2..2,10);
> LeftInt(f(x),x=-2..2,5);
   99.52163404
> LeftInt(f(x),x=-2..2,10);
   89.62938709
```

By default, the **revolve** and **LeftDisk** commands put caps on the ends of the solids. However, to speed up the display, these can be omitted by using the **nocap** option. These two procedures can also handle a set of two functions, in which case the solid, or its approximation by washers, is that given by rotating the region between the two curves. It is also possible to rotate about an arbitrary axis parallel to the x axis.

In the lab dealing with solids of revolution, the students are asked the following question.

> Design a drinking glass by revolving a suitable function about the x-axis. In your report, give the function, a three dimensional plot of your glass, and determine the volume of the liquid-filled part. You will be graded on the utility and elegance of your design.

Students seem to really enjoy this question and come up with very nice designs, one of which is shown below in Figure 3. The function used to generate the glass design shown in Figure 3 is

$$g(x) = \begin{cases} -x^3 + 0.15 & \text{if } x < 0 \\ 0.15 & \text{if } 0 \leq x < 10 \\ \frac{\sqrt{x-10}}{2} + 0.15 & \text{if } 10 \leq x < 20 \\ \frac{\sqrt{10}}{2} + 0.15 & \text{if } x \geq 20 \end{cases}$$

Figure 2: Solid of revolution and ten disk approximation.

The following Maple session shows the steps needed to define this function and plot the solid of revolution. Notice the quotes in the `revolve` call to delay evaluation.

```
> g := proc(x) if x<0 then -x^3+0.15
```

```
      elif x < 10 then 0.15
      elif x < 20 then sqrt(x-10)/2+0.15
      else sqrt(10)/2+0.15 fi
end:
> revolve('g(x)',x=-1..30,nocap);
```

Figure 3: Student glass design.

Taylor polynomials

The standard Maple `taylor` procedure includes an order term and is not directly suitable for computations. Converting to a polynomial is straightforward, but is a complication that we wanted our students to avoid so we wrote a wrapper function `Taylor` that handles the conversion. We also wrote a procedure `TayPlot` that plots the function on the same graph as a set of Taylor polynomial approximations. Examples are shown in the Maple session below and Figure 4, which shows the output of the `TayPlot` command in the Maple session. Notice the use of the `numpoints` parameter. The `TayPlot` command will accept any of optional arguments to the `plot` command.

```
> Taylor(sin(x),x=0,5);
            3         5
    x - 1/6 x  + 1/120 x
> Taylor(exp(x),x=1,3);
    exp(1) + exp(1) (x - 1) +
                          2
        1/2 exp(1) (x - 1)  +
                          3
        1/6 exp(1) (x - 1)

> TayPlot(sin(x),x=0,{13,31},
x=0..6*Pi,y=-2..2,numpoints=100);
```

In the labs, the students use these two commands to investigate convergence of Taylor polynomials and the use of known Taylor series to generate related Taylor series via multiplication, division, or substitution. For example, the series for $\exp(x^2)$ is obtained by substituting $u = x^2$ in the series for $\exp(u)$, as is partially shown in the Maple example below.

```
> subs(u=x^2,Taylor(exp(u),u=0,4));
         2        4        6         8
    1 + x  + 1/2 x  + 1/6 x  + 1/24 x

> Taylor(exp(x^2),x=0,8);
         2        4        6         8
    1 + x  + 1/2 x  + 1/6 x  + 1/24 x
```

By using Maple, it is easy for the students to experiment with Taylor polynomials and discover for themselves, for example, the reason why the orders had to be different in the two commands above to obtain the same order Taylor polynomial.

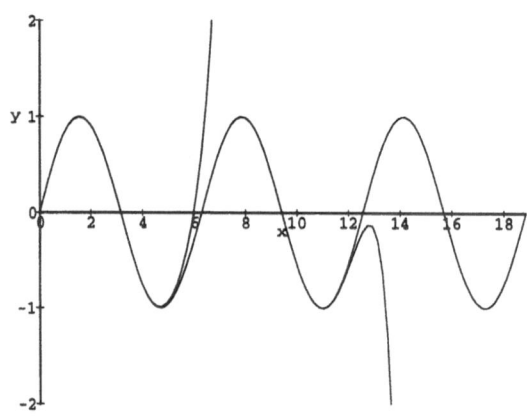

Figure 4: Plot of $\sin(x)$, along with the Taylor polynomial approximations of orders 13 and 31.

Vector calculus of parametric curves

This topic often seems to be given short shrift in typical calculus courses. With help from Maple, it isn't too hard to give this the attention we feel it deserves. As described in an earlier section, the `CalcP` package contains quite a few procedures having to do with motion on curves, ranging from the curve computation routines `Curvature`, `normalvect`, and `tanvect` to the procedures `ParamPlot` and `ParamPlot3d` which use

the Maple animation commands to show the actual motion along a parametric curve in two and three dimensions. Some simple examples are shown in the Maple session below. Several of the routines use the `linalg` package, though all will accept arguments that are lists of functions rather than vectors. Notice that the curve computation routines return either an expression if the third argument is a name, or a numerical value if the third argument assigns a value to the independent variable.

```
> tanvect([cos(t),sin(t)],t);
  [ - sin(t), cos(t) ]
> r := proc(t) [t,t^2,t^3] end;
r := proc(t) [t,t^2,t^3] end
> TanAccel(r(t),t);
                   3
      4. t + 18. t
   ------------------------
              2       4 1/2
   (1. + 4. t  + 9. t )
> TanAccel(r(t),t=1);
   5.879747321
> Curvature(r(t),t);
          4     2     1/2
     (9 t  + 9 t  + 1)
   2 -------------------
             2     4 3/2
     (1 + 4 t  + 9 t )
```

With Maple to do the tedious manipulations, the students can be given exercises to work on that would probably not be possible by hand. Alternatively, the students can shift their focus away from the manipulations to the results. Below, we list two sample exercises that have appeared in our labs.

1. Consider the helix

$$\mathbf{r}(t) = A\cos(\omega t)\mathbf{i} + A\sin(\omega t)\mathbf{j} + bt\mathbf{k}.$$

   (a) Show that the speed is constant. How does it depend on the parameters?

   (b) How do the curvature and the normal acceleration depend on the parameters?

2. A hairpin turn on a roadway can be approximated as half of an ellipse, including the major (longer) axis. If the major axis is 100 feet and the minor axis is 80 feet, what is the maximum constant speed at which a car can go through the turn while keeping the centripetal acceleration less than 0.3g?

The Maple sessions below demonstrate how the CalcP procedures can be used to do the computations needed to answer these questions.

```
# exercise 1
> helix := proc(t)
  [A*cos(w*t),A*sin(w*t),b*t] end;
helix := proc(t)
  [A*cos(w*t),A*sin(w*t),b*t] end
> Speed(helix(t),t);
     2  2     2 1/2
   (A  w  + b )
> Curvature(helix(t),t);
```

```
       2
      A w
   ----------
    2  2    2
   A  w  + b
> NormAccel(helix(t),t);
     2  2 4    4 6 1/2
   (b  A  w  + A  w )
   --------------------
       2  2    2 1/2
     (A  w  + b )
> simplify(NormAccel(helix(t),t));
     2
    A w
# exercise 2
> ellipse := proc(s)
            [50*cos(s),40*sin(s)] end;
ellipse := proc(s)
            [50*cos(s),40*sin(s)] end
> VPlot(ellipse(s),s=-Pi/2..Pi/2);
> Curvature(ellipse(s),s);
              2000
   ----------------------------
                 2        3/2
   abs(- 900 cos(s)  + 2500)
```

The maximum curvature occurs at s=0. Since the normal acceleration is given by kappa*v^2, where kappa is the curvature and v is the speed, the result can be obtained by solving the equation

```
0.3 *g= kappa_max*v_max^2
```

for the maximum speed v_max.
```
> kappa_max:=
  Curvature(ellipse(s),s=0);
  kappa_max := 1/32
> v_max :=
   [solve(0.3*32=kappa_max*v^2,v)];
  v_max := [-17.52712185, 17.52712185]
> speed_max := v_max[2]*3600/5280;
  # speed in miles per hour
  speed_max := 11.95031035
```

The hairpin curve produced by the VPlot command is shown below in Figure 5.

Conclusions

The use of Maple labs and projects in calculus at WPI has been successful, with the department moving toward full adoption in the Fall of 94. The students struggle at first, but most eventually come around. With the aid of Maple, students can certainly tackle problems that could not be done by hand, but we have found that there is another aspect of using Maple that we did not really anticipate. Simply stated, to use Maple to solve problems the students first have to understand what they are doing and second to know how to translate what they want to do into Maple commands. For some students, both of these steps are real impediments. However, we have noticed that for most students the first problem, that of understanding, is the more difficult problem. To our way of thinking, forcing the students to understand what they are doing is not a bad thing.

Biographical information

William Farr is currently an associate professor of Mathematical Sciences at Worcester Polytechnic Institute. For the last three years he has been involved in restructuring the calculus curriculum there. His research interests are in the areas of studying how students learn and bifurcation theory with symmetry. He can be reached via e-mail at `bfarr@wpi.edu`. Michael VanGeel was a student in the new calculus course the first time it was taught. The next year he was extensively involved in writing the `CalcP` package described above. He has since moved on the the University of Illinois at Champaign/Urbana, where he will be a senior next year.

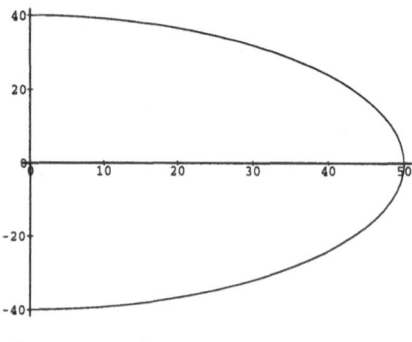

Figure 5: Hairpin curve for exercise 2.

MAPLE AT THE UNIVERSITY OF NORTH LONDON

Graham Taylor-Russell
School of Mathematical Sciences, University of North London, London, UK

1.   Background

The School of Mathematical Sciences of the University of North London has been a user of Maple for the past four years; during this time the number of students exposed to the package has grown significantly so that all degree students within the School have the opportunity to become familiar with the power and sophistication available through Maple. For the past two years the School has had a network of PC's running Maple and computer algebra is rapidly becoming central to the teaching of Mathematics in the University.

The author has been involved in promoting the use of Maple via two main avenues:

i)   the advanced level investigative unit called Computer Algebra available as an option for Second and Final Year students

ii)   the supervision of final year projects in which Maple is used as a motivator for introducing advanced concepts in areas of group theory and number theory.

This paper describes the author's experiences and details the type of assignments undertaken.

2.   Maple as an investigative tool in the unit MS217 - Computer Algebra

Students in the School of Mathematical Sciences come with a wide variety of backgrounds. Students on the BSc Mathematics pathway are expected to have studied Algebra and Calculus to the equivalent of UK A-Level standard and by and large are fairly fluent mathematically (although many require revision of some core material). On the other hand students on the BSc Mathematical Sciences pathway (including many who have come through non-traditional routes to higher education) are not required or expected to have seen material beyond the usual 16+ qualifications while in addition we have many students enrolled on joint courses. This wide variety and the presence of non-traditional students presents challenges to staff and one strategy adopted has been to develop mathematical skills using computer algebra and investigative techniques. This unit (developed by the author) is the main vehicle in this area.

The unit is available to all students who have passed the introductory calculus units in the preliminary year and hence includes students whose familiarity with the core areas of calculus and linear algebra varies widely. The unit is therefore structured as follows:

i)   Introduction to and familiarity with Maple. To service this part of the course a short guide has been produced by the author giving a brief description of the main features of the package and an introduction to writing procedures.

ii)   The use of Maple in Calculus problems. Here the intention is to underpin the students' knowledge of techniques by extended examples which might not be feasible without a computer algebra system. Students without a strong mathematical background are introduced to the Student Calculus package of Maple and are presumably given an idea of the way the techniques they may have only recently mastered can be applied. A toolbox of techniques is thus made available to them which they may use subsequently in units requiring a knowledge of calculus.

iii)   The main section of the unit aims to foster investigative skills by providing students with intensive assignments which give them a grounding in the techniques of mathematical discovery which will stand them in good stead throughout their university career and beyond. Each student undertakes assignments from a bank of investigative problems and is expected to submit a full report on at least one, the report forming a significant proportion of the marks for the unit (which is assessed by means of coursework and a practical exam). They are also encouraged to give oral presentations on their conclusions to the rest of the class.

Since the background of the students is so varied it was decided that the fairest approach was to choose assignments with which no student was expected to be familiar. The vehicle chosen for many of the assignments was number theory. This was because

a)   There was a wealth of problems which could be specified and introduced to students without the need for large amounts of background lecturing. Often all that was required was a few definitions and a motivation of why the problems might be interesting. When this was not sufficient the lecture time (which previously had been devoted to practical Maple concepts) was used to introduce theory.

b)   Many of the problems in this area lend themselves to investigation by computer techniques.

One problem that has been used in each of the last two years is to investigate the structure of perfect numbers and we now reproduce the documentation issued to students undertaking this task:

INVESTIGATION : PERFECT NUMBERS

Definition:

The number n is said to be *perfect* if the sum of all the divisors of n (including 1 but not n itself) is equal to n.

Thus 6 is a perfect number since $6 = 1 + 2 + 3$.

The following Maple procedure tests whether a given number is perfect. Note that commands such as divisors are only available after the number theory package has been loaded.

```
isperfect:=proc(j)
    local k,z,divsum;
      z:=divisors(j);
      divsum:=sum(op(k,z),k=1..nops(z));
      if divsum=2*j then
        RETURN(true) else
        RETURN(false); fi;
    od
 end;
```

Note:

The variables op(k,z) and nops(z) are Maple variables which record respectively the kth operand of z and the number of operands.

Thus   sum(op(k,z),k=1..nops(z));

computes the sum of the divisors of j.

EXERCISES:

1. Type in this procedure and check that it produces sensible answers when you try

   isperfect(6);
   isperfect(10);
   isperfect(28);

2. 6 and 28 are known to be the first two perfect numbers. In this question we want to find the next perfect number and eventually we would like to make a list of the first few perfect numbers in order to make a conjecture about their structure.

Write a procedure incorporating the test isperfect within a for loop to create a new procedure that finds the first perfect number after a given starting point.

3. Use the procedures in 2. to write down a list of the first 5 perfect numbers. For the 4th number start the search at 8000 while for the 5th start looking from 33550000.

4. Compute the numbers in the sequence

$$2^{n-1} (2^n - 1)$$

for values of n from 1 to 15. (You may find it useful to write a short Maple procedure to do this).

5. Do the results to 4. tell you anything about the possible structure of perfect numbers? If yes make a conjecture about their form otherwise fill in the table overleaf.

n	$2^{n-1} (2^n - 1)$	isperfect $(2^{n-1} (2^n - 1))$	isprime(n)
1			
2			
3			
4			
5			
6			
7			
8			
9			
10			
11			
12			
13			
14			
15			

6. Try to refine your conjecture in the light of the information contained in the table.

7. Write a procedure for testing whether the number $(2^n - 1)$ is prime or not. Run this test for all n between 1 and 15.

8. a) Try to write down a condition that describes all the perfect numbers found so far.

   b) What would be the sixth and seventh perfect numbers in the list if your condition were correct.

   c) Use the procedure isperfect to test these numbers.

9. Use Maple to find the divisor sum of a number having your general form. This will constitute a proof of a theorem having the form

"The number n is perfect if n = ? and ? " where you should replace the ? by some appropriate statement.

10. Can you prove this theorem without Maple.

MORE ON PERFECT NUMBERS

A.   Suppose we apply the following process to a number n:

1.   Add up all the digits of n. Call the result of this operation digitsum(n).

2.   Add up all the digits of digitsum(n).

3.   Continue performing this operation until you have a one digit number.

4.   Return the number found.

It can be shown that the number thus returned is 1 for all perfect numbers other than 6.

EXERCISES

a)   Write a procedure for performing this task.

[This is not easy unless you try to show that n $\equiv 1 \pmod 9$ instead]

b)   Verify the result for the first 7 perfect numbers.

B.   TRIANGULAR NUMBERS

Definition:

The number n is said to be *triangular* if

$$n = 1 + 2 + 3 + \ldots + k$$

for some number k.

EXERCISES

a)   Write a procedure to test whether a given number n is triangular or not.

b)   Show that all the perfect numbers you have found are triangular.

c)   Can you find a relationship between the structure of a perfect number and the value of k.

You should now write up a report containing Maple output as well as a description of the problem and the approach you took towards solving it. Make sure that full explanations are given of any results you are claiming and that you present all the evidence you have for any conjecture you make.

Don't be afraid to describe any wrong ideas you had while trying to find the results. Conjectures made by professional mathematicians often turn out to be false when further evidence is considered but they are still an important part of the investigative process.

Having run the course for two years using this and similar assignments (covering Fibonacci numbers, the Four Squares Theorem and prime testing) the positive conclusions that can be drawn are:

- Students are motivated to work significantly harder using this kind of approach. At times this almost became a problem as there was a danger of other courses being neglected.

- Students appreciate that mathematics is not a complete body of truths set down in stone but that mathematical discovery often proceeds by informed guesswork.

- The theoretical concepts introduced appear more relevant after students have struggled to identify patterns and derive conjectures themselves.

3. The use of Maple as a support tool in final year projects

Students enrolled on all degree courses in the School of Mathematical Sciences are required to undertake an extensive project lasting either one or two semesters in their final year. It is expected that this should involve an element of investigative work as well as the acquisition of theoretical knowledge and Maple has been a useful tool in this aspect of the work. An example of the possibilities is given below:

Using Maple in Computational Group Theory - The Reidemeister-Schreier rewriting process:

Problem : Given a presentation $\langle X \mid R \rangle$ for a group G and some generators for a subgroup H in G can we find a presentation for H.

(Note : We are trying to expand on the process used by the Maple command *pres* and to describe the algorithm involved. The theoretical justification for the process is dealt with by directed reading and is omitted from this report).

Example:

Suppose $G = \langle a, b, c \mid abc = b, bca = c, cab = a \rangle$

Calculate a presentation for the subgroup
$H = \langle ab, bc, ca \rangle$.

The problem can be broken down into four steps:

1) Find a Schreier transversal U for H in G. This is a set which satisfies the following:

   a) $1_G \in U$

   b)   U contains exactly one element from each coset of H in G.

   c)   If $w \in U$ then so do all left subwords of w.

A Schreier transversal can be found using the Maple command *cosets* or as a by-product of the Todd-Coxeter coset enumeration process for which a computer implementation is available.

2) Calculate a set of Schreier generators for H. These will consist of all distinct elements of the set

$$A = \{(ux)(\overline{ux})^{-1} : u \in U, x \in X \cup X^{-1}\}$$

where $\overline{g}$ is the coset representative of g

Complete the following table listing the elements of A and their inverses using the Maple command *cosrep*.

	a	b	c	a^{-1}	b^{-1}	c^{-1}
u_1						
u_2						
u_3						
...						
u_r						

where the u's are the elements of the transversal calculated above.

3.   Calculate the complete relation set

$$R' = \{ uru^{-1} : u \in U, r \in R\}$$

(This will have to be done by hand - though is not too tedious).

Simplify these relations if possible.

4.   Write a Maple procedure that takes the complete set of relations R' calculated above and rewrites them in terms of the Schreier generators. The algorithm for doing this is as follows:

Suppose $r' = x_1 x_2 \ldots x_n \in R'$ and define elements u_i and a_i inductively as follows

$$u_1 = 1$$
$$u_{i+1} = cosrep(u_i x_i) \quad (1 \leq i \leq n)$$
$$a_i = u_i x_i u_{i+1}^{-1}$$

The rewritten version of r' is then the relation $a_1 a_2 a_3 \ldots a_n$ and each a is an element of the generating set A.

Having completed these steps we have a presentation for H with generators A and relations as calculated in 4. Use the Maple command *pres* to check your answers to this and other examples. You will probably find redundant relations in your presentation which can be eliminated using Tietze transformations.

Reference: Presentations of Groups. D.Johnson ; Cambridge University Press.

The use of Maple in projects such as this would appear to have the following educational benefits:

- Using Maple as a computational device allows students to encounter more complicated examples than would otherwise be possible.

- Through implementing an algorithm of this type students obtain a better understanding of the processes involved.

In conclusion I have no doubt that Maple will continue to be a central tool in the final year projects   in the area of pure mathematics and students will continue to acquire benefit from its use.

Graham Taylor-Russell is currently a lecturer in the School of Mathematical Sciences at the University of North London. He received a BSc in Mathematics from the University of York and a PhD from London University in 1990. His research interests are in combinatorial group theory and the solution of equations over groups. He may be contacted at

School of Mathematical Sciences
University of North London
Holloway Road
London N7 8DB
U.K.

INTRODUCING MAPLE TO FIRST YEAR ENGINEERING STUDENTS VIA PROJECT STUDY

Ulf Rønnow
Department of Mathematics and Computer Science,
Institute for Electronic Systems, Aalborg University, Aalborg, Denmark

Introduction

The fundamental educational model at Aalborg University is project-based study. Under this form of study, the students, every semester during their entire course of study, work with specific projects within a given theme. The students are usually organized in groups of four to eight. Every group chooses a subject relevant to the given semester's theme for their project work, and under the guidance of a teacher elaborate a project report which at the end of the semester becomes the subject of an examination.

Parallel to the project work, the students follow courses in various subjects, some of which are related to the current project theme, whereas others are more general. The latter include the mathematics and physics courses ([2]).

In the first year, the project-based form of study is practiced within some broad theoretic frame, for example, Systems of Energy, Urbanization, the Work Environment, Systems of Environmental Technology. Typical project subjects for the Basic year might therefore be

Energy Transmission by Wind Turbines.

Artificial Neural Nets Applied to the Sorting of Letters.

Applying Global Positioning Systems in Agriculture.

I will here describe an experiment in introducing Maple V for first year engineering students through a project instead of, as traditionally, doing it in relation to the courses in Calculus and Linear Algebra. As a point of departure I have chosen the subject, Artificial Neural Nets Applied to Sorting.

Artificial Neural Nets

An artificial neural net model is a non-programmed, self-adapting computer system primarily developed for classification purposes (fig. 1). It is composed of many non-linear computational elements operating in parallel and arranged in patterns reminiscent of biological neural nets. The computational elements or nodes are connected via weighted links that are typically adjusted during use to improve performance.

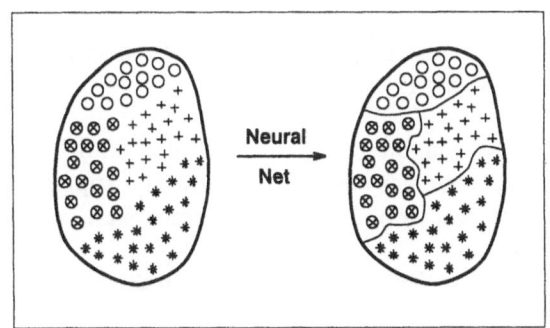

Figure 1: The Neural Net as classifier.

Before a neural net can be used for a particular classification job it must be "trained", i.e. must be presented with a series of test data, each one of which consists of an element from the set of objects to be classified together with information about the *correct* classification of the given element. Via this presentation the internal structure of the neural net is adjusted, and when the quality of the classification performance has reached the desired level, the neural net is ready to do its classification job.

There are several types of neural nets [1]. In the present paper we will take a closer look at the neural net models of the Multi-Layer Perceptron type (MLP), the general structure of which is shown in fig. 3.

Mathematically, a trained net can be considered as a function mapping the set of objects to be classified into the set of classifiers. However, the construction and mode of operation presuppose that the domain of objects to be classified, as well the range of classifiers, are represented as vectors, which means that a neural net can be considered as a function $\boldsymbol{\Psi}$ defined on a vector space $\mathbb{R}^\mu$ into another vector space $\mathbb{R}^\nu$ (fig. 2).

$$\boldsymbol{\Psi} : \mathcal{D} \mapsto \mathbb{R}^\nu, \quad \mathcal{D} \subseteq \mathbb{R}^\nu$$

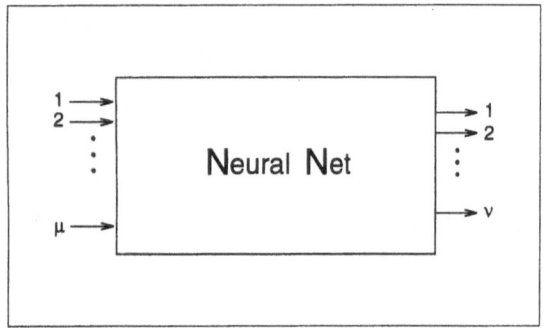

Figure 2: The external structure of a neural net of the type MLP.

Since the internal structure of a Multi-Layer Perceptron net has the following appearance (fig. 3), $\boldsymbol{\Psi}$ can obviously be represented as

$$\boldsymbol{\Psi} = \boldsymbol{\psi}_M \circ \boldsymbol{\psi}_{M-1} \circ \ldots \circ \boldsymbol{\psi}_0$$

where

$$\boldsymbol{\psi}_k : \begin{cases} \mathcal{D} \mapsto \mathbb{R}^{N_1} & (\mu = N_0) \quad k = 0, \\ \mathbb{R}^{N_k} \mapsto \mathbb{R}^{N_{k+1}} & k = 1, ..., M-1, \\ \mathbb{R}^{N_M} \mapsto \mathbb{R}^{N_M} & (\nu = N_M) \quad k = M. \end{cases}$$

M is the number of layers whereas N_k is the number of neurons in the k'th layer.

Regarding the structure of a single neuron in a Multi-Layer Perceptron net we see (fig. 4) that

$$\boldsymbol{\psi}_k = \begin{cases} \boldsymbol{u}_1, & k = 0, \\ \boldsymbol{u}_{k+1} \circ \boldsymbol{f}_k & k = 1, ..., M-1, \\ \boldsymbol{f}_M & k = M. \end{cases}$$

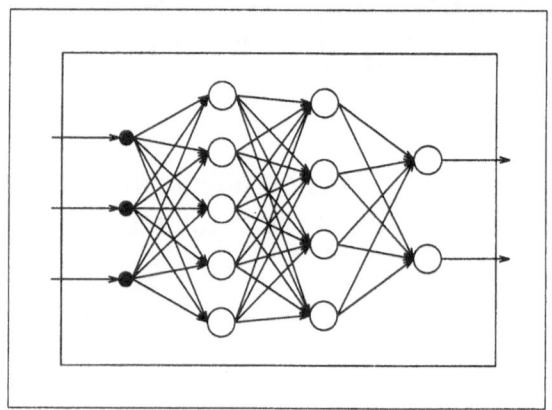

Figure 3: Internal structure of a neural net of the type MLP.

where

$$\boldsymbol{f}_k = \boldsymbol{x} \mapsto \left[f_i^k(\boldsymbol{x}) \right]_{N_k}$$
$$\boldsymbol{u}_k = \boldsymbol{y} \mapsto \boldsymbol{w}_k \boldsymbol{y} - \boldsymbol{\theta}_k, \quad k = 1, 2, \ldots, M.$$

The variables $\boldsymbol{x}$ and $\boldsymbol{\theta}_k$ (thresholds) are N_k dimensional vectors, $\boldsymbol{y}$ is an N_{k-1} dimensional vector, whereas $\boldsymbol{w}_k$ (weights) is an $N_k \times N_{k-1}$ matrix. The neuron functions $f_i^k(\boldsymbol{x})$, $i = 1, \ldots, N_k$ are (mathematical) expressions φ_i^k in the scalar variable $\boldsymbol{x}^T \boldsymbol{e}_i$, where

$$\boldsymbol{e}_i^T = [\, 0 \; 0 \; \cdots \; 1 \; 0 \; \cdots \; 0 \,]$$
$$\underset{i}{\uparrow}$$

i.e.

$$\boldsymbol{f}_k(\boldsymbol{x}) = \begin{bmatrix} f_1^k(\boldsymbol{x}) \\ f_2^k(\boldsymbol{x}) \\ \vdots \\ f_i^k(\boldsymbol{x}) \\ \vdots \\ f_{N_k}^k(\boldsymbol{x}) \end{bmatrix} = \begin{bmatrix} \varphi_1^k(\boldsymbol{x}^T \boldsymbol{e}_1) \\ \varphi_2^k(\boldsymbol{x}^T \boldsymbol{e}_2) \\ \vdots \\ \varphi_i^k(\boldsymbol{x}^T \boldsymbol{e}_i) \\ \vdots \\ \varphi_{N_k}^k(\boldsymbol{x}^T \boldsymbol{e}_{N_k}) \end{bmatrix}.$$

The net is trained by initially selecting small random weights and thresholds and then presenting all test data repeatedly. Weights and thresholds are adjusted after each trial using the information about the correct class. The adjustment is accomplished by a gradient search technique to minimize a performance function $\mathcal{P}$ equal to the

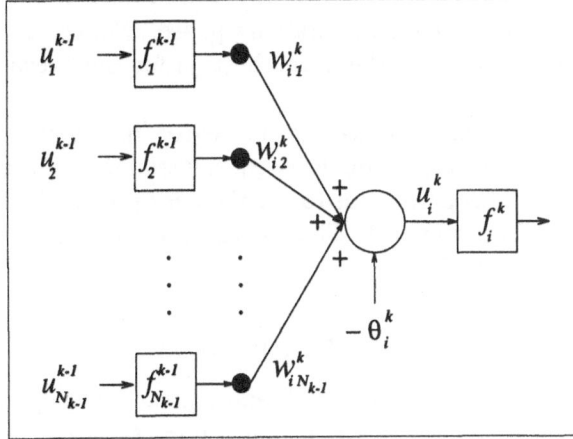

Figure 4: Structure of the i'th neuron in the k'th layer.

mean square difference between the desired and the actual net outputs, i.e. between $\boldsymbol{y}_D$ and $\boldsymbol{y}_M$

$$\mathcal{P} = (\boldsymbol{w}, \boldsymbol{\theta}) \mapsto \frac{1}{2}\big[\boldsymbol{y}_D - \boldsymbol{y}_M\big]^T \big[\boldsymbol{y}_D - \boldsymbol{y}_M\big],$$

where

$$\boldsymbol{y}_M = \boldsymbol{\Psi}(\boldsymbol{u}_0), \quad \boldsymbol{u}_0 \in \mathcal{D}.$$

Therefore the adjustment quantities are given by

$$\Delta \boldsymbol{w}_k = \nabla \mathcal{P}\big[\boldsymbol{w}_k\big]\Big|_{(\boldsymbol{w}, \boldsymbol{\theta}) = (\boldsymbol{w}, \boldsymbol{\theta})_{old}}$$

$$\Delta \boldsymbol{\theta}_k = \nabla \mathcal{P}\big[\boldsymbol{\theta}_k\big]\Big|_{(\boldsymbol{w}, \boldsymbol{\theta}) = (\boldsymbol{w}, \boldsymbol{\theta})_{old}}$$

and thus the adjusted weights and thresholds

$$\boldsymbol{w}_{k_{new}} = \boldsymbol{w}_{k_{old}} - \alpha \Delta \boldsymbol{w}_k$$

$$\boldsymbol{\theta}_{k_{new}} = \boldsymbol{\theta}_{k_{old}} - \alpha \Delta \boldsymbol{\theta}_k,$$

where α is a scaling factor to be adjusted during the training phase.

Using the chain rule we have

$$\frac{\partial \mathcal{P}}{\partial w_{ij}^k} = \frac{\partial \mathcal{P}}{\partial \boldsymbol{u}_M} \frac{\partial \boldsymbol{u}_M}{\partial \boldsymbol{u}_{M-1}} \cdots \frac{\partial \boldsymbol{u}_{k+1}}{\partial \boldsymbol{u}_k} \frac{\partial \boldsymbol{u}_k}{\partial w_{ij}^k}.$$

Setting

$$\delta_k = \begin{cases} \dfrac{\partial \mathcal{P}}{\partial \boldsymbol{u}_M}, & k = M \\[2mm] \dfrac{\partial \mathcal{P}}{\partial \boldsymbol{u}_M} \dfrac{\partial \boldsymbol{u}_M}{\partial \boldsymbol{u}_{M-1}} \cdots \dfrac{\partial \boldsymbol{u}_{k+1}}{\partial \boldsymbol{u}_k}, & \\ & k = M-1, \ldots 1, \end{cases}$$

we have

$$\delta_k = \delta_{k+1} \frac{\partial \boldsymbol{u}_{k+1}}{\partial \boldsymbol{u}_k}, \quad k = M-1, \ldots, 1,$$

and thus

$$\frac{\partial \mathcal{P}}{\partial w_{ij}^k} = \delta_k \frac{\partial \boldsymbol{u}_k}{\partial w_{ij}^k}.$$

Similarly

$$\frac{\partial \mathcal{P}}{\partial \theta_{ij}^k} = \delta_k \frac{\partial \boldsymbol{u}_k}{\partial \theta_{ij}^k}.$$

Setting up and running a Neural Net

A Maple program can now be constructed for this mathematical structure. It consists essentially of two parts, a Classification Routine and a Back Propagation Algorithm.

The Classification Routine, which presents the function $\boldsymbol{\Psi}$, is from a mathematical point of view a piece of matrix algebra, whereas the Back Propagation Algorithm, which is used to adjust the weights and thresholds, combines functional matrices and matrix multiplication.

To adapt oneself to the variable convention of Maple, the program is built up backwards, which means that the two programming parts are woven together (see the box on the next page). The program structure is applicable in all three phases of the program running

- The Setting Up, the Training and the Execution phase.

In the *setting up* phase the various mathematical expressions are (internally) developed i.e. calculated symbolically (which is the special feature of Computer Algebra systems). In the *training* phase, the Classification Routine and the Back

Propagation Algorithm are run over and over, once for each element of the test data, until the desired level of performance is reached. And finally, in the *execution* phase, only the Classification Routine is used.

The pseudo code of the central part of the program for the neural net is presented below in the box. A few words of comments is needed to explain the use of symbols and terms.

The expression $[\ldots]_k$ means a $k \times 1$ dimensional array; in particular the symbol $[a_i]_k$ means the array $[a_1, a_2, \ldots, a_k]$. The expression $\text{jacobian}(v, z)$ is presented in the text as $\frac{\partial v}{\partial z}$; the term 'assign φ_k' means that the function φ_k is assigned to the k'th neuron function.

By *input/output* at the end of the code is meant input/output for the training phase, whereas the input/output in the parenthesis is for the execution phase.

Even though there are no neuron functions in the 0'th layer, the quantity φ_0 is introduced in the code: This permits us to use the back propagation loop down to $n = 1$. The crucial programming line in the code is

$$u_n \leftarrow w_n f_{n-1}(u_{n-1}) - \theta_n$$

for $n = 1$. It becomes a valid instruction only if φ_0 is defined and the elements in φ_0 are assigned to the identity '$x \mapsto x$'.

Written out as Maple code the program looks as follows.

The Multi-Layer Perceptron Net

input M, N_i, $i = 0, 1, \ldots, M$

$\mathcal{P} \leftarrow \left[\frac{1}{2}[y_D - y_M]^T[y_D - y_M]\right]_1$

$\delta \leftarrow \text{jacobian}(\mathcal{P}, y_M)$

$y_M \leftarrow f_M(u_M)$

$f_M \leftarrow \left\{x \mapsto \left[\varphi_i^M(x^T e_i)\right]_{N_M}\right\}$

$\delta_M \leftarrow \delta \, \text{jacobian}(f_M, u_M)$

\# Start of back propagation loop.

for $n = M, M-1, \ldots, 1$ **do**

    assign φ_n

    $u_n \leftarrow w_n f_{n-1}(u_{n-1}) - \theta_n$

    for $i = 1, 2, \ldots, N_n$ **do**

        $\Delta \theta_i^n \leftarrow \delta_n \, \text{jacobian}(u_n, \theta_i^n)$

        for $j = 1, 2, \ldots, N_{n-1}$ **do**

            $\Delta w_{ij}^n \leftarrow \delta_n \, \text{jacobian}(u_n, w_{ij}^n)$

        end

    end

    $f_{n-1} \leftarrow \left\{x \mapsto \left[\varphi_i^{n-1}(x^T e_i)\right]_{N_{n-1}}\right\}$

    $\delta_{n-1} \leftarrow \delta_n \, \text{jacobian}(u_n, u_{n-1})$

end

\# End of back propagation loop.

    assign φ_0

    initialize w_k, θ_k, $k = 1, 2, \ldots, M$

input u_0, y_D, (or u_0)

output $y_M, \mathcal{P}, \Delta w_k, \Delta \theta_k$,

        $k = 1, 2, \ldots, M$, (or y_M)

```
> with(linalg):

Read in M, N[i]
Set up neuronfct[i]

> P := array(1..1,
[1/2*dotprod(y(D)-y(M),y(D)-y(M))]);
> delta := jacobian(P,y[M]);
> y[M] := f[M];
> f[M] :=
array(1..N[M],[seq(phi[M][i](u[M][i]),
i=1..N[M])]);
> delta[M] := evalm(delta & *
jacobian(f[M],u[M]);
> for n from M by -1 to 1 do
> phi[n] := neuronfct[n]; #assign the
n'th neuron function
> delta[n] := map(eval,delta[n]);
> u[n] := evalm(w[n] & * f[n-1] -
Theta[n]);
>   for i to N[n] do
>   Delta[Theta][n][i] := evalm(delta[n]
& * jacobian(u[n],[Theta[n][i]]))[1,1];
>     for j to N[n-1] do
>     Delta[w][n][i,j] := evalm(delta[n]
& * jacobian(u[n],[w[n][i,j]]))[1,1];
>       od;
>     od;
>   f[n-1] :=
array(1..N[n-1],[seq(phi[n-1][i]
(u[n-1][i]),i=1..N[n-1])]);
>   u[n] := map(eval,u[n]);
>   delta[n-1] := evalm(delta[n] & *
jacobian(u[n],u[n-1]));
> od:
```

```
> phi[0] := neuronfct[0]:  #assign the
0'th neuron function
> randomnumber := rand(-100..100)/100.:
> for l to M do
>  for i to N[l] do
>  Theta[l][i] := randomnumber():
>   for j to N[l-1] do
>   w[l][i,j] := randomnumber():
>   od;
>  od;
> od:
```

Input: u[0] and y[D]; (or u[0])
Output: y[M], P, Delta[w][k],
Delta[Theta][k]; (or y[M])

The Setting Up phase can be rather (cpu-)time intensive, mainly because of the jacobian-terms. However this phase can be speeded up considerably if we make use of the following calculations

$$\frac{\partial \mathcal{P}}{y_M} = -(y_D - y_M)$$

$$\frac{\partial u_{k+1}}{\partial u_k} = w_{k+1}\frac{\partial f_k}{\partial u_k}$$

where

$$\frac{\partial f_k}{\partial u_k} = \begin{bmatrix} \varphi_1^k(u_1^k) & & & \\ & \varphi_2^k(u_2^k) & & \mathbf{0} \\ & & \ddots & \\ \mathbf{0} & & & \varphi_{N_n}^k(u_{N_n}^k) \end{bmatrix},$$

and

$$\frac{\partial u_k}{\partial w_{ij}^k} = \begin{bmatrix} 0 \\ 0 \\ \vdots \\ 0 \\ y_j \\ 0 \\ \vdots \\ 0 \end{bmatrix} \leftarrow i \quad \text{and} \quad \frac{\partial u_k}{\partial \theta_i^k} = \begin{bmatrix} 0 \\ 0 \\ \vdots \\ 0 \\ -1 \\ 0 \\ \vdots \\ 0 \end{bmatrix} \leftarrow i$$

as

$$\begin{bmatrix} u_1^k \\ \vdots \\ u_i^k \\ \vdots \\ u_{N_k}^k \end{bmatrix} = \begin{bmatrix} w_{11}^k & & w_{1N_{k-1}}^k \\ \vdots & & \vdots \\ w_{i1}^k & & w_{iN_{k-1}}^k \\ \vdots & & \vdots \\ w_{N_k1}^k & & w_{N_kN_{k-1}}^k \end{bmatrix} \begin{bmatrix} y_1 \\ \vdots \\ y_j \\ \vdots \\ y_{N_{k-1}} \end{bmatrix} - \begin{bmatrix} \theta_1^k \\ \vdots \\ \theta_j^k \\ \vdots \\ \theta_{N_k}^k \end{bmatrix}.$$

Educational benefits

When evaluating the educational benefits in relation to the sketched project one question is crucial.

> – Does Maple enhance students' *understanding* of the significance of Mathematics, i.e. mathematical concepts, notations, definitions and theorems, etc. as a *tool* whereby concrete, practical, technical problems are solved?

It is less important if Maple improves the students' ability to solve their problems from the courses in Calculus and Linear Algebra, or even if Maple helps students to improve their general understanding of mathematics.

What really matters here is the *relationship* between mathematics and the applications. Or better, the *understanding* of this relationship.

I will explain a bit more. In using Maple to construct a neural net, the students experience how the matrix notation, the matrix algebra and the matrix formulation of the chain rule for multivariate functions facilitate their programming activities. They see that these are not just gimmicks interesting to mathematicians, but rather tools whereby they really can operate. This means that explaining mathematical rules, proving theorems etc. now become interesting and even compelling for the students.

Concluding remarks

The purpose of the above considerations is to show another way to introduce a Computer Algebra system to engineering students, that is via project study.

Our experience is that Maple in this respect is received positively by the students. In my opinion it is important *not* to integrate Maple – or any other Computer Algebra system – into the mathematical education proper, where mathematical *concepts*, *definitions* and *methods* are central.

Maple is better introduced as an useful, but independent tool, when working with mathematical *models*.

With our point of departure in this view, it is rather a question of inventing projects, such as the above, which demand the construction of mathematical models. And in this respect the possibilities are immense; besides technical subjects, areas such as physics and economics offer a wealth of subjects.

References

[1] Richard P. Lippmann. *An Introduction to Computing with Neural Nets*. IEEE ASSP *Magazine*, pages 4-20, april 1987.

[2] Ulf Rønnow. *Strategies for Future Education in Mathematics to Engineering Students*. In *Industrial Mathermatics Week* (Proceedings of the 9th Nordic Conference on Mathematical Education at the Technical Universities. Norwegian Institute of Technology.) p. 42-46, 1992.

The author Ulf Rønnow received his Ph.D. degree in Mathematics at University of Copenhagen in 1963, and is currently Associate Professor of Mathematics at Aalborg University. He can be reached at the following address:

Ulf Rønnow
Dpt. of Mathematics and Computer Science
Aalborg University
DK-9220 Aalborg
Denmark

E-mail: ugr iesd.auc.dk

ANOTHER LOOK AT LEARNING THE TECHNIQUES OF ELEMENTARY INTEGRATION

Joseph A. Pavelcak
Department of Mathematics and Computer Science, Merrimack College,
North Andover, MA

Synopsis

Pattern recognition plays a central role in discovery and progress in both science and mathematics. Integration involves the ability to recognize when a particular method should be applied. Every CAS utilizes pattern matching procedures in the algorithms for integration. Emphasis is placed on this aspect of the technique, thus developing, at least, the awareness of the importance of pattern recognition and presumably, leading to some facility in it. Use the openness of Maple to expose a pattern in the techniques of integration.

The Way It Is

The standard calculus textbook used in the Freshman course, intersperses integration techniques over a number of chapters which have other goals. Examples of such chapters would be those on the introduction of circular (trig) and transcendental functions, and applications such as the calculation of surface area, volumes, moments etc. Though the major goal is integration, any underlying organized approach to integration is obscured by these subgoals. It is argued that the myriad applications serve to motivate the learning of the techniques of integration. There is validity in this argument; however, the other concepts covered in these chapters form a dense underbrush through which the student is expected to observe some order in integration, without a clear delineation of that order. Nowhere is there a mention of pattern or of how to recognize or search for any.

Yet proficiency in integration is the aim, so every section in these chapters contains a large dose of exercises used to give the student the drill/practice needed to attain that proficiency. In general, these exercises receive a less than enthusiastic response, motivational problems not-with-standing.

This 'traditional' drill/practice/problem approach, which seemed to work in the past, is now considered a failure ostensibly because too many students fail the course. Current remedies for this contain a mixture of graphics and real-world (realistic) problem projects. This is done by reprinting those traditional texts with more graphs and inserting problems formulated to use graphing calculators or computer graphics packages. Associated with these new versions are supplemental publications composed of 'realistic' projects designed for laboratory sessions.

Both of these supplements are good. Graphics is an excellent vehicle for pattern recognition. Practice in problem solving, whether project size or textbook size, is important and expected of a course in mathematics.

The Way It Might Become

One of the perennial difficulties within the calculus course sequence is the shortage of time necessary for devoting sufficient attention to the requisite topics. In order to introduce new tools or topics, time must be taken from another topic or some topic(s) dropped out of the sequence - which leads, of course, to considerable debate. Thus when real-world project problems are introduced, a laboratory session is added, increasing the number of contact hours, but usally leaving the number of credits unchanged.

Impetus for such changes to the calculus comes from the emerging school of thought that places practice in problem solving at the head of the list of goals for the course - ahead of the acquisition of basic skills. The rationale seems to be that basic skills are a spin-off from working through the project problems. One might be tempted, following this reasoning, to introduce, along with the commands needed for plotting, the upper level Maple commands diff, int, solve,fsolve, et. al. With these powerful tools, a student has little immediate need for basic skills, but this path bodes disaster for learning calculus or any mathematics.

The Way It Should Be

Problem solving facility is an ideal difficult to reach and impossible to teach. To get started in the formulation and subdivision of a problem, leading toward a (possible) solution, the student must be comfortable with and capable of symbolic manipulation i.e. basic skills. Mathematics is a language. As is characteristic of languages, it takes a lot of practice before one can write in a language - problem solving requires the ability to write mathematics.

Exposing the student to problem solving methods and practices is generally recognized as being an expected and important role of math courses, even elementary math courses. Thus time must be found for this tedious task, whether lab sessions are scheduled or not. More importantly, the order of topics comes into play, in particular, basic skills must be taught (practiced/learned) before and concurrently with problem solving techniques, else this venture will move too slowly and not have the proper intellectual effect on the student.

Many instructors of the Calculus II integration segment, observe that introducing (trying to teach) techniques of integration, imbedded as it is among other topics, usurps a considerable amount of time and, for the majority of students, is barely successful. Clearly, a more methodical approach to integration must be set forth. Here is where Maple comes in.

Much effort, indeed man-years, has gone into the development of the processes which are invoked with the simple command int. Since Maple is open, in the sense that the code of the procedures is accessible, we may exploit the work and results of the designers. The developers of the integration procedures in CAS's have imposed an order upon the integration algorithm which is efficient. An excellent outline of the approaches to symbolic integration used in CAS's is given by Joel Moses in Communications of the ACM August 1971, Vol. 14 Number 8.

Following the same or similar approach in the classroom presentation of the calculus, due to the efficiency of the method, will reduce the time currently needed to cover the topic, as well as bring order to the confusion experienced by many students in this part of the course.

The student is not expected to learn another language (Maple), else we would be guilty of clouding the issue, just as we have accused the current Calculus textbooks of doing. However, in this proposal for teaching the integral calculus, the student will be given illustrations, written in Maple, of the "recognize, branch and perform" loop, used in the search to classify the integrand. The student is then expected to write and test short sequences of similar Maple commands for special cases. One of the pedagogical pluses of programming a computer is that it recognizes no hand waving and scorns fuzzy thinking. To 'teach' a computer to execute an algorithm correctly, the student must know how to perform that algorithm correctly, thus leading to learning.

Outline for Calculus II - Integration

The order for introducing the techniques, is that used by the standard text on Calculus. Only a subset of the integration procedures in Maple are used since the problems from the text don't require all the power. The student can experiment with that power and is urged to do so.

Four distinct techniques are covered. In each, the student is expected to use a pattern matching procedure and/or the Boolean 'type' construct, identifying the integrand, in order to determine the technique to be applied. The student will need to learn the use of a small number of Maple commands; here, the illustrations will serve as models.

See appendixes for illustrations for each of the following algorithms.

I - The first formula introduced is :

$$\int x^n dx = \frac{x^{n+1}}{n+1} \quad n \neq -1$$

For this case the 'match' procedure is used to determine the coefficients and exponent. With these values, a <u>subs</u> command can be used to perform the integration.

II - Knowledge of the integrals of elementary functionals is assumed. Here the student should generate a table of antidifferentiation "formulas" which he/she knows from the differentiation part of the course. If the table is incomplete, he/she will be unable to perform some of the integrations, thus leading him/her to extend that table with more functionals.

III - Substitution. This technique is based upon the chain rule for differentiation. N.B. Change of variable substitution or trigonometric substitution is not covered here. In fact, some of the substitutions in Calculus books are given with no explanation, leaving the student mystified. Such esoteric procedures should be removed from the textbooks, since they add little to the mathmatics of integration, but take up time.

$$\int f(u)\,du, \; u = g(x), \; du = g'(x)dx$$

In order to decide whether the integrand fits this model, the student finds the functional f, extracts the function u, performs (via Maple) the differentiation, then divides the integrand by the differential. If a quotient is a constant, then the integration is a table lookup with f as index.

IV - Integration by parts. This technique is based upon the derivative of a product. The procedure <u>intparts</u> from the <u>student</u> package is used. The student determines the factors of the integrand, selecting one of the factors for differentiation, i.e., u in :

$$\int u\,dv = uv - \int v\,du$$

Invoking <u>intparts</u> returns the right-hand side of the above equation. The student must decide how to proceed. With the aid of Maple, trial and error tests can be performed rapidly. This experimentation is an important part of the learning process - repetition.

V - Integration of rational integrand by partial fractions. Investigation of the Maple procedures is recommended to the student, since by this time he/she is familiar with the procedures used above and should be comfortable with the idiosyncrasies of Maple. Through experimentation the student is expected to find that the integration of rational functions is a difficult problem, involving factoring procedures.

All of the above require some pencil and paper work. The algorithms are presented with examples then the student uses Maple.

Conclusion

Extracting the integration from the chapters in which it is imbedded and presenting the process for study in a unified manner will improve the learning of the algorithms both in time and retention, because the student will see the pattern, which leads to understanding. That pattern is a decision tree. A test is made on the integrand to determine its type which determines the next step (branch) in the process.

Not all of the special integration 'tricks' should be considered, since the text gives no rationale, simply presents these as complicated mechanical procedures. These procedures require time, time which can be spent more productively on topics such as numerical techniques and curve fitting. This permits the introduction of problems, absent from the integration chapters, for which there is no method in the integration process tree. Such problems need more attention since textbooks tend to leave the impression that closed forms exist for all integrals.

The tools available in Maple will facilitate the learning process, but must be used judiciously. The hope is that the interest of the student will be tweaked to the point that he/she will investigate the procedures of Maple more thoroughly.

Appendix I

Implementation overview.

In the illustrations that follow it is assumed, for purposes here, that the student knows that the integration operator is linear. Therefore only products are used.

Clearly there needs to be far more explanation of the Maple operators than is given in the appendix. However, I believe it would not be very extensive. Practice is what students need and with Maple taking care of the mechanics, the methods as methods will be seen more easily. However, a major concern with this approach is the fact that our students are weak in algebraic manipulation.

It is my hope that this approach will save some time which can be used in V (partial fractions) to clear up some of those difficulties with algebra. There again, Maple can be used as a tool for drill.

Illustrative Examples

I: Integration of the form c*x^n, n <> -1. The two commands introduced are 'match' and 'subs' (substitute). The first Maple command below is 'Int' (Inert integration). It is not used in the procedure, it simply displays the the integral.

> Int(c*x^n, x);

$$\int c\, x^n\, dx$$

> match(x^2*sqrt(x) = C*x^Ex, x, V);
$$\text{true}$$

> V;
$$\{C = 1, Ex = 5/2\}$$

> Ans := subs(V, C*x^(Ex+1)/(Ex+1))

$$Ans := \frac{2}{7} x^{7/2}$$

A procedure can be written to generalize the above. The students are expected to improve that procedure, e.g. include the case for n = -1. A starter is "Integrate".

> Integrate := proc(Integrand, X) local V;
        if match(Integrand = C*X^Ex, X, V)
            then subs (V, C*X^(Ex+1)/(Ex+1))
        fi;
    end;

> Integrate(3*x^2, x);

$$x^3$$

> Integrate(6*Y^(1/2), Y);

$$4Y^{3/2}$$

> Integrate(X^Y, X);

$$\frac{X^{(Y+1)}}{Y+1}$$

> Integrate(X^Y, X);

$$\frac{1}{2} Y\, X^2$$

II: Here the student uses his/her knowledge of differentiation to generate a table of antidifferentiation formulas. The table can be as comprehensive as the student wants to make it. Below is an example table and its use.

```
>      IntSet :=  table( [     (sin)  = -cos,
                               (cos) =  sin,
                               (sec * tan) = sec,
                               (1/sqrt(a^2 - u^2) ) = arcsin,
                               (tan) =  -ln @ abs ,
                               (sec^2) = tan ,
                               (sqrt) = 2/3*sqrt^3 ,
                               (ln) = proc(u)  u*(ln(u) - 1)  end,
                               (sec) = proc(u)  ln(abs (sec(u) + tan(u) ) )  end  ] );
```

Integration is done via table lookup.

It is pointed out that an 'else' clause can be included in the "Integrate" procedure, which, along with a table similar to this one, would give the student a fairly good integration routine.

Some examples.

```
>      IntSet[sin](x);
```
$$- \cos(x)$$

```
>      IntSet[sec*tan](u);
```
$$\sec(u)$$

```
>      IntSet[tan](X);
```
$$- \ln(sbs(X))$$

```
>      IntSet[sqrt](x-1);
```
$$\frac{2}{3}(x-1)^{3/2}$$

```
>      IntSet[sec](x+4);
```
$$\ln(abs(sec(x+4) + tan(x+4)))$$

Questions for the student: Is it correct to use any argument in the above integration by table lookup ? What, if any, are the restrictions ? Use Maple 'int' to check.

III: Substitution is based upon the chain rule in differentiation. The student is asked to perform some pencil and paper examples in a stepwise fashion, then do the same with the help of Maple.

Stepwise example with Maple.

>   Integrand := 5*x^2 * sin(x^3);

$$\text{Integrand} := \quad 5\,x^2\,\sin(x^3)$$

>   match(Integrand = C*x^Ex * sin(x^Exx), x, V);

$$\text{true}$$

>   V;

$$\{\text{Exx} = 3,\ \text{C} = 5,\ \text{Ex} = 2\}$$

>   Pow := op(2, op(1, V));

$$\text{Pow} := 3$$

>   Div := diff(x^Pow, x);

$$Div := 3\,x^2$$

>   match(Integrand/Div = K * sin(x^Pow), x, Vs);

$$\text{true}$$

>   Ans := subs(Vs, K * IntSet [sin](x^Pow));

$$Ans := -\frac{5}{3}\cos(x^3)$$

To write a general Maple procedure for this, requires learning more of the language than we expect of the student. Perhaps some will be interested enough to try.

The student must specify the conditions under which the above algorithm will work.

IV: Integration by parts is based upon the derivative of a product. The student will perform trial and error experimentation using the Maple 'intparts' to do the mechanical work.

$$\int u\,dv = uv - \int v\,du$$

An approach usually suggested is to chose that part of the integrand which you know how to integrate for dv, the remainder is then u. Alternatively, chose that part of the integrand for u , which, upon differentiation becomes simpler.

Some examples.

>   Integrand := x*exp(x);

$$Integrand := x\ exp(x)$$

>   intparts(Int(Integrand, x), exp(x));   #   exp(x) is u.

$$1/2\exp(x)\,x^2 - \int 1/2\exp(x)\,x^2 dx$$

Poor choice for u , since the remaining integral is more complicated than the original.

>   intparts(Int(Integrand, x), x);

$$x\exp(x) - \int\exp(x)\,dx$$

>   Integrand := x^2 * ln(x);

$$Integrand := x^2 \ln(x)$$

It is easier to differentiate ln than integrate ln, so let u = ln(x).

>   intparts(Int(Integrand, x), ln(x));

$$1/3\ln(x)\,x^3 - \int 1/3 x^2 dx$$

>   Integrand := arcsin(x);

$$Integrand := arcsin(x)$$

Let u = arcsin(x), since the dervative is known.

>   intparts(Int(Integrand, x), arcsin(x));

$$\arcsin(x)\,x - \int\frac{x}{(1-x^2)^{1/2}}dx$$

The student will need to apply these methods to perform the remaining integrations.

Joseph A. Pavelcak received a BA in mathematics from the College of St. Thomas ,St. Paul Minnesota 1951 and a MA in mathematics from University of Massachusetts, Amherst in 1957. He is currently Associate Professor in the department of Mathematics/Computer Science at Merrimack College, North Andover Mass. Tel. (508) 837-500 X4301 or X4202 or JPavelcak@Merrimack.Edu .

WORKSHEETS: CAN WE TEACH MATHEMATICAL ALGORITHMS WITH THEM?

Michael Monagan
Departement Informatik, ETH Zurich, Switzerland

Introduction

What is the traditional way we teach mathematics? There are essentially three activities. Students read textbooks, they hear and see an instructor give explanations, and they do exercises. Some students prefer to read textbooks. I prefer to listen to an instructor. Most students need to work through the exercises. Where does a computer algebra system (CAS) like Maple fit into this picture?

In this article I am interested in showing how a CAS, specifically worksheets (notebooks), can be used to PRESENT worked examples which demonstrate mathematical procedures or algorithms. These worksheets could be used by the instructor during the lecture to present worked examples instead of using the blackboard in order to save time and to present more realistic examples. Secondly, they could replace and/or supplement worked examples in a textbook or laboratory exercise book. There are two modes in which this presentation can be made. One is "live", using the computer; the other is "dead", on transparencies or on paper.

There are essentially two ways in which one might use Maple in teaching. One is as a support tool, a calculator, to SHOW examples or phenomenon. The purpose might be to motivate a discussion. Here the instructor is using Maple to show an application. Maple is used to DO the mathematics. The second way is to use Maple to TEACH a method or algorithm for solving a problem. We can present the method as a textbook would, as a sequence of steps, and show the method with worked examples. Unlike a book, we are free to execute larger, more realistic examples. We also include the capability of DOING the computations interactively.

Let me clarify the difference between the two uses. In the first approach, the instructor is not interested in how Maple gets the answer to a given command. Of course, the instructor may be curious as to how Maple did it. However, the primary purpose is to obtain a result, THEN study it. In the second way, the instructor does care about those intermediate steps because he is teaching the students how to solve the problem. He is teaching the students a mathematical algorithm and the primary purpose of using Maple is to help the student understand the algorithm through seeing worked examples. In both cases, the possibility of doing computations live makes it possible to study what happens when we change the problem slightly. In the first way, the instructor may want to demonstrate what happens to a solution when a parameter is changed. In the second way, one may want to see what happens to an algorithm when the size of the problem is increased, perhaps to compare two different algorithms.

How do mathematics students learn algorithms? Traditionally, an instructor presents, in English, the key ideas behind the algorithm then presents the algorithm as a sequence of steps to be performed. The steps may not be written down explicitly, as many instructors proceed directly from the ideas and theorems to demonstrate the method with examples. Some instructors prefer to give a more detailed (theoretical) explanation of the method, leaving examples mainly for the exercises. Most instructors do not give a rigorous proof of an algorithm like they do a theorem.

The question being asked in this paper is: Is there any role for a CAS in teaching students mathematical procedures (algorithms)? Can this be done effectively through worksheets? It appears to be considerably more difficult to use a CAS to teach an algorithm, than just to 'do' a calculation and look at the answer. Perhaps the answer is: yes, good students would profit from this exercise, but average students would find it difficult.

I am not proposing that we teach a mathematical algorithm by asking students to write a program that implements the algorithm. Though I do think that it would be a good exercise for every student learning mathematics to do, at least once, programming is not sufficiently trivial for students to do this all the time. Instead I believe that with worksheets we have a medium that permits us to present an algorithm as an interactive sequence of steps. This is not the same as would be found in a textbook. Nor is it the same as presenting a computer program. It is more like seeing the execution of a program.

The best way to show what I mean is to present some carefully chosen examples from algebra and calculus. Four Maple worksheets follow. You will notice two fonts used for the commentary within the worksheets. Comments in the Times Roman font (this font) are for you, the reader. Comments in the Helvitica font are part of the worksheet.

Eigenvalues Worksheet

We teach our students that the eigenvalues of an n by n square matrix A are the roots of the characteristic polynomial p(lambda), a polynomial of degree n. And we give them a few examples to compute, perhaps as part of an application. The examples we give them are necessarily small because of the algebra involved. The first example appeared as a question on an assignment. The second example shows the power of a CAS permitting us to show an example too complicated to do by hand. The third example shows a real application taken from organic chemistry. All three examples involve parameters, so numerical software is not applicable.

Here are some worked examples for how

to compute the eigenvalues of a square matrix

```
> restart; # clear all variables
> with(linalg,matrix,det,nullspace);
```

$$[det, matrix, nullspace]$$

```
> A := matrix([[a,b],[b,a]]);
```

$$A := \begin{bmatrix} a & b \\ b & a \end{bmatrix}$$

Recall that the eigenvalues are the roots of the characteristic polynomial which is the determinant of the characteristic matrix lambda I - A where I is the identity matrix. In Maple, we need to construct this matrix, then compute its determinant. This requires that we either type it in again (which is most likely what an instructor would do on the blackboard) or we have to learn a new command.

```
> C := evalm( lambda-A );
```

$$C := \begin{bmatrix} -a + \lambda & -b \\ -b & -a + \lambda \end{bmatrix}$$

Note: Maple understands the scalar lambda here to mean lambda I. Next, we compute the determinant.

```
> p := det(C);
```

$$p := a^2 - 2\,a\,\lambda + \lambda^2 - b^2$$

Next we factor the polynomial p(lambda) and solve for the roots.

```
> p := factor(p);
```

$$p := (\lambda + b - a)(\lambda - b - a)$$

```
> solve(p=0, lambda);
```

$$-b + a,\ b + a$$

Let us try the same procedure on a more complicated example, which we cannot reasonable do by hand.

```
> A := matrix([[a,b,c],[b,a,b],[c,b,a]]);
```

$$A := \begin{bmatrix} a & b & c \\ b & a & b \\ c & b & a \end{bmatrix}$$

```
> C := evalm( lambda - A );
```

188

$$C := \begin{bmatrix} -a + \lambda & -b & -c \\ -b & -a + \lambda & -b \\ -c & -b & -a + \lambda \end{bmatrix}$$

```
> p := det(C);
```

$$p := -a^3 + 3\,a^2\,\lambda - 3\,a\,\lambda^2 + 2\,a\,b^2 + \lambda^3$$
$$- 2\,\lambda\,b^2 - 2\,c\,b^2 + c^2\,a - c^2\,\lambda$$

```
> p := factor(p);
```

$$p := (-a + \lambda + c)$$
$$\left(\lambda^2 - 2\,a\,\lambda - c\,\lambda + a^2 - 2\,b^2 + c\,a\right)$$

```
> solve(p=0, lambda);
```

$$a - c,\; a + \frac{1}{2}c + \frac{1}{2}\sqrt{c^2 + 8\,b^2},$$
$$a + \frac{1}{2}c - \frac{1}{2}\sqrt{c^2 + 8\,b^2}$$

We should remind our students that for large matrices, we may not be able to solve the characteristic polynomial exactly. In this case a numerical approach may be the only possibility. Note that this method for computing the eigenvalues via the characteristic polynomial is very poor numerically. Thus numerical methods use a completely different approach which cannot be explained in passing. The advantage of a CAS here is that the method an instructor would show on the blackboard can be reproduced on the computer. Once the student has seen the method executed interactively, it is reasonable that the student desire to code it.

Here is a program to compute the eigenvalues followed by an example of a 4 by 4 matrix

```
> eigenvalues := proc(A)
>    local lambda,C,p;
>    C := evalm( lambda-A );
>    p := det(C);
>    p := factor(p);
>    solve(p=0,lambda);
> end:
```

Unfortunately, one must explain the concept of local variables. Everything else in this program is trivial in the sense that there is a one to one correspondence between the worked examples and the program. The issue of local variables should not be avoided by not declaring them to be local. That would be bad programming. Since the concept of local variables is present in every programming language, and will have been met by every student who has taken an introductory course in programming, this level of programming language knowledge should be considered acceptable for a mathematics class.

The following example arises in the study of Huckel molecular orbitals in organic chemistry.

```
> A := matrix([[a,b,0,0],[b,a,b,0],[0,b,a,
  b],[0,0,b,a]]);
```

$$A := \begin{bmatrix} a & b & 0 & 0 \\ b & a & b & 0 \\ 0 & b & a & b \\ 0 & 0 & b & a \end{bmatrix}$$

```
> eigenvalues(A);
```

$$a - \frac{1}{2}b + \frac{1}{2}\sqrt{5}\,b,\; a - \frac{1}{2}b - \frac{1}{2}\sqrt{5}\,b,$$
$$a + \frac{1}{2}b + \frac{1}{2}\sqrt{5}\,b,\; a + \frac{1}{2}b - \frac{1}{2}\sqrt{5}\,b$$

There are some interesting points that should be mentioned. Factoring the characteristic polynomial is redundant as the solve command does this anyway. It is included to emphasize what the solve command does. It essentially tries to factor the polynomial before applying formulae for the roots. Neither the intructor nor the students will likely know how to factor multivariate polynomials. Does it matter? No since it is clearly what factoring polynomials means.

Partial Fraction Decomposition Worksheet.

We normally introduce partial fractions when teaching integration of rational functions. The idea is conceptually simple. To compute a partial fraction decomposition of a rational function $F(x)$, we write $F(x)$ in the form

```
> restart; # clear all variables
> F(x) = P(x)+Sum( a[i](x)/d[i](x), i=1..n
  );
```

$$F(x) = P(x) + \left(\sum_{i=1}^{n} \frac{a_i(x)}{d_i(x)} \right)$$

where P(x) is a polynomial and d[i](x) are the factors of d(x), the denominator of F(x), where deg(a[i](x)) < deg(d[i](x)). The factors d[i](x) must be pairwise relatively prime, i.e. GCD(d[i](x),d[j](x)) = 1 for all i<>j.

The problem is the mechanics and algebra needed to DO a partial fraction decomposition. Even a relatively simple problem, such as this one

> (2*x^4-4*x^3+3*x^2+1)/(x^3-2*x^2+x
) = 2*x+1/x+2/(x-1)^2;

$$\frac{2\,x^4 - 4\,x^3 + 3\,x^2 + 1}{x^3 - 2\,x^2 + x} =$$

$$2\,x + \frac{1}{x} + 2\,\frac{1}{(x-1)^2}$$

would be a real challenge for most students to get right. Consequently, many instructors don't cover partial fractions very well. And there is no way that we can ask students to tackle a realistic problem like this one

> f := (2*x^4-4*x^3+3*x^2+1)/(x^3-x^2*
b-2*x^2+x*b*2);

$$f := \frac{2\,x^4 - 4\,x^3 + 3\,x^2 + 1}{x^3 - x^2\,b - 2\,x^2 + 2\,x\,b}$$

> f = convert(f,parfrac,x);

$$\frac{2\,x^4 - 4\,x^3 + 3\,x^2 + 1}{x^3 - x^2\,b - 2\,x^2 + 2\,x\,b} = 2\,x + 2\,b + \frac{1}{2}\,\frac{1}{b\,x}$$

$$- \frac{13}{2}\,\frac{1}{(-2+b)(x-2)}$$

$$+ \frac{3\,b^2 - 4\,b^3 + 1 + 2\,b^4}{b\,(-2+b)(x-b)}$$

I believe we can teach students more effectively how to do this using a CAS. Of course, I am assuming we consider it useful to teach the students the method. Many students will tell us that if the system can do it, why should I bother to learn how to do it? To this I reply: if you learn how to do it, you will

understand it better, and you will be able to use the same technique to integrate a special function g(x) multiplied by any rational function once you know how to integrate g(x)/x and g(x)*x.

Recall that the sequence of steps is
1: Split f(x) into a polynomial p(x) and proper rational function r(x)/d(x) with deg(r(x)) < deg(d(x)).
2: Factor the denominator d(x) into linear or perhaps quadratic factors with real coefficients.

> d(x) = product(d[i](x)^e[i], i=1..n);

$$d(x) = \prod_{i=1}^{n} d_i(x)^{e_i}$$

3: Write down the form r(x)/d(x) for the partial fractions

> r(x)/d(x) = Sum(Sum(a[i,j](x) / d[i](x)
^j, j=1..e[i]), i=1..n);

$$\frac{r(x)}{d(x)} = \sum_{i=1}^{n} \left(\sum_{j=1}^{e_i} \frac{a_{i,j}(x)}{d_i(x)^j} \right)$$

where deg(a[i,j](x)) < deg(d[i](x)).
4: Construct a system of linear equations to solve. Solve for the unknowns a[i] and substitute the solutions into (3).

For example, let us step through this problem

> f := (2*x^4-4*x^3+3*x^2+1)/(x^3-x^2*
b-2*x^2+x*b*2);

$$f := \frac{2\,x^4 - 4\,x^3 + 3\,x^2 + 1}{x^3 - x^2\,b - 2\,x^2 + 2\,x\,b}$$

> n := numer(f); d := denom(f);

$$n := 2\,x^4 - 4\,x^3 + 3\,x^2 + 1$$

$$d := x\,(x^2 - x\,b - 2\,x + 2\,b)$$

The rem and quo functions do polynomial long division. They compute the quotient and remainder.

> p := rem(n,d,x);

$$p := (3 + 2\,b^2)\,x^2 - 4\,x\,b^2 + 1$$

> q := quo(n,d,x);

$$q := 2x + 2b$$

> **d := factor(d);**

$$d := x(x-2)(x-b)$$

Thus to integrate f(x) we have

> **Int(f,x) = Int(q,x)+Int(p/d,x);**

$$\int \frac{2x^4 - 4x^3 + 3x^2 + 1}{x^3 - x^2 b - 2x^2 + 2xb}\, dx =$$

$$\int 2x + 2b\, dx$$

$$+ \int \frac{(3 + 2b^2)x^2 - 4xb^2 + 1}{x(x-2)(x-b)}\, dx$$

The answer that we are looking for must be of the form

> **ANSWER := A/x + B/(x-2) + C/(x-b);**

$$ANSWER := \frac{A}{x} + \frac{B}{x-2} + \frac{C}{x-b}$$

Equating the answer with p/d, multiplying through by the denominator we have

> **e := A*(x-2)*(x-b)+B*x*(x-b)+C*x*(x-2) = p;**

$$e := A(x-2)(x-b) + Bx(x-b)$$
$$+ Cx(x-2) =$$
$$(3 + 2b^2)x^2 - 4xb^2 + 1$$

Grouping the all coefficients of x^i together, we have

> **e := collect(lhs(e)-rhs(e), x);**

$$e := (A + C - 3 - 2b^2 + B)x^2$$
$$+ (-Bb + 4b^2 - 2A - Ab - 2C)x$$
$$+ 2Ab - 1$$

This must be identically zero. Thus each coefficient must be zero.

> **eqns := { coeffs(e,x) };**

$$eqns := \{ -Bb + 4b^2 - 2A - Ab - 2C,$$
$$A + C - 3 - 2b^2 + B, 2Ab - 1 \}$$

> **sols := solve(eqns, {A,B,C});**

$$sols := \left\{ A = \frac{1}{2}\frac{1}{b}, B = -\frac{13}{2}\frac{1}{-2+b}, \right.$$
$$\left. C = \frac{3b^2 - 4b^3 + 1 + 2b^4}{b(-2+b)} \right\}$$

> **subs(sols, ANSWER);**

$$\frac{1}{2}\frac{1}{bx} - \frac{13}{2}\frac{1}{(-2+b)(x-2)}$$
$$+ \frac{3b^2 - 4b^3 + 1 + 2b^4}{b(-2+b)(x-b)}$$

To do this example in Maple we have to learn several Maple commands and be able to use them effectively. This is an impediment. But once the set of commands is put down on paper, the student and instructor can take this example and solve their own problems.

How does this activity of working through the steps of an algorithm differ from writing a program? If we want to write a program, we would have to program the construction for the formula for the ANSWER and then the construction of the equations. This would have to contemplate an arbitrary number of factors d[i]. This is where the programming is tricky. We did this by hand in the worksheet. We just wrote down the form of the ANSWER. We didn't have to think about how to write a program to construct it. This is the principle technical reason why I feel that the worksheet is a better medium for teaching an algorithm than a program. It is simpler. It avoids the tricky parts of programs because it provides the flexibility of mixing steps done by hand, and computations done by Maple. From the student's point of view, this approach is better than seeing a program because the student always sees the steps being done. The student is free to step through them interactively. This reinforces the algorithm without hiding the algorithm in a black box which happens when you write a program.

Series Solutions of Linear ODE's Worksheet

Suppose we are given the following ODE.

> **restart; # clear all variables**
> **ode := diff(y(x),x$2) + x*diff(y(x),x) +**

x^2*y(x) = 0;

$$ode := \left(\frac{\partial^2}{\partial x^2} y(x) \right) + x \left(\frac{\partial}{\partial x} y(x) \right)$$

$$+ x^2 y(x) = 0$$

How can we compute a Taylor series solution? Let s(x) be the Taylor series solution for the ODE.

> s(x) = Sum(a[k]*x^k, k=0..infinity);

$$s(x) = \sum_{k=0}^{\infty} a_k x^k$$

One approach that will give us the first few terms of s(x) is to truncate the series s(x) and substitute it into the equation symbolically

> s := sum(a[k]*x^k, k=0..5);

$$s := a_0 + a_1 x + a_2 x^2 + a_3 x^3 + a_4 x^4$$
$$+ a_5 x^5$$

> r := subs(y(x)=s, ode);

$$r := \left(\frac{\partial^2}{\partial x^2} \%1 \right) + x \left(\frac{\partial}{\partial x} \%1 \right) + x^2 \%1 = 0$$

$$\%1 := a_0 + a_1 x + a_2 x^2 + a_3 x^3 + a_4 x^4$$
$$+ a_5 x^5$$

We evaluate the derivatives and group like powers of x together

> r := eval(r);

$$r := 2 a_2 + 6 a_3 x + 12 a_4 x^2 + 20 a_5 x^3 + x$$
$$\left(a_1 + 2 a_2 x + 3 a_3 x^2 + 4 a_4 x^3 \right.$$
$$\left. + 5 a_5 x^4 \right) + x^2 \left(a_0 + a_1 x + a_2 x^2 \right.$$
$$\left. + a_3 x^3 + a_4 x^4 + a_5 x^5 \right) = 0$$

> r := collect(lhs(r),x);

$$r := a_5 x^7 + a_4 x^6 + \left(5 a_5 + a_3 \right) x^5$$
$$+ \left(4 a_4 + a_2 \right) x^4$$
$$+ \left(20 a_5 + 3 a_3 + a_1 \right) x^3$$

$$+ \left(12 a_4 + 2 a_2 + a_0 \right) x^2$$
$$+ \left(6 a_3 + a_1 \right) x + 2 a_2$$

We then discard terms of degree higher than 3. They are not correct as they depend also on higher order terms.

> r := rem(r,x^4,x);

$$r := \left(20 a_5 + 3 a_3 + a_1 \right) x^3$$
$$+ \left(12 a_4 + 2 a_2 + a_0 \right) x^2$$
$$+ \left(6 a_3 + a_1 \right) x + 2 a_2$$

Solving for the unknown coefficients a[2], a[3], a[4], and a[5] in terms of a[0] and a[1] (which themselves depend on the initial conditions of the ODE)

> eqns := {coeffs(r,x)};

$$eqns := \left\{ 6 a_3 + a_1, 12 a_4 + 2 a_2 + a_0, \right.$$
$$\left. 20 a_5 + 3 a_3 + a_1, 2 a_2 \right\}$$

> sols := solve(eqns,{a[2],a[3],a[4],a[5]});

$$sols := \left\{ a_5 = -\frac{1}{40} a_1, a_3 = -\frac{1}{6} a_1, \right.$$
$$\left. a_4 = -\frac{1}{12} a_0, a_2 = 0 \right\}$$

leads to the solution

> s := subs(sols, s);

$$s := a_0 + a_1 x - \frac{1}{6} a_1 x^3 - \frac{1}{12} a_0 x^4 - \frac{1}{40} a_1 x^5$$

The key to finding the coefficients a[k] of the series s(x) is that for a power series to vanish identically over any interval, each coefficient in the series must be zero. Let us verify that the solution computed satisfies the ODE to order O(x^4)

> r := eval(subs(y(x)=s, ode));

$$r := -a_1 x - a_0 x^2 - \frac{1}{2} a_1 x^3$$
$$+ x \left(a_1 - \frac{1}{2} a_1 x^2 - \frac{1}{3} a_0 x^3 - \frac{1}{8} a_1 x^4 \right)$$

$$+ x^2\left(a_0 + a_1 x - \frac{1}{6}a_1 x^3 - \frac{1}{12}a_0 x^4\right.$$

$$\left. - \frac{1}{40}a_1 x^5\right) = 0$$

> **expand(r);**

$$- \frac{1}{3}a_0 x^4 - \frac{7}{24}a_1 x^5 - \frac{1}{12}x^6 a_0 - \frac{1}{40}x^7 a_1 =$$

$$0$$

This gives a method for computing the first few terms. To find a formula for the k'th coefficient, a[k], in terms of the previous coefficients, we use the following method. Since the ODE is of order 2, we must look at the term x^(k+2) because under differentiation it will become a term in x^k. And, since the coefficients of our ODE are polynomials of degree 2, we must consider the term of order x^(k-2) as well. Hence we need only consider the terms

> **s := sum(a[i]*x^i, i=k-2..k+2);**

$$s := a_{k-2}\,x^{(k-2)} + a_{-1+k}\,x^{(-1+k)} + a_k\,x^k$$

$$+ a_{1+k}\,x^{(1+k)} + a_{k+2}\,x^{(k+2)}$$

Substituting these terms into the ODE, evaluating the derivatives, then simplifying, we obtain

> **s := eval(subs(y(x)=s,ode)):**
> **s := simplify(lhs(s));**

$$s := -2\,a_{k-2}\,x^{(k-2)} - a_{-1+k}\,x^{(-1+k)}$$

$$+ a_{1+k}\,x^{(1+k)} + 2\,a_{k+2}\,x^{(k+2)}$$

$$+ a_k\,x^{(k-2)}k^2 - a_k\,x^{(k-2)}k + x^k\,a_{k-2}$$

$$+ x^{(1+k)}\,a_{-1+k} + x^{(k+2)}\,a_k$$

$$+ x^{(k+3)}\,a_{1+k} + x^{(4+k)}\,a_{k+2}$$

$$+ x^{(k-2)}\,a_{k-2}\,k + x^{(-1+k)}\,a_{-1+k}\,k$$

$$+ x^k\,a_k\,k + a_{k-2}\,x^{(k-4)}k^2$$

$$- 5\,a_{k-2}\,x^{(k-4)}k + a_{-1+k}\,x^{(-3+k)}k^2$$

$$- 3\,a_{-1+k}\,x^{(-3+k)}k + a_{1+k}\,x^{(-1+k)}k^2$$

$$+ a_{1+k}\,x^{(-1+k)}k + a_{k+2}\,x^k\,k^2$$

$$+ 3\,a_{k+2}\,x^k\,k + 2\,a_{k+2}\,x^k$$

$$+ 6\,a_{k-2}\,x^{(k-4)} + 2\,a_{-1+k}\,x^{(-3+k)}$$

$$+ x^{(1+k)}\,a_{1+k}\,k + x^{(k+2)}\,a_{k+2}\,k$$

> **termk := coeff(s,x^k);**

$$termk := a_{k-2} + a_k\,k + a_{k+2}\,k^2 + 3\,a_{k+2}\,k$$

$$+ 2\,a_{k+2}$$

Since this coefficient must be zero, we obtain a formula for a[k+2] in terms of a[k] and a[k-2], namely

> **rec := solve(termk,a[k+2]);**

$$rec := -\frac{a_{k-2} + a_k\,k}{k^2 + 3\,k + 2}$$

The purpose of this example is to show clearly that such examples cannot reasonably be attempted by hand. Also, like the partial fraction example, writing a program to do it is tricky. So we don't attempt to do it. Note: this example is taken from the Maple V Flight Manual. There the example is given but only the output of the last step, the result, is shown. To understand the steps, it is vital that the student can see what each command is doing. It is also important to state what each command does if it is not clear, just as one would explain steps in a textbook. If you are preparing worksheets, always show the output of all steps.

The Euclidean Algorithm Worksheet

There are several reasons why I want to show this example. Firstly, it is a very nice example for comparing recursive procedures versus procedures which have a loop. Secondly, it is the oldest procedure known in Mathematics. It is due to the Greek mathematician Euclid and dates back to circa 300 BC. So it is of important historical interest. Thirdly, it is important because Maple uses it whenever it does arithmetic with fractions. That makes it relevant to the student. Do you remember how to add two fractions? For example
You add two fractions using the following

formula

```
> restart; # clear all variables
> a/b+c/d = (a*d+b*c)/(b*d);
```

$$\frac{a}{b} + \frac{c}{d} = \frac{a\,d + b\,c}{b\,d}$$

If we apply this formula to 3/4 + 5/6 we get 3/4 + 5/6 = (3x6+4x5)/(4x6) = 38/24. The answer is correct, but it is not simplified. We can cancel out a factor of 2 from the the numerator 38 and the denominator 24 to get 38/24 = (2x19)/(2x12) = 19/12. Maple does this simplification automatically.

```
> 3/4 + 5/6;
```

$$\frac{19}{12}$$

The problem Euclid was trying to solve is the following. How can we find the "greatest" integer that divides both the numerator and the denominator of a fraction? Alternatively, how can we find the "greatest common divisor" of two integers a and b ? We denote this quantity by GCD(a,b).

One way to compute the greatest common divisor of two integers a and b is to factor both integers into prime factors. It is easy to see what the GCD is when you have the factorizations. Unfortunately, the problem of factoring integers is known to be a hard problem, that is, an efficient method is not known, and is believed by many experts not to exist. About the best we can today on the fastest computers using the best known methods for factoring integers is to factor 100 digit integers. But we need not factor two integers if we want to compute their GCD! Euclid has a better method. His method or algorithm is based on the following three observations

1: GCD(a,b) = GCD(b,a), e.g. GCD(4,6) = GCD(6,4) = 2
2: GCD(0,a) = a, e.g. GCD(0,2) = 2
3: GCD(a,b) = GCD(a-b,b), e.g. GCD(6,4) = GCD(2,4) = 2

You may want to try to prove fact (3).

Here is how we can use these three facts to compute the GCD of 38 and 24.

```
GCD(38,24) = GCD(14,24) = GCD(24,14)
GCD(24,14) = GCD(10,14) = GCD(14,10)
GCD(14,10) = GCD(4,10) = GCD(10,4)
GCD(10,4) = GCD(6,4) = GCD(2,4) = GCD(4,2)
GCD(4,2) = GCD(2,2) = GCD(0,2) = GCD(2,0) = 2
```

When Euclid originally wrote down his method, he did so in words, because he didn't have a better language in which to express his method. It took him about a page of Greek to describe his method. We can do better with a programming language like Maple

```
> GCD := proc(a,b)
>     if  a<b then GCD(b,a)
>     elif b=0 then  a
>     else  GCD(a-b,b)
>     fi
> end:
> GCD(38,24);
```

$$2$$

Now there are three things that one must think about when writing a recursive program like this. The first is, will it stop? Clearly this program will stop only if b becomes 0. Can you argue that b must eventually become 0? What is Euclid's key idea here? The second is, if the procedure does stop, is the output correct? In this example, this is clear because each branch of the if statement simply implements one of the stated identities.

The final thing to consider is whether the program is efficient. Whereas a mathematician might be happy with the above program, a computer scientist would not. What happens when we run the program on GCD(10^10,1)? It repeatedly subtracts 1 from 10^10 until it reaches zero. This is a lot of steps! What Euclid's algorithm is really doing is computing the remainder of the integer a divided by b by repeated subtraction instead of long division. Let us modify fact (3) so we can use long division

3: GCD(a,b) = GCD(irem(a,b),b), e.g. GCD(10,4) = GCD(2,4) = 2

We arrive at the normal version of Euclid's algorithm that is used in Maple and many other computer algebra systems. The irem function in Maple computes the remainder of two integers using long division

```
> GCD := proc(a,b)
>     if a<b then GCD(b,a)
>     elif b=0 then a
>     else GCD(irem(a,b),b)
>     fi
> end:
```

This Maple procedure is acceptable. It allows us to compute efficiently the GCD of integers of hundreds of digits in size. However, it is recursive, and because of this it keeps all the intermediate integers computed in the remainder sequence around before it can finally return the GCD. This requires a lot of memory. In this case, however, this problem can be eliminated by rewriting the program in a loop. This is in fact how it would be implemented in most systems.

```
> GCD := proc(a,b) local c,d,t;
>     c := a; d := b;
>     while d <> 0 do
>         t := irem(c,d); c := d; d := t;
>     od;
>     c
> end:
> GCD(10^60,2^200);
```

$$1152921504606846976$$

As an exercise, show that the new program is correct. Hint: try to argue that when the condition d <> 0 is tested each time round the loop, that GCD(c,d) = GCD(a,b) holds.

Discussion

We list a number of problems, limitations and recommendations that we have found in presenting algorithms using worksheets. Some are problems with Maple, not worksheets, which can be repaired. Others are problems inherent with using worksheets.

1: Maple worksheets are presently limited in that we cannot write mathematical formulae in the text regions, and secondly, the formulae

that Maple can display are limited. Maple does not support a rich enough set of mathematical characters to cover all mathematical disciplines. These limitations, though annoying, will eventually be removed as Maple's mathematical typesetting capabilities are improved.

2: A serious limitation is the difficulty of manipulating parts of formulae. Suppose we are given the expression

```
> e := 1 + alpha^2/sqrt(Pi)*x - alpha/sq
rt(Pi)*x + alpha^3/sqrt(Pi)*x^2 - alph
a/sqrt(Pi)*x^2;
```

$$e := 1 + \frac{\alpha^2 x}{\sqrt{\pi}} - \frac{\alpha x}{\sqrt{\pi}} + \frac{\alpha^3 x^2}{\sqrt{\pi}} - \frac{\alpha x^2}{\sqrt{\pi}}$$

How would you simplify the above expression as a polynomial in x ? In this case we can use the collect command to group together all coefficients powers of x

```
> e := collect(e,x);
```

$$e := \left(\frac{\alpha^3}{\sqrt{\pi}} - \frac{\alpha}{\sqrt{\pi}} \right) x^2 + \left(\frac{\alpha^2}{\sqrt{\pi}} - \frac{\alpha}{\sqrt{\pi}} \right) x + 1$$

Collect does not work, though, if we want to collect in sqrt(Pi). To simplify each coefficient, however, the collect function provides the option of allowing us to apply any function to the coefficients.

```
> collect(e,x,simplify);
```

$$\frac{\alpha \left(\alpha^2 - 1 \right) x^2}{\sqrt{\pi}} + \frac{\alpha \left(\alpha - 1 \right) x}{\sqrt{\pi}} + 1$$

The following I find particularly difficult to do using Maple. Select the coefficient alpha of P*V/T in the expression

```
> 1+alpha*P*V/T - beta*P^2*V^2/T^2;
```

$$1 + \frac{\alpha P V}{T} - \frac{\beta P^2 V^2}{T^2}$$

Alternatively, how do you replace all occurrences of P*V/T by R, say, in this expression to obtain

```
> 1 + alpha*R - beta*R^2;
```

$$1 + \alpha R - \beta R^2$$

It is these manipulations that often limit the usefulness of worksheets. Note, even if Maple provided a way to select and edit parts of expressions with the mouse, this is not helpful. A worksheet must consist of only Maple commands. Fortunately, in a worksheet you can type in the answer you need, if necessary. That is exactly what a textbook does. And then you can continue. This is what we did in the first example for the partial fraction example. We didn't know how to create the equation to be solved using a Maple command. We just typed it in. In the collect example above, it doesn't matter that we don't know how to extract the coefficient alpha, we can simply type it in.

3: Another difficulty is not being able to get Maple to produce the answer that YOU want. For example, Maple will simplify (-8)^(1/3) to be a complex number, and refuse to simplify sqrt(x^2*y) to be x*y.

```
> simplify( (-8)^(1/3) );
```
$$1 + I\sqrt{3}$$
```
> simplify( sqrt(x^2*y) );
```
$$\sqrt{x^2\,y}$$

How can I get Maple to do what I want here, namely, to return -2 and x*sqrt(y)? The commands, if they exist, may not be easy to find. This is an inherent problem. Can Maple provide a command for everything manipulative that you might want to do? The argument that the programming language provides the flexibility to effect transformations not provided is simply not valid. Many are simply difficult to program. Thus I expect difficulties like this to remain for many years. Incidentally, you can get around these two common difficulties in Maple V Releae 3 as follows:
```
> readlib(surd):
> convert( (-8)^(1/3), surd );
```
$$-8^{1/3}$$
```
> simplify( " );
```
$$-2$$
```
> simplify( sqrt(x^2*y), symbolic );
```
$$x\sqrt{y}$$

4: Avoid the op command in worksheets. The op command (for extracting the ith operand of an expression) depends on the order of the terms in sums and factors in products. Since

Maple doesn't use a canonical order, what is op(1,f) in your Maple session might not be so when your students run your worksheet. Note: you can use the sort command to order the terms of a polynomial.

Conclusion

The proposal in this article is to use worksheets to present mathematical algorithms. The algorithm would be presented in English, as in a textbook, together with a worked example where all steps are shown. This can be done either live on a computer, or dead, on transparencies or paper. The proposal not does insist that an algorithm be presented as a program. There is a big advantage in simply presenting an algorithm by showing the sequence of calculations done for one or two examples. In cases where the algorithm is simple, as in the eigenvalues and Euclidean algorithm examples, it is usually very good to present a program. In other cases, such as the partial fraction example, it would be an impediment.

Finally, one word of advice for those writing worksheets. Always show the output of all commands if you intend to put the worksheets on paper or transparencies. It is difficult to follow the method if you can't see what those commands are doing, especially if the commands are manipulative and not part of the actual algorithm.

The author Michael Monagan did his Ph.D. at the University of Waterloo in computer algebra in 1989. He is presently Oberassistent at ETH Zurich. He is a member of the Maple group, an author of the Maple books, and author of the column ''Tips for Maple Users'' in the Maple Technical Newsletter.

Dr. Michael Monagan
Institute for Scientific Computing
ETH Zentrum,
CH 8092 Zurich
Switzerland
monagan@inf.ethz.ch

IV B. MAPLE IN SCIENCE AND THE APPLICATIONS

COMPUTER ALGEBRA AS A TOOL FOR ANALYZING NONLINEAR SYSTEMS

Diana Murray
Physics Department, SUNY, Stony Brook, Stony Brook, NY

ABSTRACT

We have studied a variety of weakly perturbed nonlinear dynamical systems using the method of normal forms, a reduction scheme, introduced by Poincare in the late nineteenth century. The method was formalized by Birkhoff who applied it extensively to Hamiltonian mechanics. By invoking a near-identity coordinate transformation, the method of normal forms converts the nonlinear differential equations into simplified equations of motion for the zeroth-order approximation to the true solution [1,2,3]. These coordinate transformations, nonlinear functions of the zeroth-order approximations, are found by solving a sequence of linear equations which are determined by the spectrum of the operator associated with the linear, unperturbed motion. The algebra related to these calculations is intensive and well-suited to the symbolic and programming capabilities of Maple. Through the use of procedures written in Maple we have performed high-order normal form computations and have investigated the exploitation of an inherent nonuniqueness of the zeroth-order approximation [4,5]. We will show how this nonuniqueness can be utilized to obtain transformed equations of motion that can be tailored to the needs of the investigator. Coupled with computer algebra, the method of normal forms is a potentially powerful tool for the examination of nonlinear systems [6].

Nonlinear differential equations in the form of linear systems with small nonlinear perturbations are useful in modeling a wide range of physical phenomena. Of interest in the design of charged particle accelerators is to understand the conditions for resonant oscillations of betatron motion due to the multipole components in the guiding magnetic fields. In this example, the unperturbed motion is simple harmonic. Studies in celestial dynamics and of electrical circuits have also led to weakly perturbed oscillatory systems. Predator-prey systems in ecology and convection in fluid media such as the atmosphere can be modeled as multi-dimensional systems of nonlinear ordinary differential equations.

We have applied normal form perturbation expansion methods to two general classes of weakly perturbed nonlinear problems:
I. oscillatory conservative, dissipative and parametric systems where the unperturbed motion is simple harmonic:

$$\frac{dx}{dt} = -y, \quad \frac{dy}{dt} = x + \varepsilon\, F(x, w, w^*; \varepsilon),$$
$$\frac{dw}{dt} = -i\, w, \quad \frac{dw^*}{dt} = i\, w$$

where x, y, w, w^* are scalar variables, F is a scalar function and ε is a small parameter;
II. n-dimensional first order ODE systems:

$$\frac{dx}{dt} = A\, x + \varepsilon\, G(x; \varepsilon)$$

where x and G are n-dimensional vectors and A is a constant nxn matrix. The method of normal forms invokes a change of the dependent variables to transform the original equations of motion into simpler ones that more clearly depict the interaction between the linear aspect and the nonlinear perturbations and the resonances which arise due to this interaction. The transformations are generated in the neighborhood of a fixed

---

199

point of the system. These normal form calculations are algebraically severe and extremely difficult and tedious to perform (at least correctly) by hand. Maple, with its symbolic, numerical and graphical capabilities. is a natural and indispensible tool in such investigations. This paper will illustrate through simple examples the method of normal forms (NF) and its capacity to reduce dynamical systems to compact forms. We will also discuss the application of procedures written in Maple to perform high-order perturbation calculations and to analyze the results numerically and graphically.

We will explain the essential elements of NF by analyzing the unforced Duffing equation:

$$\frac{dx}{dt} = y, \quad \frac{dy}{dt} = -x - \varepsilon\, x^3 \, ; \quad |\varepsilon| < 1. \quad (1)$$

The frequency of oscillation of this conservative system is a function of the small parameter ε and is determined from the first integral. the energy:

$$E = \frac{1}{2}\dot{x}^2 + \frac{1}{2}x^2 + \frac{1}{4}\varepsilon\, x^4 = \text{constant}.$$ For $\varepsilon > 0$ there is one stationary point, $x = 0$, which is a center. Real motions exist only when the energy is positive and these are periodic. For $\varepsilon < 0$ there is a center at $x = 0$ and 2 saddle points at

$x = \pm 1 / \sqrt{|\varepsilon|}$. Real motions exist for all energy values. Separatrices, trajectories through the saddle points, separate regions of periodic and aperiodic motion. A first "conditioning" step is to transform the system to the diagonalized coordinates associated with its linear part. This reduces the amount of computation involved because now the determination of the coordinate transformation is uncoupled. For complicated and higher-dimensional systems, the diagonalizing transformation is easily derived with the aid of Maple's linear algebra package. Introducing $z = x + i\, y$, eq (1) becomes

$$\frac{dz}{dt} = -i\, z - \varepsilon\, \frac{3\,i}{8}(z + z^*)^3 \quad (2)$$

where the asterisk denotes complex conjugation. In this case the diagonalization provides a further simplification as the results for the second transformed variable, z^*, are simply the complex conjugation of those for z.

With the method of normal forms one seeks a polynomial near-identity coordinate transformation

$$z = u + \varepsilon\, T_1(u, u^*) + \varepsilon^2\, T_2(u, u^*) + O(\varepsilon^3) \quad (3)$$

in which the dynamical system takes a simpler form. The variable u is the zeroth-order approximation to z and satisfies the normal form equation:

$$\frac{du}{dt} = -iu + \varepsilon\, U_1(u. u^*) + \varepsilon^2\, U_2 + O(\varepsilon^3) \quad (4)$$

As we will demonstrate. there is a (well-known) lack of uniqueness associated with the zeroth-order approximation that can be exploited to construct the resulting normal form in a number of ways. Substituting eqs (3&4) into eq (2), and collecting terms in each order of the expansion parameter. ε. convert the nonlinear differential equation into a series of equations that can be solved recursively for the polynomials, T_i, and the normal form terms, U_i:

$$U_1 = -i\, T_1 + i\, u \frac{\partial T_1}{\partial u} - i\, u \frac{\partial T_1}{\partial u^*} - \frac{i}{8}(u + u^*)^3 \quad (5a)$$

$$U_2 = -i\, T_2 + i\, u \frac{\partial T_2}{\partial u} - i\, u \frac{\partial T_2}{\partial u^*}$$
$$-\frac{3\,i}{8}(u + u^*)^2 (T_1 + T_1{}^*) - U_1 \frac{\partial T_1}{\partial u} - U_1{}^* \frac{\partial T_1}{\partial u^*} \quad (5b)$$

Note that each expansion equation is of the form $U_i = L \circ T_i + \text{known term}$ where

$$L \equiv -i\,(1 - u\frac{\partial}{\partial u} + u^*\frac{\partial}{\partial u^*})$$ is the homological equation [1] which operates on the T_i, and also that the U_i and T_i for each order p can be expressed in terms of monomials: $C_{j,k}\, u^j\, u^{*k}$; $j + k = 2\,p + 1$.

These observations will help in constructing a straightfoward Maple procedure to solve eqs (5a-b).

Taking a closer look at the action of the operator L, it is seen that the transformation polynomial associated with order p cannot be used to annihilate a monomial of the form $u^{p+1}\, u^{*p}$ in the order p calculation. These monomials are called "resonant" monomials; they have the same phase as the zeroth-order approximation u:

$$L \circ u = L \circ (u^{p+1} u^{*p}) = 0.$$

Herein lies the origin of the nonuniqueness of the normal form [4,5]. Although a resonant

monomial in T_p cannot affect the order p calculation, it will stategically appear in higher orders offering the investigator the opportunity to endow the normal form with one of several desirable properties. The idea of a "tailor-able" zeroth-order approximation will become clear by examining the results of the Duffing calculation:

$$U_1 = -\frac{3i}{8} u^2 u^*,$$

$$T_1 = \frac{1}{16} u^3 - \frac{3}{16} u u^{*2} - \frac{1}{32} u^{*3} + F_1(u u^*) u$$

where $F_1(u u^*) = f_1 u u^*$, $f_1 = $ constant, is the undetermined resonant term associated with the first-order expansion polynomial T_1. By definition $L \circ F_n(u u^*) u = 0$. In each order the $F_n(u u^*) u$ are chosen here to have the same monomial character as T_n. Although f_1 does not affect U_1, it appears in U_2 (and all U_i, $i > 1$):

$$U_2 = (\frac{51}{256} - \frac{3}{4} f_1) i u^3 u^{*2}$$

$$U_3 = (-\frac{1419}{8192} + \frac{51}{64} f_1 - \frac{3}{8} f_1^2 - \frac{3}{4} f_2) i u^4 u^{*3}$$

$$T_2 = \frac{3}{1024} u^5 + (-\frac{15}{256} + \frac{13}{16} f_1) u^4 u^*$$

$$+ (\frac{69}{512} - \frac{9}{16} f_1) u^2 u^{*3} + (\frac{21}{1024} - \frac{3}{32} f_1) u u^{*4}$$

$$- \frac{1}{512} u^{*5} + f_2 u^3 u^{*2}$$

As can be seen, the f_i can be chosen specifically to alter either the NF or the near-identity transformation.

One possible choice for the undetermined terms, the "usual" choice, is $f_i = 0$ for all i. This

choice, referred to by Bruno as the "distinguished" transformation [4], results in the fewest number of monomial terms in the near-identity transformation and, as will be shown below, smooth and uniform convergence of the expansion.

Since the goal of performing a normal form analysis is to obtain simplified equations of motion, one may choose each f_i to annihilate the corresponding U_{i+1} resulting in a "minimal" normal form (MNF) [7]. Choosing

$$f_1 = \frac{17}{64}, \quad f_2 = \frac{131}{8192}, \quad \text{etc., results in the}$$

following normal form for an order n calculation:

$$\frac{du}{dt} = -i u - \varepsilon \frac{3i}{8} u^2 u^* + O(\varepsilon^{n+1}).$$

Writing $u = \rho \exp(-i \phi)$ where ρ is the amplitude of the perturbed oscillation and ϕ is the phase, we obtain

$$\frac{d\rho}{dt} = 0; \quad \omega = \frac{d\phi}{dt} = 1 + \varepsilon \frac{3}{8} \rho^2.$$

The amplitude of oscillation is constant as expected for a conservative system. The key consequence of the minimal form choice is that the full functional update of the fundamental frequency of the perturbed oscillation is obtained in a first-order calculation. We will show that this choice also leads to the "best" (as defined below) approximation to the true solution from among a number of other traditional choices:
1) choosing the undetermined terms to be zero (the "usual" choice),
2) making the normalizing transformation canonical thus preserving Hamilton's equations in each order,
3) requiring that the initial conditions be satisfied by the zeroth-order approximation.
For the details concerned with these choices see ref. [8].

The minimal normal form is the simplest, most compact equation of motion. A notable consequence of the normal form analysis of the Duffing equation is the numerical superiority of the MNF choice over the other options enumerated. For this example, the MNF choice leads to a better approximation of the exact solution. The criteria for this conclusion have been graphically tested with Maple. When a calculation is performed to a prescribed order, n, it is found that the mismatch of the MNF choice is the least. The exact solution $x(t)$ of the differential equation in eq (1) is replaced by the various normal form approximations which obey the equation through order n. What is left over, the mismatch, is of order n+1:

$$M = \frac{d^2 x_{approx}}{dt^2} + x_{approx} + \varepsilon\, x^3_{approx} = O(\varepsilon^{n+1}).$$

Fig. 1 depicts the mismatch for several orders of calculation. Although the MNF approximations are an order of magnitude better than the usual and canonical approximations, the errors in each case are quite small. In fact, each of the approximations plotted along with a numerical solution shows excellent agreement, but the MNF approximation has the longest time-validity.

Another graphical test addresses the issue of convergence of the perturbation expansion. There is no unique convergence criterion. We have studied the ratios between consecutive terms in the near-identity transformation [8]:

$$R_{n+1} = \frac{1}{\rho^2} \sqrt{\frac{T_{n+1} T_{n+1}^*}{T_n T_n^*}}.$$

The advantage of this definition is that the ratio is independent of the amplitude and hence of ε.

R_1, the ratio of T_1 and $T_0 = u$ and R_2 are plotted in Fig. 2 for the choices discussed. As can be seen, MNFs represents the largest correction in first order and the smallest correction in second order (and the next several higher orders as well). By introducing a large correction in first order, MNFs captures most of the effect of the perturbation early on.

Fig. 3a displays the ratios for usual normal forms to very high order (n =15). The asymptotic convergence of the usual expansion seems to be uniform and bounded. In the MNF case, the higher order ratios, beginning with R_7, intermittently display large values indicating that perhaps the high order expansion series is divergent (Fig. 3b). Further investigation shows that this is not the case. Fig. 4 is a graph of

$$R_{n/0} = \frac{1}{\rho^{2n}} \sqrt{\frac{T_n T_n^*}{T_0 T_0^*}}$$

for n = 3,....,8. The magnitude of each successive transformation polynomial decreases, indicating that for $\varepsilon < 1$ and for intitial conditions sufficiently small [9], the series will indeed converge.

The determination of the normalizing transformation, although straightfoward, is algebraically intensive and involves symbolic quantities (the undetermined terms). Since the expansion equations in each order have well-defined structure and involve monomials that are easily enumerated, a simple Maple procedure can be defined to execute these calculations. A comprehensive program performs the following functions:
1) accepts as input the order of the perturbation calculation, the way in which the nonuniqueness of the zeroth-order approximation will be utilized, the initial condition and the value of the expansion parameter ε.
2) determines the normalizing coordinate transformation, the normal form, the undetermined terms and the initial value of the zeroth-order approximation by inverting the normalizing transformation,
3) calculates the exact frequency of oscillation (if possible),
4) determines a numerical approximate solution using a fourth-order Taylor series procedure,
5) graphically illustrates the efficacy of the approximate solutions.

When is it important to be able to perform efficiently very high order ($n \geq 10$) calculations of this type? We recently contributed to an investigation [9] of the ability of MNFs to obtain accurate solutions to non-integrable Hamiltonians of the type encountered in accelerator physics in comparison to the more widely used method of Lie transformations [10]. As a particle will revolve around a circular accelerator on the order of 10^9 times it is necessary to track its motion with an approximate solution that has the proper time validity (i.e. to high order in the expansion parameter). Table I contains a comparison of exact and MNF results for the frequency of oscillation for motion subject to the following Hamiltonian:

$$H = \frac{1}{2}(x^2 + p^2) + \frac{1}{3}(x^3 + p^3) + x^4 + p^4 + x\,p^2$$

where x is the position and p is the momentum of the particle. Our collaborator performed the Lie transform calculations on a Cray supercomputer. We were able to keep pace, running our Maple procedures on a Sun workstation. For these types of problems, MNFs is as effective as Lie transforms.

In dissipative systems [11], there is amplitude as well as frequency updating. Consider the simple system with cubic damping:

$$\frac{dx}{dt} = y, \quad \frac{dy}{dt} = -x - \varepsilon\, y^3.$$

Performing an analysis similar to the one outlined above for the Duffing equation results in the following normal form:

$$\frac{du}{dt} = -i\,u - \varepsilon\,\frac{3}{8}\,u^2\,u^* + \varepsilon^2\,(\frac{3}{8}(f_1 - f_1^*)$$

$$+ \frac{27\,i}{256})\,u^3\,u^{*2} + \varepsilon^3\,(-\frac{3}{8}(2\,f_1^2 + f_1\,f_1^* - 3f_2 + f_2^*).$$

$$+ \frac{27\,i}{128}(f_1 + f_1^*) - \frac{567}{8192})\,u^4\,u^{*3} + O(\varepsilon^4)$$

With the usual choice, through sixth order in ε, one obtains amplitude and phase equations that are updated in alternating orders and are calculationally quite unwieldy:

$$\frac{d\rho}{dt} = -\frac{3}{8}\,\varepsilon\,\rho^3 - \frac{567}{8192}\,\varepsilon^3\,\rho^7 - \frac{62721}{4194304}\,\varepsilon^5\,\rho^{11} + O(\varepsilon^7)$$

$$\frac{d\phi}{dt} = \omega = 1 - \frac{27}{256}\,\varepsilon^2\,\rho^4 - \frac{7965}{262144}\,\varepsilon^4\,\rho^8$$

$$- \frac{214083}{268435456}\,\varepsilon^6\,\rho^{12} + O(\varepsilon^7)$$

The benefit of implementing MNFs here is twofold - to all orders the fundamental frequency of oscillation has no update: $\frac{d\phi}{dt} = \omega = 1$ and the

amplitude equation $\frac{d\rho}{dt} = -\frac{3}{8}\,\varepsilon\,\rho^3$ has a simple closed-form solution:

$$\rho(t) = \frac{\rho_0}{\sqrt{1 + \frac{3}{4}\,\varepsilon\,\rho_0^2\,t}}.$$

The method of normal forms is similarly applied to systems whose perturbations have Taylor series expansions. The pendulum with cubic damping

$$\frac{d^2x}{dt^2} + (\frac{dx}{dt})^3 + \sin x = 0$$

is also dramatically simplifed under MNFs. Considering only small amplitudes, we expand the sine function and introduce the scaling

$$x \longrightarrow \sqrt{\varepsilon}\ x \text{ to obtain}$$

$$\frac{d^2x}{dt^2} + x + \varepsilon\,[(\frac{dx}{dt})^3 - \frac{1}{6}\,x^3] + \sum_{n=2}^{\infty} \varepsilon^n\,(-1)^n\,\frac{x^{2n+1}}{(2n+1)!} = 0.$$

We find that the usual choice leads to amplitude and phase equations that are both updated in each order, whereas MNFs yields the greatly simplified equations:

$$\frac{d\rho}{dt} = -\varepsilon\,\rho^3 + \frac{1}{64}\,\varepsilon^2\,\rho^5, \quad \frac{d\phi}{dt} = 1 - \frac{1}{16}\,\varepsilon\,\rho^2 \quad \text{exact}$$

to all orders.

For perturbed harmonic oscillator equations such as those discussed, the reducing power of MNF's is ubiquitous. Further examples are listed in Table 2.

The next example has an exact solution and was concocted to exhibit behavior expected to be "troublesome" for the method of normal forms:

$$\frac{dx}{dt} = -\frac{1}{10}\,x - \frac{1}{10}\,x\,y + \frac{1}{4}\,x^3$$

$$\frac{dy}{dt} = -\frac{1}{5}\,x + \frac{1}{2}\,x^2 - \frac{1}{5}\,y^2 + \frac{1}{2}\,x^2\,y\ ;$$

$$x(0) = 1,\ y(0) = \frac{1}{10}.$$

Fig. 5a depicts the phase portrait associated with the exact system. Although the components initially increase despite their negative eigenvalues, they eventually decay towards the stable fixed point (the origin). The initial condition is outside the basin of attraction of the stable fixed point. As seen in Fig. 5b, a low-order normal form result is a poor approximation to the exact solution. Using computer algebra it is simple to perform an "infinite" order calculation and actually determine the solutions to the original system in closed form:

$$x = \frac{u}{\sqrt{1-v}}; \quad y = \frac{v}{1-v} \quad \text{where u and v are given}$$

by the normal form equations:

$$\frac{du}{dt} = -\frac{1}{10}u; \quad \frac{dv}{dt} = -\frac{1}{5}v + \frac{1}{2}u^2.$$

The hope in performing an analytical perturbation expansion calculation is to derive a qualitative understanding of the dynamics of the original system from the transformed, simplified equation of motion (the normal form). For example, one may seek to construct a phase portrait for the system or to perform a bifurcation analysis [12]. We have shown that by exploiting the nonuniqueness associated with the method of normal forms we are able to handle a wide variety of weakly perturbed nonlinear systems in a number of ways. In particular, implementing minimal normal forms yields strikingly simple, compact equations of motion. Conceptually and computationally it is desirable to have the amplitude and phase equations fixed early in the calculation. This simplicity coupled with the computational power of Maple allows one to investigate a number of complex, but important issues including
- convergence properties of the near-identity transformation
- dynamical systems for which one needs to develop an approximate solution valid for long times
- the transient motion in multidimensional dissipative systems (see ref. 11 for an example)
- the effect of truncating modes in a multidimensional system.
This last point and the extension of the analysis to higher dimensional problems will be described elsewhere.

Acknowledgement: D.M. is grateful for the financial support of the Institute for Pattern Recognition at SUNY Stony Brook.

References

1. V.I. Arnold, Geometrical methods in the theory of ordinary differential equations, Springer-Verlag, NY (1988)
2. J. Guckenheimer and P.J. Holmes, Nonlinear oscillations, dynamical systems and bifurcations of vector fields, Springer-Verlag, NY (1983)
3. S. Wiggins, Introduction to applied nonlinear dynamical systems and chaos, Springer-Verlag, NY (1990)
4. A.D. Bruno, Local methods in nonlinear differential equations, Springer-Verlag, Berlin (1989)
5. M. Kummer, How to avoid secular terms in classical and quantum mechanics, Nuovo Cimento, **1B**, 123 (1971)
6. R.H. Rand and D. Armbruster, Perturbation methods, bifurcation theory and computer algebra. Springer-Verlag, NY (1987)
7. P.B. Kahn and Y. Zarmi, Minimal normal forms in harmonic oscillations with small nonlinear perturbations. Physica D **54**, 65-74 (1991)
8. P.B. Kahn, D. Murray and Y. Zarmi, Freedom in small parameter expansions for nonlinear perturbations, Roy. Soc. Proc. A **443**, 83-94 (1993)
9. E. Forest and D. Murray, Freedom in minimal normal forms, Physica D, to be published (1994)
10. A.J. Dragt and J.M. Finn, Lie series and invariant functions for analytic symplectic maps, J. Math. Phys. **17**, 225 (1976)
11. D. Murray, Normal form investigations of dissipative systems, Mechanics Research Communications, to be published (1994)
12. J.D. Crawford, Bifurcation analysis, Rev. Mod. Phy. **63**, no.4, 991-1037 (1991)

Diana Murray (murray@mathlab.sunysb.edu) is expected to receive her PhD in physics from SUNY Stony Brook in August 1994. Her research interests include the application of perturbation expansion methods to nonlinear dynamical systems and nonlinear time series analysis.

TABLE 1

Comparison of exact and MNF results for the frequency of oscillation due to the Hamiltonian of eq. 6.

x(0)	w (exact)	w (MNF)	error
0.1	1.02382272107350180	1.02382221705492299	$4.9\ 10^{-7}$
0.125	1.03838122227850189	1.03837742078126191	$3.7\ 10^{-6}$
0.15	1.05706875210537283	1.05704910131313645	$1.9\ 10^{-5}$
0.175	1.08025281094985680	1.08017486853108192	$7.2\ 10^{-5}$
0.2	1.10827235537607029	1.10801860918391330	$2.3\ 10^{-4}$
0.225	1.14141494048353416	1.14070667845660645	$6.2\ 10^{-4}$
0.25	1.17990185548137750	1.17815520027277578	$1.5\ 10^{-3}$
0.275	1.22388247066031169	1.21999328548711406	$3.2\ 10^{-3}$
0.3	1.27341938452659925	1.26548913825049846	$6.2\ 10^{-3}$

TABLE 2

Some simple perturbed harmonic oscillator systems and their minimal normal forms (exact to all orders):

$\ddot{x} + x + \varepsilon\, x^2 = 0$	$\dot{u} = -i\, u + \varepsilon^2\, 5i/12\, u^2\, u*$
$\ddot{x} + x + \varepsilon\, \dot{x}^2 = 0$	$\dot{u} = -i\, u + \varepsilon^2\, i/6\, u^2\, u*$
$\ddot{x} + x + \varepsilon\, x^3 = 0$	$\dot{u} = -i\, u - \varepsilon\, 3i/8\, u^2\, u*$
$\ddot{x} + x + \varepsilon\, \dot{x}^3 = 0$	$\dot{u} = -i\, u - \varepsilon\, 3/8\, u^2\, u*$
$\ddot{x} + x + \varepsilon\, x^2 + \varepsilon^2\, x^3 = 0$	$\dot{u} = -i\, u + \varepsilon^2\, i/24\, u^2\, u*$
$\ddot{x} + x + \varepsilon\, \dot{x}^2 + \varepsilon^2\, x^3 = 0$	$\dot{u} = -i\, u - \varepsilon^2\, 5i/24\, u^2\, u*$
$\ddot{x} + x + \varepsilon\, \dot{x}^4 = 0$	$\dot{u} = -i\, u + \varepsilon^2\, 7i/40\, u^4\, u*^3$
$\ddot{x} + x + \varepsilon\, x^5 = 0$	$\dot{u} = -i\, u - \varepsilon\, 5i/16\, u^3\, u*^2$
$\ddot{x} + x + \varepsilon\, \dot{x}^5 = 0$	$\dot{u} = -i\, u - \varepsilon\, 5/16\, u^3\, u*^2$
$\ddot{x} + x + \varepsilon\, (\dot{x}^3 + x^3) = 0$	$\dot{u} = -i\, u - \varepsilon\, (1+i)\, 3/8\, u^2\, u* - \varepsilon^2\, 3/32\, u^3\, u*^2$
$\ddot{x} + x + \varepsilon^2\, x^3 + \varepsilon^3\, x^4 = 0$	$\dot{u} = -i\, u - \varepsilon^2\, 3i/8\, u^2\, u*$
$\ddot{x} + x + \varepsilon\, x^2 + \varepsilon^2\, x^3 + \varepsilon^3\, x^4 + \varepsilon^4\, x^5 = 0$	$\dot{u} = -i\, u + \varepsilon^2\, i/24\, u^2\, u*$
$\ddot{x} + \sin(x) = 0$ for small amplitude	$\dot{u} = -i\, u + \varepsilon\, i/16\, u^2\, u*$

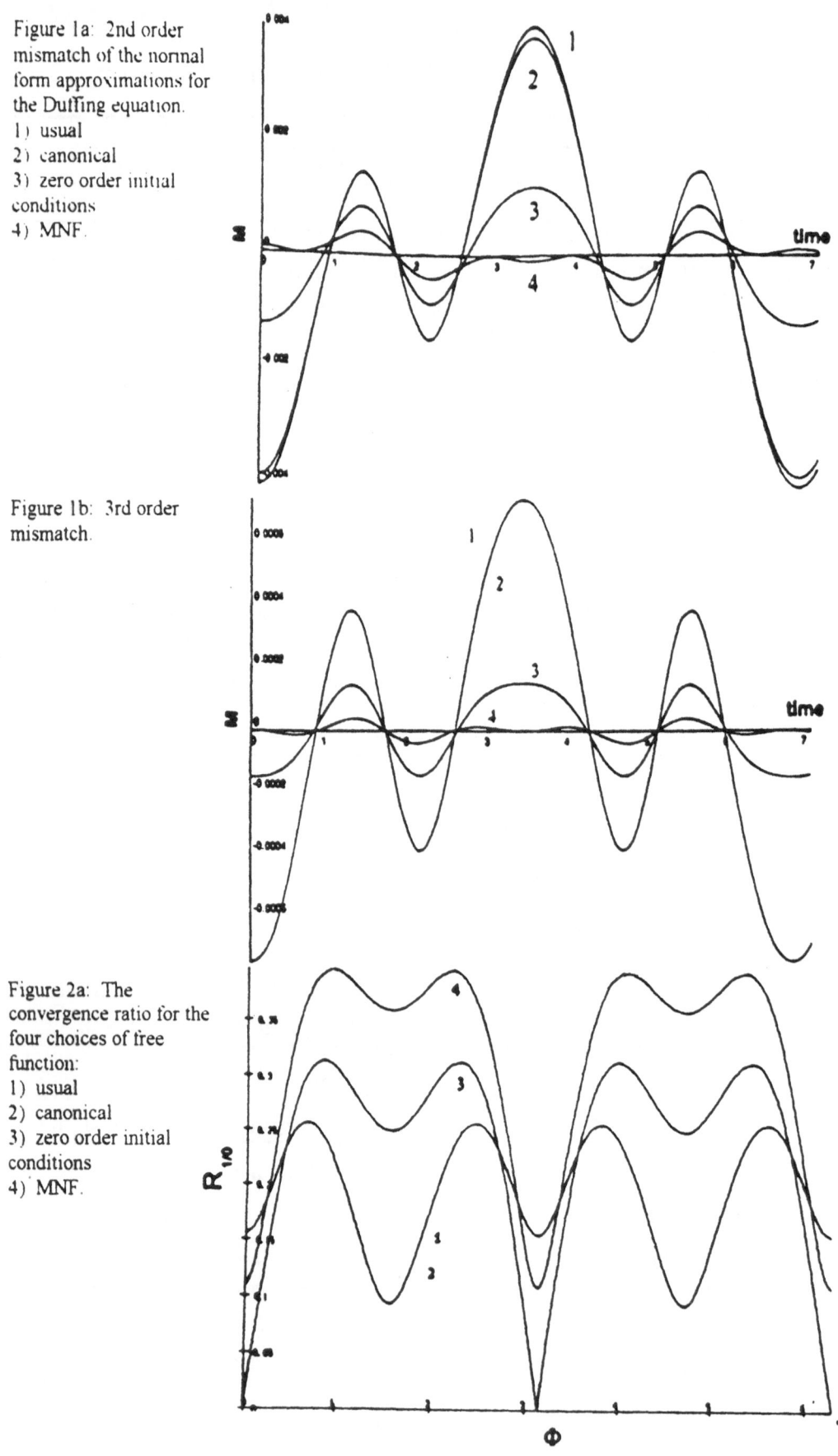

Figure 1a: 2nd order mismatch of the normal form approximations for the Duffing equation.
1) usual
2) canonical
3) zero order initial conditions
4) MNF.

Figure 1b: 3rd order mismatch.

Figure 2a: The convergence ratio for the four choices of tree function:
1) usual
2) canonical
3) zero order initial conditions
4) MNF.

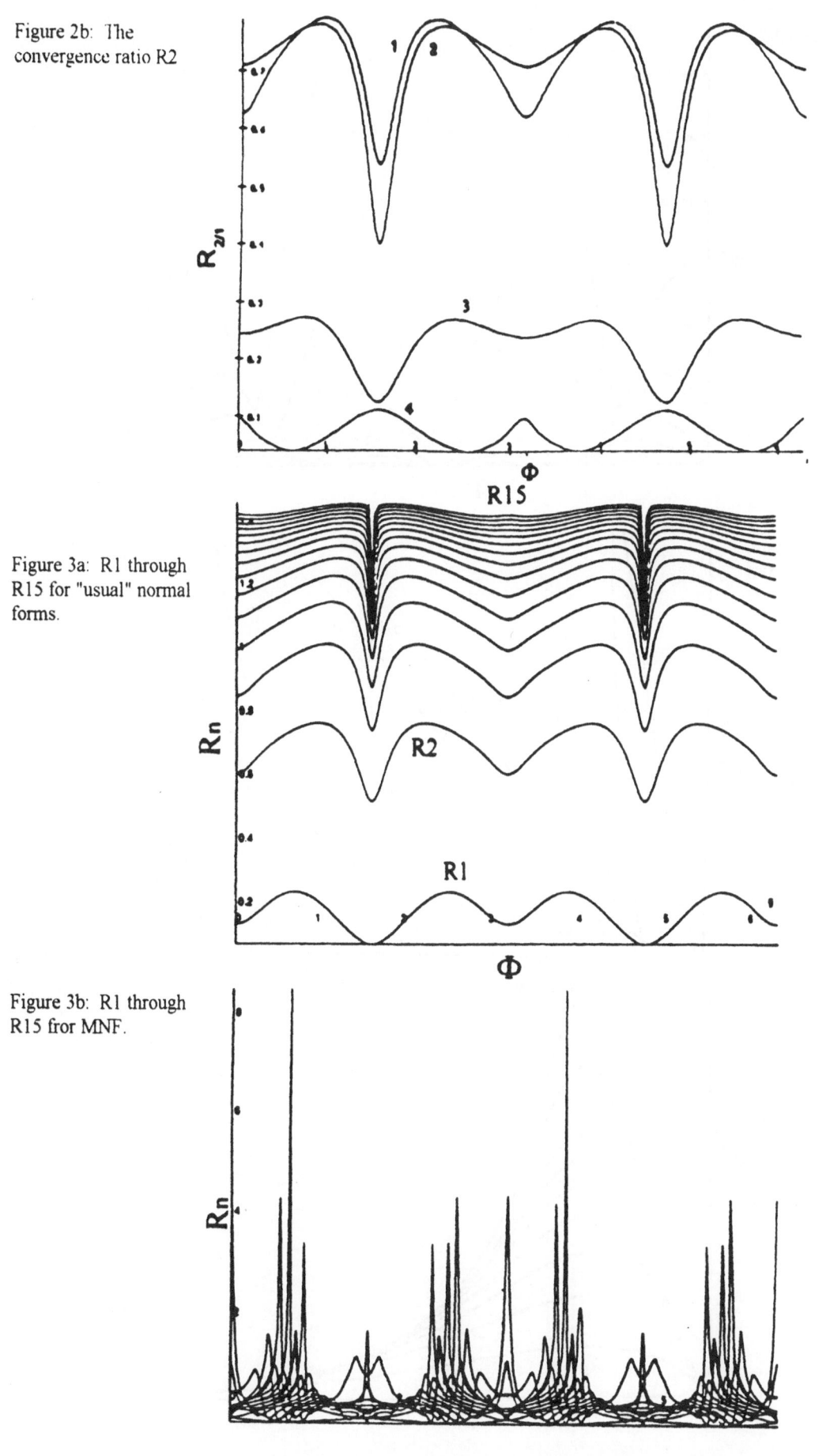

Figure 2b: The convergence ratio R2

Figure 3a: R1 through R15 for "usual" normal forms.

Figure 3b: R1 through R15 fror MNF.

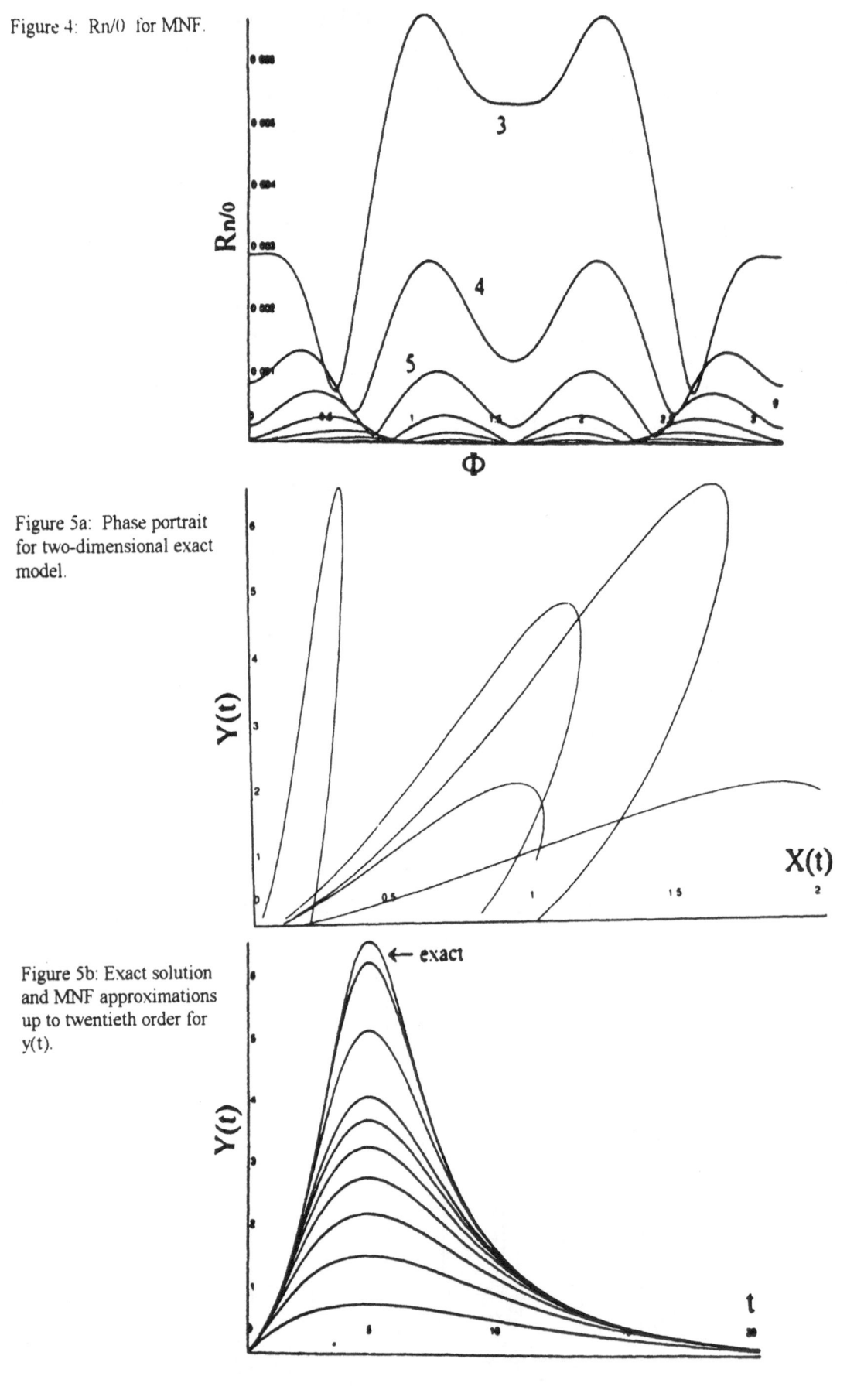

Figure 4: Rn/0 for MNF.

Figure 5a: Phase portrait for two-dimensional exact model.

Figure 5b: Exact solution and MNF approximations up to twentieth order for y(t).

MAPLE V AND GEOMETRICAL OPTICS: ABERRATION COEFFICIENTS IN ARBITRARY OPTICAL SYSTEMS

Eisso Atzema
CAN/RIACA, Amsterdam, The Netherlands

Abstract

Traditionally, ray tracing methods are used to obtain numerical approximations for the aberration coefficients of a fully specified optical system. In this paper, it will be shown how computer algebra can be used to compute the aberration coefficients in a completely symbolic way.

1 Introduction

In recent years, it has been shown that computer algebra can play an innovative rôle both in science and in industry (see [3]). An example of a technical discipline to which computer algebra can be successfully applied is the field known as either technical or geometrical optics. In particular, computer algebra is of assistence in the design of optical instruments.

From a purely theoretical point of view, it is not too difficult to formulate criteria for the performance of optical instruments. The basic concept here is that of aberration, i.e. the deviation from a perfect image. The criteria for the performance of a system are then expressed in terms of the coefficients of particular Taylor expansions related to the systems. These coefficients are known as aberration coefficients.

In practice, the computations that are required to determine the aberration coefficients soon become far too involved to be performed by pad and pencil and approximating methods have to be resorted to. With the help of computer algebra, however, the exact values of the coefficients can be determined without much ado. Besides, by allowing one or more of the specifications of the system to go undeclared it is possible to find expressions for the aberration coefficients in terms of the undeclared variables. This opens up the possibility of hitherto unavailable means of optimalisations of the optical system. In this way, for instance, the effect of translation of one single lens in an optical system can be fully determined in one go.

A possible drawback for any package based on symbolic software could be the speed of the computations. When written in symbolic form, the expressions one has to deal with soon become enormous and very difficult to manipulate, even for a computer algebra system. Therefore, it is of the utmost importance to find algorithms that allow computer algebra systems to work their way through the formulae in the most efficient way possible. As it is, it looks as if for the case of geometrical optics at least two algorithms aspiring to this status are available. Ultimately, both algorithms go back to the fact that the transition of a system of rays gives rise to a so-called *symplectic transformation*, but they make use of different consequences of this property. One of these algorithms is based on the theory of so-called *eikonals*, the other makes use of so-called *Lie transformations*. In the former case, the coefficients of the Taylor expansion of the eikonal with respect to all its variables can be taken as the aberration coefficients. In the latter case, Lie transformations are used to compute efficiently a Taylor series that has the aberrations coefficients for its coefficients.

The application of computer algebra to the eikonal approach has recently been considered by André Heck and Marc Biemond at **CAN** (see [8] & [9]). The other approach has been discussed in, among others, [11] and [15]. In this paper, some further applications of computer algebra to the computation of the optical map and the use of Lie transformations will be studied. Whereas Dragt and Wolf concentrate on the optical transformation for one surface, the emphasis in this paper will be on the use of Lie transformations for the optical transition through several surfaces. As far as we know this topic has not been systematically considered yet.

In the following, first the basic concepts in geometrical optics and the ideas underpinning aberration theory will be sketched. After this, an outline will be given of the principal ideas underlying the Lie approach to geometrical optics. Also, its implementation in computer algebra software will be discussed. Finally, some illustrations of the possible applications of the software to geometrical optics will be provided. All programming is done in *Maple*. It has not yet been attempted to write software that is maximally efficient. The principal goal of this paper is to give a first impression of how computer algebra can be used in the case of geometrical optics.

2 What is geometrical optics about?

Classically, geometrical optics is divided into the fields of the optics of reflection and that of the optics of refraction. Underlying all theory in these two fields are the laws of reflection and refraction, respectively. According to the former, any light ray that is reflected at a surface is reflected under an angle of excidence which is equal to the angle of incidence. The angle of incidence could be defined here as the angle of the incident ray with the normal to the surface at the point of incidence. In a similar way, the angle of excidence can be defined. Both the case of refraction and that of reflection are covered by

Theorem 1 (Snell's law) *Suppose we have a refracting surface* $\mathcal{F}$*. Let us denote the medium in front of* $\mathcal{F}$ *by* $\mathcal{M}_0$*. Likewise, the medium behind the surface is denoted by* $\mathcal{M}_1$*. We assign so-called indices of refraction* μ_0 *and* μ_1 *to* $\mathcal{M}_0$ *and* $\mathcal{M}_1$*, respectively. Furthermore, let us denote the angle of incidence of an incident ray by* ι_0*. Similarly,* ι_1 *denotes the corresponding angle of excidence. Then for the case of refraction, one has*

$$\sin(\iota_0) : \sin(\iota_1) = \mu_0 : \mu_1.$$

In the case of reflection, the same law applies, where one has to choose μ_i *such that* $\mu_0 + \mu_1 = 0$

In this paper, we will deal with the refraction of systems of rays at a series of surfaces. Such a series is called an *optical system*. We will denote the

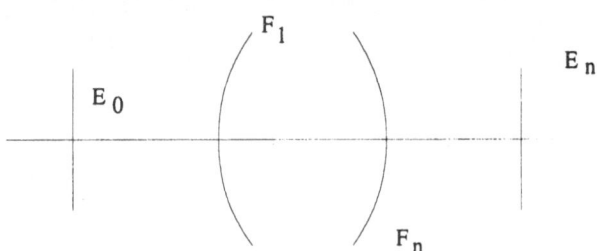

Figure 1: An optical system

surfaces of this system by $\mathcal{F}_1$ through $\mathcal{F}_n$. The medium between $\mathcal{F}_i$ and $\mathcal{F}_{i+1}$ will be denoted by $\mathcal{M}_i$. The medium in front of the first refracting surface is $\mathcal{M}_0$. $\mathcal{M}_0$ will be referred to as the *object space*. Likewise, $\mathcal{M}_n$ is the medium behind the last refracting surface and will be referred to as the *image space*. The index of refraction of a medium $\mathcal{M}_i$ is μ_i. Furthermore, we assume that refracting surfaces are intersected perpendicularly by one single line, the optical axis. The distances between surfaces $\mathcal{F}_i$ and $\mathcal{F}_{i+1}$ are denoted by d_i. For every $\mathcal{F}_i$, there are two reference planes with respect to which the incident and the excident rays are defined. We will denote the distances of these reference planes for a surface $\mathcal{F}_i$ as measured along the optical axis by e_i' and e_i'', respectively. An exception will be made for the distances of the most outlying reference planes. The distance from the reference plane before the system to $\mathcal{F}_1$ will be denoted by e_0, that of the plane behind the system to $\mathcal{F}_n$ by e_n. All lengths are measured in the direction from E_0 to E_n. It will be assumed that the reference planes between two surfaces coincide. We will denote the coinciding planes between $\mathcal{F}_i$ and $\mathcal{F}_{i+1}$ by E_i (see fig.(1)).

Finally, we will coordinatise the space of the optical system with orthogonal coordinates (x, y, z), with the z-axis coinciding with the optical axis. In all reference planes, we will use the x,y-coordinates induced by the x, y, z-coordinates of the ambient space. The direction of a light ray departing from a plane E_i will be indicated by a vector $\bar{s}_i = (\alpha_i, \beta_i, \gamma_i)$ with $|\vec{s}_i| = 1$.

With this notation, any incident ray is characterised by a pair $(\bar{x}_0, \bar{\sigma}_0)$ with respect to E_0, where $\bar{x}_0 = (x_0, y_0)$ are the coordinates of the point of intersection of the ray with E_0 and the vector $\bar{\sigma}_0$ is formed by the first two components of the vector $\bar{s}_0 = (\alpha_0, \beta_0, \gamma_0)$ indicating the direction of the light ray. In order to emphasise the link between geometrical optics and mechanics, we will write $\bar{q}_0$ for $\bar{x}_0$; to indicate the direction of the light ray, we will use $\bar{p}_0 = \mu_0 \bar{\sigma}_0$. Thus, any incident light ray will be given by a pair $(\bar{q}_0, \bar{p}_0)$. In the same way, any excident ray is characterised with respect to E_n by a pair $(\bar{q}_n, \bar{p}_n)$. We will denote the space of incident directions $\bar{p}_0$ by F_0. The space of excident directions will be referred to as F_n. Thus, the transition of the incident light rays through the optical system gives rise to a map

$$M : E_0 \times F_0 \to E_n \times F_n.$$

This map will be called the *optical map* for the system. Ultimately, all properties of an optical system can be expressed in terms of the properties of its optical map.

2.1 Paraxial Optics

For small angles of incidence at an optical system, the angles of excidence will be small as well. This gives rise to the so-called theory of *paraxial optics*. The basic idea here is the following. Suppose we have a system of refracting surfaces centred around an axis, such that the axis intersects all these surfaces perpendicularly. Then, all incident paraxial rays, i.e. all rays very near and almost parallel to this axis, will still be paraxial after transition through the system. For the refraction of these rays at a surface $\mathcal{F}$ of the system, this means that we can now approximate Snell's law by a more convenient expression. Essentially, we consider the linear part of the optical map only.

Let us restrict to spherical refracting surfaces. In this case, we may assume that the axis of the system is the z-axis. The tangent plane at the vertex of the refracting surface $\mathcal{F}$ can be taken as the x, y-plane. Up to the first order, the equation of the refracting surface will now be of the form

$$z = \frac{x^2 + y^2}{2R},$$

where R is the radius of curvature of $\mathcal{F}$. For an incident paraxial ray with optical direction $\bar{\sigma}_0$ and corresponding excident ray with direction $\bar{\sigma}_1$, both $\bar{\sigma}_i$ will be very small. Likewise, the lengths of the vectors $\bar{x}_i$ will be small. It is now not too difficult to check that, up to the first order, Snell's law is equal to the condition

$$\mu_0(\bar{\sigma}_0 + \bar{x}_0/R) = \mu_1(\bar{\sigma}_1 + \bar{x}_0/R).$$

Writing $\bar{q}_i := \bar{x}_i$ and $p_i := \mu_i \sigma_i$, we thus obtain

$$\bar{q}_1 = \bar{q}_0, \qquad \bar{p}_1 = \bar{p}_0 + \frac{(\mu_0 - \mu_1) \cdot \bar{q}_0}{R}.$$

These linear expressions describe the refraction of paraxial rays at a surface. Put geometrically, these equations describe how the coordinates $\bar{q}_1, \bar{p}_1$ of a refracted ray with respect to the tangent plane of the refracting surface relate to the coordinates $\bar{q}_0, \bar{p}_0$ of the incident ray with respect to the same tangent plane.

Note that it is also possible to describe the relation between the coordinates of the incident and the excident rays with respect to planes other than the tangent plane. Indeed, suppose we have a reference plane E_0 parallel to the tangent plane with a distance e_0 to $\mathcal{F}$. Then, it is easily verified that for one and the same ray

$$\bar{p}_0 = \bar{p}, \qquad \bar{q}_0 = \bar{q} + (e_0/\mu_0) \cdot \bar{p}.$$

A similar relation applies to the relation between the coordinates of the refracted ray with respect to an arbitrary reference plane E_1 and its coordinates $(\bar{q}_1, \bar{p}_1)$ with respect to the tangent plane at the vertex of the refracting curve.

The above relations are linear. Therefore, the relations between the coordinates of an incident ray with respect to a reference plane E_0 and those of the excident ray with respect to a reference plane E_1 will be linear as well. In fact, we have

Theorem 2 *Suppose we have an incident system of paraxial rays which are refracted at one surface. Let the coordinates of an incident ray with respect to a reference plane E_0 at distance e_0 from the refracting surface be $(\bar{q}_0, \bar{p}_0)$, with $(\bar{q}_1, \bar{p}_1)$ those of the corresponding excident ray with respect to a plane E_1 at distance e_1. Then, up to the first order,*

$$\begin{pmatrix} \bar{q}_1 \\ \bar{p}_1 \end{pmatrix} =$$

$$\begin{pmatrix} 1 & \tilde{e}_1 \\ 0 & 1 \end{pmatrix} \cdot \begin{pmatrix} 1 & 0 \\ 1/\tilde{R} & 1 \end{pmatrix} \cdot \begin{pmatrix} 1 & -\tilde{e}_0 \\ 0 & 1 \end{pmatrix} \cdot \begin{pmatrix} \bar{q}_0 \\ \bar{p}_0 \end{pmatrix},$$

where $\tilde{e}_i = e_i/\mu_i$ and $\tilde{R} = R/(\mu_0 - \mu_1)$.

It is now easily verified that for all reference planes E_0 there is a reference plane E_1 such that pencils of rays departing from a point on E_0 will unite on a point of E_1 after refraction. Such pairs of planes are called conjugate planes. The kind of transformation between the reference planes induced by transition through the optical system is a dilatation and the dilatation factor is called the *magnification* of the optical system with respect to the reference planes.

2.2 Aberrations

Note that the preceding only applies to the case of paraxial optics. In general, the rays departing from one point will not unite into one point again after refraction. Indeed, it can be shown that if this can be achieved, then the magnification of the system has to be either 1 or −1 for all conjugate planes (see [2]). This situation can, for instance, be achieved by using mirrors only. In the case of general systems we have to count on deviations from the perfect image. These deviations are called the *aberrations* of a system. Essentially, the principal task of geometrical optics is to find means to measure the deviation from the perfect image and to reduce these aberrations to a minimum while retaining the desired magnification of the system.

A good measure for these aberrations is provided by the coefficients of the higher-order terms of the optical map of the system. For this reason, these coefficients are also called aberration coefficients. Note that in the case of rotation symmetric systems we only have aberration coefficients of odd order. In fact, it can be easily shown that because of the rotation symmetry, the $(2n+1)$th-order part of the Taylor expansion of M is of the form

$$\sum_{a+b+c=n,} \langle \zeta_{abc}, (\bar{q}_0, \bar{p}_0) \rangle U^a V^b C^c,$$

where a, b and c are integers ≥ 0, $\langle .. \rangle$ denotes the standard inner product, ζ_{abc} is a 4-tuple of numbers and $U = \langle \bar{q}_0, \bar{q}_0 \rangle$, $V = \langle \bar{q}_0, \bar{p}_0 \rangle$ and $W = \langle \bar{p}_0, \bar{p}_0 \rangle$ are the rotation invariants.

In the following, we will discuss how these aberration coefficients may be computed for a given system with given reference planes.

3 The computation of aberrations

The most direct procedure to determine the aberration coefficients is to determine the optical map up to a certain order. A procedure to do so is easily indicated. In the case of one refraction, the determination of the optical map is an elementary application of perturbation theory. Once we have obtained the optical map for this case, the optical map for a compound system simply is the composition of the optical maps for the constituent refracting surfaces.

The procedure to determine the optical map for one refracting surface is the following. For the sake of practicality, we will only consider the case of spherical refracting surfaces. Note, however, that the extension to more general cases is straightforward. Suppose we have a refracting sphere $\mathcal{F}$ given by the equation

$$z = P(x,y) = \frac{1}{2R}(x^2 + y^2) + \dots \qquad (1)$$

The normal at a point $\bar{x}$ of this surface will be indicated by $\bar{s} = (-P_x, -P_y, 1) = (s_1, s_2, -1)$, where P_x and P_y are short for the derivatives of P with respect to x and y, respectively. Furthermore, we assume that the reference planes E_0 and E_1 both lie at the top of $\mathcal{F}$. The coordinates of the incident light rays will be indicated by $(\bar{q}_0, \bar{p}_0)$, those of the excident rays by $(\bar{q}_1, \bar{p}_1)$. Finally, we will indicate the point where the incident ray impinges on $\mathcal{F}$ by the coordinates $\tilde{q} = (q_1, q_2, q_3)$. It now is not difficult to see that the coordinates of $\tilde{q}$ can be determined up to any order with the help of basic perturbation theory. In fact, let us parametrise an incident ray ℓ by

$$p_\ell : \tilde{q} \mapsto (\bar{q}_0, 0) + \lambda(\bar{p}_0, \sqrt{\mu_0^2 - \langle \bar{p}_0, \bar{p}_0 \rangle}), \qquad \lambda \in I\!R.$$

After substitution of this parametrisation into the equation (1) for the refracting surface, we now obtain an equation of the form

$$\lambda \sqrt{\mu_0^2 - \langle \bar{p}_0, \bar{p}_0 \rangle} = Z_1(\bar{q}_0, \bar{p}_0) + \lambda \cdot Z_2(\lambda, \bar{q}_0, \bar{p}_0),$$

where Z_1 is a function of $\bar{q}_0$ and $\bar{p}_0$ only and Z_2 is quadratic in $\bar{q}_0$ and $\bar{p}_0$. Rewriting gives us

$$\lambda = Z_3(\bar{q}_0, \bar{p}_0) + \lambda \cdot Z_4(\lambda, \bar{q}_0, \bar{p}_0),$$

where Z_3 is a function of $\bar{q}_0$ and $\bar{p}_0$ only and Z_4 is quadratic in $\bar{q}_0$ and $\bar{p}_0$. Consequently, let us denote the Taylor expansion of λ around the origin and up to order m by λ_m, while λ^m denotes the mth-order part of λ. Then,

$$\lambda^m \overset{m}{=} Z_3 + \lambda_{m-1} \cdot Z_4(\lambda_{m-1}, \bar{q}_0, \bar{p}_0),$$

where $\overset{m}{=}$ means that equality only applies for the mth-order terms. This means that if we know λ up to order $m-1$, then we are able to compute λ up to order m as well.

In order to compute the coordinates of the excident rays, we have the following two relations:

$$\bar{p}_1 - \frac{\langle \bar{p}_1, \tilde{s} \rangle}{\langle \tilde{s}, \tilde{s} \rangle} \tilde{s} = \bar{p}_0 - \frac{\langle \bar{p}_0, \tilde{s} \rangle}{\langle \tilde{s}, \tilde{s} \rangle} \tilde{s} \qquad (2)$$

and

$$\bar{q}_1 = \tilde{q} - \frac{q_3 \bar{p}_1}{\sqrt{(\mu_1)^2 - \langle \bar{p}_1, \bar{p}_1 \rangle}}, \qquad (3)$$

212

where $\tilde{s}$ is the normal at the point of impingement of the ray. The first of these equations is a vector formulation of Snell's law. The second relates the point where the incident light ray impinges on $\mathcal{F}$ to the point of excidence. Together with the expressions for $\tilde{q}$ up to a given order, these two relations suffice to determine the optical map up to one order less. In fact, suppose we have determined $(\bar{q}_1, \bar{p}_1)$ up to order n and $\tilde{q}$ up to order $n+1$. Then, the first equation yields $\bar{p}_1$ up to order $n+1$, while the second one gives $\bar{q}_1$ up to order $n+1$. Some care is needed for the lowest order cases, but even in the first-order case, everything turns out to work properly. Thus, it is not difficult to write a short computer algebra procedure which computes the optical map for one refraction up to arbitrary order. (For a similar approach, see [5], pp.124-133.)

Once the optical map is obtained for one refraction, there should not be any difficulty in obtaining the optical map for more refractions. In fact, we only need to compose the individual optical maps and truncate the resulting map after the required order. In practice, however, the relative substitutions give rise to expressions which contain a large number of terms higher than the one required. Only at the last moment these terms are discarded. The result is that the internal memory of Maple is used up long before the program has arrived at the final expressions (so-called *intermediate swell*).

One way to obviate this problem is to make use of the fact that M can be written as a Lie transformation. In order to see what this means, we first need

Definition 1 *Let g be an infinitely differentiable function acting on a space $\mathbb{R}^{2n} = \{(\bar{q}, \bar{p})\}$. Then, the 'adjoint operator' $ad(g)$ is a function that maps any infinitely differentiable function f to another one according to*

$$ad(g)(f) := \sum_{i=1}^{n} \frac{\partial g}{\partial p_i}\frac{\partial f}{\partial q_i} - \frac{\partial g}{\partial q_i}\frac{\partial f}{\partial p_i}.$$

The adjoint operator can now be used to define a Lie transformation.

Definition 2 *Let C^∞ be the space of infinitely differentiable functions. Suppose g is a C^∞-function acting on a space $\mathbb{R}^{2n}$. Then, the Lie transformation $e^{ad(g)}$ of g working on C^∞, the space of infinitely differentiable functions, is defined as*

$$e^{ad(g)} := \sum_{i=0}^{\infty} \frac{1}{i!} ad^i(g),$$

where $ad^0(g)$ is the identity map.

In general, the series on the right-hand side cannot be formally computed, neither can an arbitrary function on an even-dimensional space always be written as a Lie transformation. However, for our purposes the following suffices. Suppose we have a Lie transformation. We can extend the domain of this transformation to a $2n$-tuple of functions by applying the Lie transformation to the components of the tuple. Then, we have

Theorem 3 *Let M be a symplectic diffeomorphism from $\mathbb{R}^{2n}$ to $\mathbb{R}^{2n}$, such that its Jacobian $Jac(M)$ can be written in the form $Jac(M) = exp(JS)$, where J is the antisymmetric $2n \times 2n$ matrix defined by*

$$\begin{pmatrix} 0 & I_n \\ -I_n & 0 \end{pmatrix}$$

and S is a symmetric $2n \times 2n$-matrix. Then, for any given order m, there is a function $\widetilde{M} : \mathbb{R}^{2n} \to \mathbb{R}$ such that, up to order m,

$$M(\bar{q}, \bar{p}) = e^{ad(\widetilde{M})}(\bar{q}, \bar{p}).$$

Proof see [4],p.2217 □

Since the optical map is symplectic (see [7]) and since it can be checked that it has the required properties, we have

Corollary 1 *The optical map for an optical system can be expressed in terms of a Lie transformation. In particular, for any optical map $M : E_0 \times F_0 \mapsto E_n \times F_n$, there is a function $\widetilde{M}$ on $E_0 \times F_0$ such that*

$$e^{ad(\widetilde{M})}(\bar{q}_0, \bar{p}_0) = M(\bar{q}_0, \bar{p}_0)(= (\bar{q}_n, \bar{p}_n)).$$

For a general optical transformation, $\widetilde{M}$ will be hard to compute. For the optical map of one refracting surface, however, the Lie transformation can be computed rather easily. Indeed, we already know how to compute q_{11}, the first component of $\bar{q}_1$, up to order m. Now, suppose we have computed $\widetilde{M}$ up to order m. Let us denote this part of $\widetilde{M}$ by $\widetilde{M}_m$. We will denote the m-order part by $\widetilde{M}^m$. For other series, the same notation will apply. Then, we have

$$e^{ad(\widetilde{M}_m + \widetilde{M}^{m+1})}(q_{01}) \overset{m}{=} (q_{11}).$$

Since $[\widetilde{M}_2, q_{01}] = 0$, it follows that

$$\sum_{i=0}^{\infty} \frac{1}{i+1!} ad^i(\widetilde{M}^2)([\widetilde{M}^{m+1}, q_{01}]) \overset{m}{=}$$

$$q_{11} - e^{ad(\widetilde{M}_m)}(q_{01}).$$

This allows us to compute $[\widetilde{M}^{m+1}, q_{01}]$. In fact, the left-hand side of this equation may be viewed as a map ϖ applied to $\widetilde{M}^2$, which in turn is applied to $[\widetilde{M}^{m+1}, q_{01}]$. In other words, we have

$$\varpi(\widetilde{M}^2)([\widetilde{M}^{m+1}, q_{01}]) = (q_{11})_{m-1} - e^{ad(\widetilde{M}_{m-1})}(q_{01}).$$

The map $\varpi(\widetilde{M}^2)$ is called the *derivative of the Lie transformation* of $\widetilde{M}^2$. Derivatives of Lie transformations are invertible and we have

$$\varpi(\widetilde{M}^2)^{-1} = \sum_{i=0}^{\infty} \frac{B_i}{i!} ad^i(\widetilde{M}^2),$$

where B_i denotes the ith Bernoulli number (See [1], p.90, ex.3b with $V = Id$.). Upon closer inspection, we see that

$$[\widetilde{M}^{m+1}, q_{01}] = \frac{\partial \widetilde{M}^{m+1}}{\partial p_{01}}.$$

Thus, making use of the symmetry of the optical map, we can determine $\widetilde{M}.1^{m+1}$, i.e. the part of $\widetilde{M}^{m+1}$ that does not depend on $\bar{q}_0$ only. We deduce the remaining part $\widetilde{M}.2^{m+1}$ of $\widetilde{M}^{m+1}$, from the property that

$$p_{11} \overset{m}{=} e^{ad(\widetilde{M}_{m-1} + \widetilde{M}.1^{m+1} + \widetilde{M}.2^{m+1})}(p_{01}).$$

It follows that

$$\frac{\partial \widetilde{M}.2^{m+1}}{\partial q_{01}} \overset{m}{=} e^{ad(\widetilde{M}_{m-1} + \widetilde{M}.1^{m+1})}(p_{01}) - p_{11}.$$

This suffices to determine $\widetilde{M}.2^{m+1}$ and hence the full expression $\widetilde{M}^{m+1}$.

In general, $\widetilde{M}$ will be a more manageable expression than M itself. This will prove to be essential if we want to find the optical map for a compound system in a more efficient way than by straightforward substitution. The crucial point is the following

Theorem 4 *Suppose we have a set of Lie transformations* $f_i = e^{ad(\tilde{f}_i)}$ *($i \in \{1, ..n\}$). Then*

$$f_n \circ ... \circ f_1 = e^{ad(\tilde{f}_1)}...e^{ad(\tilde{f}_n)}.$$

Proof This directly follows from the properties of Lie transformations given in [6], p.13 or [13], p.55. $\square$

For an optical transformation M induced by n refractions M_i, it follows that

Corollary 2 *For a sequence of optical transformations* $M_1, ..., M_n$ *with* $M := M_n \circ ... \circ M_1$, *we have*

$$M = e^{ad(\widetilde{M}_1)}...e^{ad(\widetilde{M}_n)}.$$

Thus, we have obtained an analogue of the matrix formulation of paraxial optics. As in the case of paraxial optical, it is better to split up the refraction at one surface in three parts. First we consider the transformation induced by the "flow" of the rays from an arbitrary reference plane E_0 to the reference plane E tangent to the refracting surface. Secondly, we compute the transformation induced by the refraction with reference to E and finally we consider the flow from E to an arbitrary reference plane E_1.

All three of these transformations can be written as a Lie transformation. In fact, we already know that the actual refraction can be written as a Lie transformation. Furthermore, we easily verify that a Lie series of the form

$$e^{l \cdot ad(\sqrt{\mu_0^2 - \langle p_0, p_0 \rangle})} : (\bar{q}_0, \bar{p}_0) \mapsto (\bar{q}_0', \bar{p}_0')$$

describes an optical flow over a distance l in a medium $\mathcal{M}_0$ with index of refraction μ_0. In the same way, a flow in any medium $\mathcal{M}_i$ can be described. Therefore, we have

Theorem 5 *Suppose M is an optical transformation with respect to two arbitrary reference planes corresponding to refraction at one surface. Then, M is described by the composition of Lie transformations*

$$e^{e_0 \cdot ad(\sqrt{\mu_0^2 - \langle p_0, p_0 \rangle})} \cdot e^{ad(\widetilde{M}_1)} \cdot e^{-e_1 \cdot ad(\sqrt{\mu_1^2 - \langle p_0, p_0 \rangle})}.$$

This provides us with a full generalisation of the formulae for the paraxial case to the general optical map.

Compared to straightforward substitution, this formulation offers various advantages. First of all, the expressions for one surface are more compact and easier to deal with. Secondly, in order to compute one component of the optical map up to order m with the help of the Lie transformation, it is not necessary to compute the other components up to order $m - 2$, as is the case for substitution. Furthermore, since the concatenation of Lie transformations uses no term of order higher than the ones we are interested in, the Lie approach is more efficient and uses less space than substitution.

Finally, one more edge over straightforward substitution should be mentioned. In this paper we

214

have only discussed the geometrical optics in optically homogeneous media. Clearly, it is also possible to develop a geometrical optics of non-homogeneous media. For this kind of geometrical optics, however, the rays are no longer straight lines and the straightforward approach sketched in the preceding will no longer work. In contrast, the Lie approach can be easily generalised to non-homogeneous media.

4   The Computer Algebra

In this section, we will say something more about the implementation of the methods described above to compute the optical map for an optical system formed by several spherical surfaces in a modern computer algebra system. We have chosen to use Maple V Release 2 for programming because the package is easy to use, efficient, and sufficiently powerful to do all the required computations. All calculations were done on a Silicon Graphics Challenge L with two 150 Mhz R4400 MIPS processors, 128 MB internal memory, and 256 MB swap space; computer times are in seconds. However, many of these computations are still feasible on less powerful computers as well. Because of lack of space, we will concentrate on the general design of the program and the results obtained so far. We only give very few details concerning the actual code.

4.1   A Single Refracting Surface

The optical map M for a single refracting sphere $\mathcal{F}$ can be relative easily computed up to a rather high order. Once M is obtained, there is no problem in obtaining the corresponding corresponding function $\widetilde{M}$ in the exponent of the Lie transformation.

A minor difficulty resides in the symbols we can use. Because of the limited range of characters in Maple, we have to be careful in choosing our notations, so as not to loose all relation to the standard mathematical notations. In the program, we refer to the variables in the object space by adding in (of 'incident') and by adding ex (of 'excident') to the variables of the image space. Components of an expression will be indicated by .i. Thus, we write pin.i, pex.i, muin and muex. Furthermore, we will write M[n] for M_n. The expression M[[n]] will correspond to M^n. Other notations will be explained when introduced.

We start by computing the expressions for the coordinates q_imp.i[m] of the point of impingement $\tilde{q}$ of a ray on the refracting surface up to

order m. In order to do so, we first have to expand the parametrisation of $\mathcal{F}$ up to the desired order and to compute pin.3[m] up to order m. Next, we compute lambda, the parameter in the parametrisation of ℓ, for the point of intersection of ℓ with $\mathcal{F}$. The procedure to do so has been outlined in the main text. We immediately see that lambda[[0]] and lambda[[1]], the zero-th and first-order part of λ, are equal to zero. We take these values as our point of departure. The coordinates q_imp.i[m] are now easily determined, as are the coordinates s.i[m-1] of the normal to $\mathcal{F}$ at $\tilde{q}$

Our next step is the computation of the coordinates qex.i[m] and pex.i[m] of the excident rays up to order m with the help of the two equations (2) and (3). We first use (3) to determine the qex.i[m]. After this, we use (2) to determine pex.i[m]. In order to keep these computations manageable for Maple, we have to truncate and collect the expressions we are dealing with as often as possible. Thus, we have for the computation of pex.i[m] the code

```
>for j to 2
>do
>  pex.j[i]:=  pin.j -
   ((pin.1 - pex.1[i-1])*s.1[i-1] +
   (pin.2 - pex.2[i-1])*s.2[i-1] -
   (pin.3[i-1]-pex.3[i-1]))*
   s.j[i]*s_inv[i];
>  pex.j[i]:= collect(mtaylor
   (pex.j[i],[qin.1,qin.2,pin.1,pin.2],
   i+1), [qin.1,qin.2,pin.1,pin.2],
   normal,distributed);
>od;
```

Since the expressions just computed have to be rotation-symmetric, we can write them in a more compact form with the help of the rotation invariants U, V and W (see p.4). For qex.1[3] we thus obtain

```
> qex.1[3]:= simplify(expand(qex.1[3]),
  {U=pin.1^2+pin.2^2, V=pin.1*qin.1+
  pin.2*qin.2, W=qin.1^2+qin.2^2},
  [qin.1,qin.2,pin.1,pin.2,U,V,W]);

  qex.1[3]:= qin.1 +  (- muin +  muex).


    ( R pin.1 W + muin qin.1 W )

       /  2
      / R  muex muin
     /
```

This result is in accordance with the expression found in the literature (cf. [14], p.64, where the eikonal for a spherical surface is computed up to the fourth order. Switching from the eikonal to the optical map gives the expression above.)

The Lie transformation $\widetilde{M}$ for one single refraction M can now be found by following the procedure explained in the main text. To shorten the procedure, we have defined new routines AD(f,g), EXP(f,g,m) and LIE_DERV_INV(f,g,m). It will be clear what the first of these three does. The second computes the Lie transformation of a function f applied on g up to order m in the variables qin.1, qin.2, pin.1, pin.2. The last one computes the derivative $\varpi(f)^{-1}$ of a Lie transformation applied on g up to order m in the variables qin.1, qin.2, pin.1, pin.2.

The procedure outlined in the main text requires that we already have the expression for the second-order part of $\widetilde{M}$. However, it can be easily verified with the help of the formulae on p.3 (see also [5], p.112), that

$$M_2 = \frac{\mu_1 - \mu_0}{2R}(q_1^2 + q_2^2).$$

After we have obtained Mtilde[[i-1]], we can now obtain the next-order part as follows.

```
> qex.1[1,i-1]:= select(x->
> degree(x,[qin.1,qin.2,pin.1,pin.2])<i,
  qex.1[1,m]);
> lie_derv[qin.1,i-1 ] := qex.1[1,i-1] -
  EXP(Mtilde[i-1],qin.1,i);
> lie_derv[[qin.1,i-1]]:= select(x->
  degree(x,[qin.1,qin.2,pin.1,pin.2])=i-1,
  lie_derv[qin.1,i-1]);
> DMtilde[[pin.1,i-1]]:=
  LIE_DERV_INV(Mtilde[[2]],
  lie_derv[[qin.1,i-1]],i-1);
```

The expression $\widetilde{M}.1^{m+1}$ can now be directly computed by integrating DMtilde[[pin.1,i-1]] with respect to pin.1. This, however will not give the whole of the expression and a symmetry argument has to be applied to find the terms not depending on pin.1. The expression $\widetilde{M}.2^{m+1}$ can then be determined in a similar way. Special care has to be taken to get the signs correct. In the case of $i = 4$, we thus find

```
Mtilde[4] :=
                               2
    - 1/24 (mu.0 - mu.1) W (6 R.1  U +
```

```
      2
12 mu.0 R.1  mu.1 + W mu.0 mu.1 +

      2           2
W mu.0  +  W mu.1  + 6 R.1 V mu.0 +

                     /    3
6 R.1 V mu.1)  /  (R.1  mu.0 mu.1)
                     /
```

As far as we are aware, this result has not been given in this explicit form in the literature yet.

The above procedures to compute the optical map M for one surface up to order m and its corresponding exponent $\widetilde{M}$ in the Lie transformation up to order $m + 1$ turn out to be reasonably efficient. In the following, the statistics are given up to 10th-order. Simplification with the help of U, V and W has been omitted.

run-time statistics for one refracting surface				
	Time (in seconds)		Memory (MB)	
m	M	$\widetilde{M}$	M	$\widetilde{M}$
3	0.7	1.4	0.8	0.8
5	3.3	6.6	1.1	1.6
7	23.6	64.0	2.9	4.7
9	142.1	495.3	13.6	19.5

This table shows that the computation of the function $\widetilde{M}$ for one refracting surface soon becomes very time consuming. To compensate for this, the computation of the optical map for several refractions with the help of Lie transformations turns out to be more efficient than straightforward substitution.

4.2 The case of several surfaces

A routine for straightforward composition of subsequent refractions at a number of surfaces is now no longer any problem. We start by switching to the optical map for arbitrary reference planes. It is easy to verify that the formulae we need for the transition from a reference plane $E \times F$ to a plane $E' \times F'$ are of the form

$$p_i' = p_1, \qquad q_i' = q_i + l \cdot p_i/p_3 \qquad i = 1, 2.$$

Thus the new coordinates may be computed as follows. The first loop is for the reference plane in the image space, the second for that in the object space. Note that the optical direction does not change under the first shift of reference plane. After this, we can simply compose the optical maps we have obtained.

The procedure for the Lie transformation looks similar. We start with the Lie transformation for the flow from the last refracting refracting surface to the final reference plane and then go backwards. The last transformation is that of the flow from E_0 to the first refracting surface.

It is interesting to compare the *cpu-times* and memory that both methods need. In the following table, we have listed the respective *cpu-times* for the computation of the 1st-order optical map for w refracting surfaces. Simplification with the help of U, V and W has been omitted. As before, in the case of the Lie Transformations, only one coefficient (q_1) has been computed.

run-time statistics for w refracting surfaces				
	Time (in seconds)		Memory (MB)	
w	subst.	Lie series	subs.	Lie series
3	0.6	0.8	0.1	0.1
4	1.2	1.3	0.1	0.1
5	5.0	3.3	1.2	0.8
6	28.0	10.8	3.3	1.3
7	192.4	45.1	11.8	4.3
8	–	227.9	–	17.6

Obviously, in the paraxial case considered here, we only have to do with simple linear transformations and no advanced theory seems to be needed. However, even for paraxial optics some of the advantages of the Lie method that we hope to find come to the fore. At first, the Lie transformations are slower than straightforward substitutions, but from the start they are more efficient with respect to memory. For w large enough, the Lie method becomes faster as well. For $w = 8$, the substitution program crashes because of an `object-too-large`-error. For higher-order optical maps, the computations in both cases require enormous amounts of time for $w > 2$; we have not checked which of the two methods is faster. For two surfaces and the third-order optical map, substitution requires 13 seconds, whereas the Lie transformation requires 5 seconds.

4.3 A Petzval Lens

When most of the parameters of an optical systems are specified, both the substitution method and the Lie method become much faster and we are able to deal with larger systems as well. In order to show what our software is able to achieve in this case, we have computed the optical map for a so-called Petzval lens, where we have taken the specifications from [12], p.225. The particular lens we used consists of four lenses and hence eight refracting surfaces. An element of symbolic computation is retained in that we have not specified the position of the reference planes in the object and the image space. For the other variables, we used the following specifications

```
>R.1:=  0.53000: R.2:= -4.60000:
>R.3:= -1.39700: R.4:=  2.40000:
>R.5:=  0.59500: R.6:= -0.42200:
>R.7:= -0.38000: R.8:= -1.61000:
>
>mu.0:=1: mu.1:=1.517: mu.2:=1:
>mu.3:=1.620: mu.4:=1: mu.5:=1.517:
>mu.6:=1: mu.7:=1.620: mu.8:=1:
>
>d[1]:=0.19500: d[2]:=0.02565:
>d[3]:=0.05000: d[4]:=0.37050:
>d[5]:=0.17000: d[6]:=0.00940:
>d[7]:=0.05000:
```

In the following table, we have listed the respective *cpu-times* for the computation of the *m*st-order optical map for the Petzval lens. Simplification with the help of U, V and W has been omitted. In the case of the Lie Transformations, only one coefficient has been computed.

run-time statistics for a Petzval lens				
	Time (in seconds)		Memory (MB)	
m	subs.	Lie series	subs.	Lie series
1	2.8	1.0	0.9	0.1
3	—	5.2	—	1.2
5	—	37.0	—	1.5
7	—	150.8	—	6.6
9	—	1063.1	—	12.4

Already for the first-order aberrations, the substitution method fails. The program is able to compute the transition through the first seven surfaces, for which it needs about 4000 seconds. For the eight surfaces of the Petzval lens, however, Maple crashed on an `object-too-large`-error. Clearly, the Lie method does much better here. In fact, at least for the lower-order aberrations, it still does not crash as more variables are left unspecified. This gives us, for instance, the possibility to move a lens along the optical axis. With the help of the expression thus obtained, a further optimalisation of the design of the Petzval lens can be achieved.

5 Conclusions

From the above examples it is clear that computer algebra can be used to compute the aberration coefficients for an optical system. It is too early to say anything about which of the two methods we introduced is most efficient. In both cases, the program can probably be considerably improved. In both cases, we used a step order one for our approximations. Since all even order aberrations vanish in the case of rotation-symmetrical systems, it seems plausible that a step order of two is feasible as well. Furthermore, no optimal use of the symmetries has been used yet. Finally, it still has to be investigated whether the combination of the `mtaylor` and `collect` functions are really the most efficient way to deal with Taylor series.

Regardless of the current state of our software, the computation of fully symbolic expressions turns out to pose particular problems. In the case of only a small number of unspecified variables, however, computer algebra can be fruitfully used in combination with Lie transformations in order to compute such expressions as are relevant for the design of optical instruments.

References

[1] N. Bourbaki *Eléments de Mathématiques. Groupes et Algèbres de Lie. Chap. 2 et 3* (Hermann, 1972)

[2] C. Carathéodory, "Ueber den Zusammenhang der Theorie der absoluten optischen Instrumente mit einem Satze der Variationsrechnung," *S.B. Bayerische Academie der Wissenschaften. Math.-naturwissenschaftliche Abt.* 1926, pp. 1-18

[3] A.M. Cohen, *Computer Algebra in Industry. Problem Solving in Practice* (Wiley & Sons, 1993)

[4] A.J. Dragt, J.M. Finn, "Lie series and invariant functions for analytic symplectic maps," *Journal of Mathematical Physics* 17 (1976), pp. 2214-2227

[5] A.J. Dragt, E. Forest, K.B. Wolf, "Foundations of a Lie algebraic theory of geometrical optics. In [11], pp.105-157

[6] W. Gröbner, *Lie-Reihen und ihre Anwendungen* (D.V.W., 1960)

[7] V. Guillemin & S. Sternberg *Symplectic Techniques in Physics* (C.U.P., 1984)

[8] A. Heck & M. Biemond, "Computer Algebra and Geometrical Optics I: The Eikonal of a Single Refracting Surface." In: H. Appiola, M. Laine and E. Valkeila (Eds.), *Proceedings of the Workshop on Symbolic and Numeric Computing, Helsinki 1993*. Research Reports - Rolf Nevanlinna Institute B10 (1994): 127-138.

[9] A. Heck & M. Biemond "Computer Algebra and Geometrical Optics II: The Eikonal of a Symmetric Optical System." (to appear)

[10] M.B. Monagan "The Collect Function in Maple," *Maple Newsletter* 4 (1989), pp.17-19

[11] J. Sánchez Mondragón & K.B. Wolf (eds.), *Lie Methods in Optics. Proceedings of the CIFMO - CIO workshop held at Léon* (January 7-10, 1985) (Springer, 1986) (= *Lecture Notes in Physics* 250.)

[12] W.J. Smith *Modern Lens Design. A Resource Manual* (McGraw-Hill, 1992)

[13] S. Steinberg "Lie series, Lie transformations and their applications." In: [11], pp.45-103

[14] J.L. Synge, *Geometrical Optics. An Introduction to Hamilton's Method* (Cambridge University Press, 1937) (= *Cambridge Tracts in Mathematics and Mathematical Physics*, vol. 37)

[15] K.B. Wolf (ed.), *Lie Methods in Optics II - Proceedings of the Second Workshop held at Cocoyoc* (July 19-22, 1988) (Springer, 1989) (= *Lecture Notes in Physics* 352).

Eisso Atzema (atzema@can.nl) received a M.S. degree in mathematics from the University of Nijmegen (The Netherlands) in 1988 and a Ph.D. degree in the history of mathematics from the University of Utrecht (The Netherlands) in 1993. He is currently project coordinator for a project devoted to the application of computer algebra to geometrical optics at the Research Institute for the applications of Computer Algebra (**RIACA**) in Amsterdam.

SOLUTION OF BANDED LINEAR SYSTEMS OF EQUATIONS IN MAPLE USING LU FACTORIZATION

Robert M. Corless[1] and Khaled El-Sawy[2]

Department of Applied Mathematics[1], Department of Civil Engineering[2], University of Western Ontario, London, Ontario, Canada

Abstract

We describe two Maple implementations of an LU-factorization method for the solution of column diagonally dominant banded matrices. We compare the efficiency of these programs with Maple's built-in routine linsolve. We also give an example of the new indexing functions for Maple V Release 3. The code described in this paper is available by anonymous ftp to pineapple.apmaths.uwo.ca (129.100.23.160) in the directory pub/maple/band.

Introduction

Many engineering problems require a system of linear equations $[A]\{x\} = \{c\}$ to be solved. In many applications (e.g. finite difference and finite element methods) the system of linear equations takes a banded shape (most simply, a tridiagonal shape). One can solve this system ignoring the structure of the matrix $[A]$ using ordinary Gaussian elimination. This method reduces the matrix $[A]$ to a triangular form using row operations and after that solves for the unknowns by back-substitution. No use of the structural properties of the matrix $[A]$ is made. Programmed naively, this method has the drawback of being applicable only once for each vector/matrix $\{c\}$. To solve for another vector/matrix $\{c\}$, the same process must be repeated, which is a waste of computer time (i.e. money).

The standard refinement of Gaussian elimination for efficiency in the solution of multiple linear systems is to factorize (decompose) the matrix $[A]$ into the product $[A] = [L][U]$, where $[L]$ is a lower triangular matrix and $[U]$ is an upper triangular matrix (i.e. $[A]\{x\} = [L][U]\{x\} = \{c\}$). In the case that row exchanges are necessary for numerical stability, we can write this factorization as $[P][A] = [L][U]$ where $[P]$ is a permutation matrix. In the applications of this paper, we will presume that no row exchanges are neces-

sary. It is the case that no zero pivots are encountered, for example, if the matrix $[A]$ is column diagonally dominant [4]. This often occurs in the finite-difference applications we have in mind. Note, however, that even though no zero pivots are encountered, numerical stability may still be a problem. Row-exchanges tend to increase the bandwidth of the matrix, and for efficiency we are willing to lose a bit of numerical stability by not doing the row-exchanges. Once the matrix $[A]$ is factorized, the solution to $[A]\{x\} = \{c\}$ can be found using *forward elimination* for the system $[L]\{y\} = \{c\}$ and then *back substitution* for the system $[U]\{x\} = \{y\}$ (just write $[A]\{x\} = [L][U]\{x\} = [L]([U]\{x\}) = [L]\{y\} = \{c\}$ where $[U]\{x\} = \{y\}$). This is efficient because the most expensive part of the process is the factorization, and this need only be done once.

Using the previous factorization and taking advantage of the structure of the banded matrix $[A]$, we can refine this factorization so that no zero operations are performed [3]. This will save a lot of time, especially with matrices of large dimensions and small band width. We have written a package to factorize banded matrices $[A]$.

We compared the efficiency of our package with Maple's linalg package. We expected that the method which took advantage of the banded structure of the system would be faster by a factor n^2/p^2 where p is the bandwidth of the matrix, and this turned out to be roughly true for matrices with floating point entries. However, we were surprised that if the matrices had integer entries, Maple's built-in routines were faster than our special-purpose routines. We expect that one of the reasons for this is that Maple uses a special algorithm to minimize the size of the calculations' intermediate integers.

Short Tutorial on LU Factorization

The *LU* factorization can be illustrated in the following example. Suppose $[A]$ is a 4×4 matrix and is defined by

$$[A] = \begin{bmatrix} 6 & 2 & 1 & -1 \\ 2 & 4 & 1 & 0 \\ 1 & 1 & 4 & -1 \\ -1 & 0 & -1 & 3 \end{bmatrix}.$$

Now we are looking for a factorized form $[A] = [L][U]$, where

$$[L] = \begin{bmatrix} L_{11} & 0 & 0 & 0 \\ L_{21} & L_{22} & 0 & 0 \\ L_{31} & L_{32} & L_{33} & 0 \\ L_{41} & L_{42} & L_{43} & L_{44} \end{bmatrix}$$

and

$$[U] = \begin{bmatrix} U_{11} & U_{12} & U_{13} & U_{14} \\ 0 & U_{22} & U_{23} & U_{24} \\ 0 & 0 & U_{33} & U_{34} \\ 0 & 0 & 0 & U_{44} \end{bmatrix}.$$

This can be done in Maple as follows.

```
> A := array (1..4, 1..4,
>       [ [ 6, 2,  1, -1],
>         [ 2, 4,  1,  0],
>         [ 1, 1,  4, -1],
>         [-1, 0, -1,  3] ]);
```

$$A := \begin{bmatrix} 6 & 2 & 1 & -1 \\ 2 & 4 & 1 & 0 \\ 1 & 1 & 4 & -1 \\ -1 & 0 & -1 & 3 \end{bmatrix}$$

The following is one way to define the matrices $[L]$ and $[U]$.

```
> L := array (1..4, 1..4):
> U := array (1..4, 1..4):
> for i to 4 do
>   for j to 4 do
>     if (j-i)>0 then L[i,j]:=0 fi;
>     if (i-j)>0 then U[i,j]:=0 fi;
>   od
> od;
```

Now we prepare the equations and unknowns.

```
> eqns := NULL:
> unknowns := NULL:
> LU := evalm(L &* U):
> for i to 4 do
>   for j to 4 do
>     eqns := eqns, LU[i,j] = A[i,j];
>     if not type(L[i,j], numeric) then
>       unknowns:=unknowns, L[i,j]
>     fi;
>     if not type(U[i,j], numeric) then
>       unknowns:=unknowns, U[i,j]
>     fi;
>   od
```

```
> od;
```

Now we can ask Maple to solve for the unknowns in $[L]$ and $[U]$.

```
> solve ({eqns}, {unknowns});
```

$$\left\{ L_{1,1} = 6\frac{1}{U_{1,1}}, U_{1,2} = \cdots \right.$$

three lines omitted

$$\left. U_{3,3} = U_{3,3}, U_{4,4} = U_{4,4} \right\}$$

The solution is not unique! This is because the number of unknowns is larger than the number of equations. Let us use Doolittle's assumption $L_{i,i} = 1$ to choose a unique solution.

```
> for i to 4 do
>   for j to 4 do
>     if i=j then L[i,j]:=1 fi
>   od
> od;
```

We re-prepare the equations and unknowns by executing the same commands as before. We then re-solve for the unknowns in $[L]$ and $[U]$.

```
> solve ({eqns}, {unknowns});
```

$$\left\{ U_{2,3} = \frac{2}{3}, U_{4,4} = \cdots \right.$$

one line omitted

$$\left. L_{4,3} = \frac{-9}{37} \right\}$$

```
> assign (");
> print (L);
```

$$\begin{bmatrix} 1 & 0 & 0 & 0 \\ \frac{1}{3} & 1 & 0 & 0 \\ \frac{1}{6} & \frac{1}{5} & 1 & 0 \\ \frac{-1}{6} & \frac{1}{10} & \frac{-9}{37} & 1 \end{bmatrix}$$

```
> print (U);
```

$$\begin{bmatrix} 6 & 2 & 1 & -1 \\ 0 & \frac{10}{3} & \frac{2}{3} & \frac{1}{3} \\ 0 & 0 & \frac{37}{10} & \frac{-9}{10} \\ 0 & 0 & 0 & \frac{191}{74} \end{bmatrix}$$

It is easy to check that $[A] = [L][U]$. Executing the command evalm(A - L&*U) yields the zero matrix.

Thus the 16 entries of $[A]$ can be used to determine 16 of the 20 unknowns in $[L]$ and $[U]$. If a unique solution is desired, four additional arbitrary conditions on the entries of $[L]$ and $[U]$ are needed. Different choices of the additional arbitrary conditions are due to *Crout*, *Doolittle* and *Choleski*. Crout's method uses $U_{ii} = 1$, while Doolittle's method uses $L_{ii} = 1$, and Choleski's method (for symmetric matrices) uses $L_{ii} = U_{ii}$. The method used in this study is Doolittle's factorization.

Of course, we do not wish to solve a system of nonlinear equations every time we want to factorize a matrix, and of course this solution has already been done for us. The solution process is in fact just Gaussian elimination. For a detailed description of the actual algorithm, see for example [3].

Banded Matrix Definition and Storage

If a matrix $[A]$ has nonzero entries in the qth super-diagonal, then the upper band width of $[A]$ is at least q. Similarly the lower band width p can be defined. The total band width bw can be defined as bw$= p + q + 1$.

There are at least two ways to deal with the storage of banded matrices in Maple. The first is to store them in a modified indices format to minimize the memory requirements, so that most zero terms are not stored. This method is commonly used in FORTRAN programs. For example, consider the matrix $[A]$ of dimensions $n \times n = 8 \times 8$, upper band width $p = 3$, lower band width $q = 3$ and total band width bw$= 7$ defined below.

$$
\begin{bmatrix}
A_{11} & A_{12} & A_{13} & A_{14} & 0 & 0 & 0 & 0 \\
A_{21} & A_{22} & A_{23} & A_{24} & A_{25} & 0 & 0 & 0 \\
A_{31} & A_{32} & A_{33} & A_{34} & A_{35} & A_{36} & 0 & 0 \\
A_{41} & A_{42} & A_{43} & A_{44} & A_{45} & A_{46} & A_{47} & 0 \\
0 & A_{52} & A_{53} & A_{54} & A_{55} & A_{56} & A_{57} & A_{58} \\
0 & 0 & A_{63} & A_{64} & A_{65} & A_{66} & A_{67} & A_{68} \\
0 & 0 & 0 & A_{74} & A_{75} & A_{76} & A_{77} & A_{78} \\
0 & 0 & 0 & 0 & A_{85} & A_{86} & A_{87} & A_{88}
\end{bmatrix}
$$

The modified indices format of this matrix is defined by $[\hat{A}]$ where

$$
[\hat{A}] =
\begin{bmatrix}
0 & 0 & 0 & A_{11} & A_{12} & A_{13} & A_{14} \\
0 & 0 & A_{21} & A_{22} & A_{23} & A_{24} & A_{25} \\
0 & A_{31} & A_{32} & A_{33} & A_{34} & A_{35} & A_{36} \\
A_{41} & A_{42} & A_{43} & A_{44} & A_{45} & A_{46} & A_{47} \\
A_{52} & A_{53} & A_{54} & A_{55} & A_{56} & A_{57} & A_{58} \\
A_{63} & A_{64} & A_{65} & A_{66} & A_{67} & A_{68} & 0 \\
A_{74} & A_{75} & A_{76} & A_{77} & A_{78} & 0 & 0 \\
A_{85} & A_{86} & A_{87} & A_{88} & 0 & 0 & 0
\end{bmatrix}
$$

For this example it looks as though not much space has been saved. However, that is due to the small size of the example. If we used this scheme on a 100 by 100 matrix, of bandwidth 7, then $[\hat{A}]$ would have 7×100 entries compared with 100×100 entries in the full matrix, a savings of 93%. The 12 unnecessary zeros visible in $[\hat{A}]$ are thus of no consequence for large n.

The second way to store banded matrices efficiently is to define this array using a user-defined index function. This will be discussed in the following section.

Banded Arrays and User-Defined Index Functions in Maple

The elements of arrays are stored somewhere in memory. Typically in Fortran there is a definite amount of space (say eight bytes) alloted for each matrix entry, and a contiguous block of memory is set aside at compile time containing n^2 of these eight-byte blocks for a given $n \times n$ matrix A. Elements of the matrix are stored and recovered by first computing the address of the eight-byte block corresponding to A_{ij} by an *indexing function* such as *base address* $+8(n(i-1)+j)$. This expression allows you to find the value of the matrix entry A_{ij} in storage, or to place a new value there.

In Maple, matrix entries need not take a fixed amount of space. Thus it would be very difficult for the programmer to explicitly calculate where each entry A_{ij} is. Maple provides some indexing functions built in to Maple, based on its mechanism for *tables*. This effectively separates the programmer and the user from the details of just where the matrix entry is to go. This "information hiding" allows for great ease of use and power.

Maple provides built-in indexing functions for symmetric matrices and for sparse matrices, and some others. You can simply tell Maple that a matrix is symmetric, and it will only store half the matrix (roughly), giving a great savings. If you tell Maple that your matrix is sparse, Maple will store only those entries you explicitly create, returning zero for all uninitialized entries (without actually creating storage for all those zeros). There is no built-in indexing function for banded matrices. We can, however, write our own.

The syntax for doing this in Maple V Release 2 and earlier was to create a function 'index/<blah>' which required three formal parameters:

A = array to which index function
 is to be applied
index = $[i, j]$ is the reference to the desired entry
is_LHS = boolean variable
 = true if A[i,j] is being assigned to
 = false otherwise

It is not completely straightforward to pass extra arguments such as the band width to the indexing function. A global bandwidth variable is not suitable because we may want to use different band widths for different matrices in the same session.

We used a simple trick to deal with this situation in the Maple V Release 2 implementation. The band width of the array was parsed from the index function name <blah> which must have the format "band???" where the "???" represents a number for the band width. The statement used to parse the name for the bandwidth is

```
> bw := traperror (sscanf (my_index_fcn,
>                 'index/band%i')[1]);
```

The indexing function used this statement *on its own name* to find out what the bandwidth is. This information was thus associated directly with each matrix.

We found it helpful to have a procedure BandIndexFcn, which automatically created an index function for a given bandwidth. The argument for this procedure was the index function name written in the format "band???" where ??? is the total band width bw of the matrix. An example showing the use of this procedure to create an index function for tridiagonal matrices is given below.

```
> BandIndexFcn (band3);

OK ... Band Index Function, band3,  is created

> A := array(band3, 1..6, 1..6);

        A := array(band3, 1 .. 6, 1 .. 6, [])

> print (A);
```

$$\begin{bmatrix} A_{1,1} & A_{1,2} & 0 & 0 & 0 & 0 \\ A_{2,1} & A_{2,2} & A_{2,3} & 0 & 0 & 0 \\ 0 & A_{3,2} & A_{3,3} & A_{3,4} & 0 & 0 \\ 0 & 0 & A_{4,3} & A_{4,4} & A_{4,5} & 0 \\ 0 & 0 & 0 & A_{5,4} & A_{5,5} & A_{5,6} \\ 0 & 0 & 0 & 0 & A_{6,5} & A_{6,6} \end{bmatrix}$$

The advantage of doing this over the FORTRAN-style modification discussed earlier is that Maple's other linear algebra routines work directly with this form: we can add, subtract, multiply, and divide banded matrices formed in this way (though the result is not known to Maple to be a banded matrix and hence will be full of unnecessary zeros on occasion). Utilities such as rowdim and print also work.

Indexing Functions in Maple V Release 3

The indexing functions have completely changed in Maple V Release 3, so the first indexing procedure we wrote no longer works. However, the first method, that of mimicking the FORTRAN storage of a banded matrix, had the happy advantage of being directly portable to Maple V Release 3.

We have re-written the indexing function approach, with the kind assistance of Jérôme Lang and David Clark. The indexing functions in Maple V Release 3 are far superior—we can now have matrices with more than one indexing function, so we can very simply and easily define a symmetric banded matrix, for example.

Because the new indexing functions are so different from before, we feel it is worthwhile to include the code here.

Here is the code to create an indexing function for matrices with (p, q) bandwidth (that is, p nonzero subdiagonals and q nonzero superdiagonals). It uses substitution of the global variables 'index/band/lower' and 'index/band/upper' into a *template* procedure,

and assigns the result to 'index/band<p>-<q>'. These programs are available by anonymous ftp to pineapple.apmaths.uwo.ca in the subdirectory pub/maple/band.

```
BandIndexFcn := proc(p:posint,q:posint);
 assign( 'index/band'.p.'-'.q,
   subs('index/band/lower'=p,
        'index/band/upper'=q,
    proc(indices,tabl) local i,j;
    i := indices[1];
    j := indices[2];
# The number of actual arguments
# distinguishes assignment from evaluation.
    if nargs=2 then
       if type(indices,list(integer)) and
          (i-j > 'index/band/lower' or
           j-i > 'index/band/upper') then
         0
       else
          eval(tabl[op(indices)], 1)
       fi
    elif type(indices,list(integer)) and
          (i-j > 'index/band/lower' or
           j-i > 'index/band/upper') then
       ERROR('band width','index/band/lower',
                'index/band/upper',
                'exceeded at',i,j)
    else
       tabl[op(indices)] :=  args[3..nargs]
    fi
    end));
  'indexing function band'.p.'-'.q.' created.';
 end:
```

The basic mechanism of the new indexing functions is that if they are called with three arguments, then assignment is occurring; while if they are called with only two arguments, then the value of the array entry is wanted. Recursive use of the various possible indexing functions is automatic—the table that is passed into this indexing function has had one indexing function name (band<p>-<q>) stripped from its list, and so further references are made with respect to all remaining indexing functions, if any. If there are none, the built-in table indexing is used. This provides a convenient way to define matrices with more than one property.
Remark. A matrix A with a band indexing function defined by the procedure above will return 0 if asked to evaluate A_{ij} where i or j is outside the bounds of the matrix. We could correct this by testing every time if the indices i and j are inside the bounds. For efficiency, we prefer not to do this, since any program that refers to the matrix outside the bounds is in error anyway (though it is possible that by returning zero we will allow the program to compute correct answers).

This is in keeping with the seeming philosophy of the new indexing functions, which allow the user to get him- or her-self into trouble by mixing incompatible indexing functions. For example, one can give a matrix both **symmetric** and **antisymmetric** indexing functions, and then the results of **print(A)** depend on the ordering of the indexing functions. We do not regard this as a bug, but rather as a useful freedom, which requires concomitant care on the part of the user.

It is useful in programs to determine from the matrix object what its bandwidth is. We use the sscanf trick from the previous implementation to accomplish this, by scanning the index function name.

```
bandwidth := proc(A:array) local idxfnc,n;
  idxfnc := [linalg[indexfunc](A)];
  idxfnc := select(
    (p -> substring(p,1..4)='`.band),idxfnc);
  if idxfnc=[] then
    n := linalg[coldim](A)-1;
    [n,n]
  else
    sscanf(idxfnc[1],'band%d-%d')
  fi
end:
```

Finally, explicit creation of the indexing function is a chore best done automatically. If a user wants a banded matrix, she or he simply asks for one, and the following procedure will create the necessary indexing function if it has not already been done.

```
bandmatrix := proc(p:integer,q:integer)
  local bandidx;
  bandidx := 'band'.p.'-'.q;
  if not assigned('index/'.bandidx) then
    BandIndexFcn(p,q)
  fi;
  array(bandidx,args[3..nargs])
end:
```

Here are some examples. We first create a random 5 by 5 banded matrix, with one subdiagonal and one superdiagonal (so the matrix is tridiagonal). The routine randband is a simplistic random banded matrix generator procedure contained in our package but not detailed here.

```
> A := randband(1,1,1..5,1..5);
```

$$A := \begin{bmatrix} -85 & -55 & 0 & 0 & 0 \\ -37 & -35 & 97 & 0 & 0 \\ 0 & 50 & 79 & 56 & 0 \\ 0 & 0 & 49 & 63 & 57 \\ 0 & 0 & 0 & -59 & 45 \end{bmatrix}$$

We have also written two simple routines for LU factorization and solution of banded linear systems, based on Algorithms 5.3.1–5.3.3 in [3]. The routine bandecomp computes the factorization, and returns the results in a matrix of the same banded shape as A.

```
> F := bandecomp(A);
```

$$\begin{bmatrix} -85 & -55 & 0 & 0 & 0 \\ \dfrac{37}{85} & \dfrac{-188}{17} & 97 & 0 & 0 \\ 0 & \dfrac{-425}{94} & \dfrac{48651}{94} & 56 & 0 \\ 0 & 0 & \dfrac{4606}{48651} & \dfrac{2807077}{48651} & 57 \\ 0 & 0 & 0 & \dfrac{-2870409}{2807077} & \dfrac{289931778}{2807077} \end{bmatrix}$$

Both the $[L]$ and the $[U]$ factors are stored in the one matrix, as is usual in FORTRAN and convenient here (except if you forget whether or not you have factored $[A]$ already and pass the wrong matrix to the solver, below). Note that since the diagonal of $[L]$ contains only 1's we don't need to store them at all.

Now choose a random right-hand side.

```
> b := linalg[randvector](5);
```
$$b := [-8, -93, 92, 43, -62]$$

We can solve this system by solving $Ly = b$ and $Ux = y$ in turn. This method is encoded in bandsolve.

```
> x := bandsolve(F,b);
```

$$x := \left[\frac{-97936493}{96643926}, \frac{827068483}{483219630}, \frac{-35165237}{48321963}, \right.$$
$$\left. \frac{18382928}{16107321}, \frac{17187770}{144965889} \right]$$

We verify that the solution is correct by examining the residual. This is always recommended. See the section on accuracy, below.

```
> evalm(A&*x - b);
```
$$[0, 0, 0, 0, 0]$$

Maple has some advantages over purely numerical systems. For instance, one can compute the residual at a higher precision than that used for calculation, if that is necessary (here we used exact arithmetic and so the residual is exactly zero).

Another advantage is that we can experiment with the use of ∞. In the routine bandecomp, we chose to replace zero pivots with $1/\infty$. Since we were not going to exchange rows, a zero pivot normally means termination, but we thought we might see what happened if we let Maple handle this (hopefully rare) case symbolically.

```
> B := bandmatrix(1,1,1..2,1..2,[[0,1],[2,0]]);
```
$$B := \begin{bmatrix} 0 & 1 \\ 2 & 0 \end{bmatrix}$$

This matrix obviously has a zero pivot.

```
> c := linalg[randvector](2);
```
$$c := [77, 66]$$

```
> BF := bandecomp(B);
```
$$BF := \begin{bmatrix} \dfrac{1}{\infty} & 1 \\ 2\infty & -2\infty \end{bmatrix}$$

We see that the factorization contains ∞.

```
> xb := bandsolve(BF,c);
```
$$xb := \left[\left(77 + \frac{1}{2}\frac{66 - 154\infty}{\infty} \right)\infty, -\frac{1}{2}\frac{66 - 154\infty}{\infty} \right]$$

```
> map(expand,");
```
$$\left[33, -33\frac{1}{\infty} + 77 \right]$$

Now that solution can be interpreted as $[33, 77]$, which, as it turns out, is the exact solution to the original problem. Unfortunately, the fact that that worked is because of a bug in the infinity handling in Maple—it is supposed to complain if asked to cancel ∞/∞.

```
> evalm(B &* " - c);
```
$$\left[11\frac{-3 + 7\infty}{\infty} - 77, 0 \right]$$

```
> map(expand,");
```

$$\left[-33\,\frac{1}{\infty},\ 0 \right]$$

It may be that for some examples incorrect answers will be generated—in particular if the matrix is singular we expect this to be the case. However, if we are careful to always check the solutions this should be fine. Maple should generate an error message if it is asked to subtract $\infty - \infty$, or divide ∞/∞, so if there are infinities in the calculation there should always be infinities in the answer, so we will be warned to be careful. Unfortunately as noted above this may not always hold.

It may also be that processing the symbol infinity will add a significant overhead to the efficiency of this program. If that turns out to be the case, we will remove the statement and replace it with an error message.

We are presuming that *small* pivots are not a problem, because of the arbitrary precision floating point arithmetic. If rounding errors are causing a difficulty, then the precision will be increased. This, too, can be very computationally expensive, but if it is too expensive, then likely this is the wrong approach anyway and some form of partial pivoting must be tried.

Note that we *had* to teach Maple how to efficiently solve banded systems. Just telling it how to define a banded matrix was not enough, because otherwise it would have used inappropriate algorithms to solve them. This holds true, of course, for the FORTRAN-style banded matrix as well.

Solving Linear Systems of Equations in Maple

The linear solver in Maple is designed to accept different types of input and accordingly choose the solving strategy. For floating point calculations, a Gauss-Jordan method with partial pivoting on the maximum pivot is used to minimize the round off errors, while for integer calculations the same method is used with partial pivoting on the *minimum* pivot to reduce the size of the intermediate integers produced while solving (i.e. reduce the memory and time requirements). In this study pivoting is not used because row exchanges will destroy the structure of the banded matrices, although simple extensions of the program can be made [3].

The solution method used here can be applied both to *banded matrices* and to *full matrices*. For banded matrices, the matrix can be stored in the normal format (using either the standard indexing functions or the banded indexing functions of the previous section) or in the FORTRAN-style modified indices format. The matrix is *LU* factorized first and the simple forward elimination and back- substitution are performed. A package of procedures LUpack (for the FORTRAN-style storage system) has been developed for that purpose.

Comparison Between LU Solver and Maple's 'linalg/linsolve'

Accuracy

The instability of LU factorization without pivots (which is exactly the same as Gaussian elimination without pivots or row exchanges) for general matrices is well-known [3]. What do we mean by "instability"? We mean that the *problem solved* by the algorithm will in general be *far* away from the problem you wanted it to solve.

We use here the principle that a good numerical algorithm will give you the exact solution of a nearby problem. In engineering and other physical contexts this idea makes a great deal of sense, and is both simpler and more useful than a 'forward error analysis' or comparing the computed solution with the 'exact' solution, which, after all, will usually be the exact solution of a simplified problem which is only some rough approximation to the true reality anyway. It cannot be stressed enough that we need to study the effects of perturbations in the data (which incidentally allow us to understand the effects of rounding errors in the solution, which are usually trivial compared to the effects of data error). This leads to the study of 'condition number'. This will not be gone into here, but see [1] for an introductory essay on the topic. See the LA-PACK manual for an introduction to deeper views.

Here we face a different problem: a numerical method is *unstable* if it solves a problem that is *not* close to the intended one. The simplest way we can look at this here is to compute residuals:

$$\{r\} = [A]\hat{x} - \{b\}$$

where $\hat{x}$ is our computed solution. Note that $\{r\}$ is a *computable* vector, and its size reflects the stability of our algorithm because $\hat{x}$ is the exact solution to

$$[A]\{y\} = \{b + r\}$$

and if $\{r\}$ is small compared to $\{b\}$ with respect to the physical details we have omitted in our model, then we are finished: we have the exact solution to a problem that is just as good a model of the underlying problem as was originally written down.

It is known that Gaussian elimination without row exchanges is unstable for general matrices. Thus we can expect here that the residuals $Ax - b$ will not usually be small compared to x, A, or b for the random problems examined above. In fact, they were usually acceptable for the problems we have solved, but we remark that *whenever this method is used in practice, the residual must be examined*. The accuracy in the solution, of course, is related to the residual by the condition number of A which should also be computed.

It would be relatively trivial to modify the Maple V Release 3 band indexing function to allow partial pivoting, at the cost of increasing the upper band width of the factored form of the matrix. The maximum increase in band width is known in advance, and thus the factored matrix can be created to be that size, and 'sparse'. This would cause fill-in no more than necessary. It is anticipated that this modification will be explored shortly. This modification would greatly increase the stability of the solution, and it is clear that the user should be allowed to choose whether or not to pivot.

Flop counts

Before we start the comparison of the actual performance of the different solvers, it is better to have a look at the theoretical number of operations used by each of them. The following table shows details of the different solvers' number of operations. The following table compares 'flop counts' for each algorithm. One 'flop' is one floating-point multiply or divide *plus* one floating-point addition or subtraction. Notice that this is an *inappropriate* measure for integer or rational computations, because each rational operation takes an amount of time dependent on the size of the numbers involved, while the floating-point operation takes a fixed amount of time. Of course, if the floats are many digits long, this time itself may be large.

For large values of n the order of the number of operations for both the *Gauss Elimination Solver* and *Full Matrix LU Solver* is $O\left(\frac{n^3}{3}\right)$, while the order for the *Banded Matrix LU Solver* is $O(n\beta^2)$. This shows the reduction in the number of operations achieved by the LU solver (since $\beta << n$).

To check this reduction in number of operations, we did a numerical comparison between three methods of solving. We solved a series of random n by n problems, each with a band width equal to 7, and recorded the times for each method. We took $n = 10, 20, 30, \ldots, 130$. We compared the algorithms when the problems had floating-point entries, and when they had integer entries.

The methods used were

- Maple's 'linsolve' using sparse matrix storage for the coefficient matrix $[A]$,

- a banded matrix solver and our indexing function,

- an LU solver for the FORTRAN-style modified indices storage for the coefficient matrix $[A]$.

Fig. 1 shows the results of the comparison. For floating point calculations, it is clear that the proposed LU method is suitable, especially for large matrices. The CPU time used by the LU solver is much less than the time used by the linsolve method. The method using the banded indexing function is also much faster than the linsolve method, but takes about 75% more time than the FORTRAN-style modified index method, as is shown more clearly in Fig. 2. By looking at a logarithmic plot, we see that the time taken by linsolve grows roughly as $O(n^2)$ while the band solvers each take $O(n)$ time. By fitting straight lines to the data in Fig. 2 we find that the band index function solver takes approximately $0.11n - 1.35$ seconds (on an IBM RISC 6000 running a β-version of Maple V Release 3), while the modified index method takes $0.064n - 0.87$ seconds to solve the same problem. We suspect that the approximately 75% higher cost of the Maple V Release 3

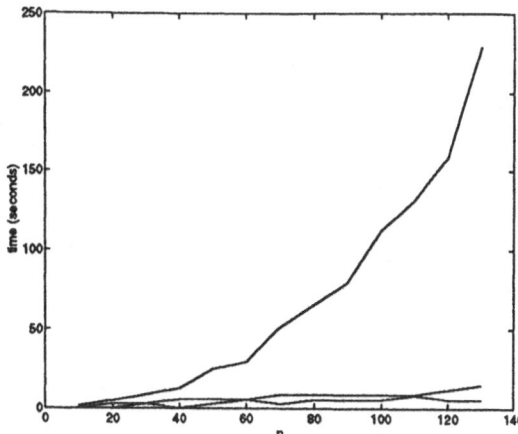

Figure 1: Computation times for the solution of random banded floating-point systems. The upper curve is the time taken by linsolve and is apparently quadratic in n, the size of the system. The lower curves are the times for the modified index method and the Maple V Release 3 indexing function method.

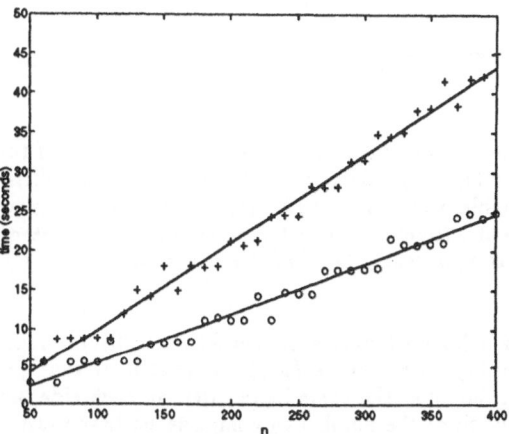

Figure 2: Execution times of the two banded matrix solvers, for solving seven banded linear systems of size n. The modified (FORTRAN-style) indexing method is the lower curve, taking approximately $0.064n - 0.87$ seconds to solve seven linear systems, while the Maple V Release 3 band indexing function method takes approximately $0.11n - 1.35$ seconds to solve the same systems.

Table 1: Algorithm complexity

	Multiplications/divisions	Additions/subtractions
Gauss Elimination No Pivoting	$\frac{n^3}{3} + n^2 - \frac{n}{3}$	$\frac{n^3 - n}{3}$
LU full matrix	$\frac{n^3}{3} + n^2 - \frac{n}{3}$	$\frac{n^3 - n}{3}$
LU banded matrix	$(\beta^2 + 3\beta + 1)n - \frac{2\beta^3}{3} - 2\beta^2 - \frac{4\beta}{3}$	$(\beta^2 + 2\beta)n - \frac{2\beta^3}{3} - \frac{3\beta^2}{2} - \frac{5\beta}{6}$

n = matrix no. of rows, $\beta = p = q$ (case where upper = lower bandwidth)

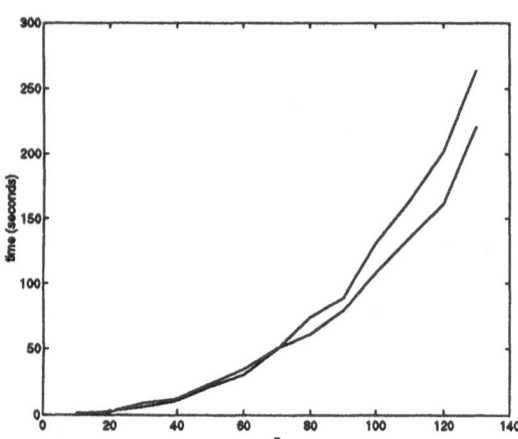

Figure 3: Computation times for the solution of random integer systems. Two of the curves are indistinguishable on this graph. All three curves are comparable, but linsolve is slightly better for larger systems. All are apparently $O(n^\alpha)$ for $2 < \alpha < 3$.

band indexing function approach is due to its greater generality—we allow a (p, q) bandwidth in that program, not just the same bandwidth on each side, and this doubles the number of comparisons that must be done in the indexing function (the actual comparisons don't take that much time, it is the overhead of interpreting the longer code that increases the cost).

The rate of growth in time in the case of integer calculations is shown on Fig. 3. Surprisingly, for integer calculations we see that the linsolve routine is roughly as suitable as either LU solver, but slightly better. This is partly because Maple chooses the *minimum* pivot to minimize intermediate expression swell, and partly because it, too, takes account of sparsity. All three sets of solutions have costs which grow roughly as $O(n^2)$, though to be more precise the linsolve cost appears to grow as $O(n^{2.1})$ while the band index function solver cost appears to grow as $O(n^{2.3})$. For some reason, the cost of the FORTRAN-style modified index method appears to be growing as $O(n^{2.6})$. We do not put much trust in these decimal places, because we have only tested these routines on a relatively small number of problems.

Conclusions

1. We have implemented a matrix factorization in Maple. There is a matrix factorization package already in the Maple Share Library, but it does not take banded matrices into account [2].

2. We have shown that for floating-point matrices, the expected efficiency gains from the use of banded matrix solvers can be realized.

3. We have given an example of the new user-defined indexing functions in Maple V Release 3.

4. We have demonstrated that Maple's built-in linear solver is more suitable for integer calculations than the proposed LU method.

References

[1] Robert M. Corless, "Six, Lies, and Calculators", American Mathematical Monthly, **100**, no. 4, 1993.

[2] Robert M. Corless, David J. Jeffrey, and M. A. H. Nerenberg, "The Row Echelon Decomposition of a Matrix", in the linalg subdirectory of the Maple Share Library, 1990.

[3] Gene H. Golub and Charles F. Van Loan, *Matrix Computations*, Johns Hopkins, 1983.

[4] William W. Hager, *Applied Numerical Linear Algebra*, Prentice-Hall, 1988.

[5] Sergio Pissanetzky, *Sparse Matrix Technology*, Academic Press, 1984.

Robert M. Corless is an Associate Professor in the Department of Applied Mathematics at the University of Western Ontario, and has been at Western since 1987. His Ph.D. was in classical applied mathematics (though it was from the Mechanical Engineering Department) under the supervision of G. V. Parkinson at the University of British Columbia in 1987. Earlier, he got an M. Math. from the University of Waterloo in 1982, again in applied mathematics. It was at Waterloo early in 1981 that he was introduced to Maple.

Professor Corless is the author of *Essential Maple*, an introductory work on Maple for scientific programmers. The book is to be published by Springer-Verlag.

He is currently on sabbatical leave, visiting the A # group at IBM T. J. Watson Research Center in Yorktown Heights, New York.

Khaled El-Sawy is a Ph.D. student in Civil Engineering at the University of Western Ontario, under the supervision of Professor Ian Moore. He received his Civil Engineering B.Sc. in 1982 from Ain Shams University in Cairo, Egypt. He then worked for several years, mainly as a design and construction supervision engineer for Mohamed Ibrahim Consulting Office in Cairo. He then came to Canada and received his M.Sc. in Civil Engineering in 1991 from the University of Western Ontario.

USING MAPLE TO DESIGN A COMPLEX MIXTURE EXPERIMENT

Richard A. Bilonick
CONSOL Inc., Pittsburgh, PA

Introduction

The design for an experiment can sometimes be taken directly from one of many books on the subject. Often, the design must be at least slightly modified. All too often, a complex experiment requires a design that cannot be found ready-made. Most experimental design software has focused on 2- and 3-level factorial designs and other specialized designs (e.g. Box-Benkin). Adapting or creating an experimental design can be a daunting task for many experimenters and even applied statisticians. Little experimental design software exists to tackle the more complex design problems. Maple can be used to produce complex designs and by using its built-in procedures the programming task is greatly simplified. In the following, Maple is used to design a complex mixture experiment consisting of seven factors.

Extreme Vertex Designs for Mixture Experiments

In a mixture experiment, the independent variables (or factors) are referred to as "components" and their levels are proportions [1]. These component levels sum to one for each trial. The measured response depends on these proportions alone and not on the absolute amounts. For example, the texture of a pizza crust will depend on the proportions of the ingredients: flour, milk, sugar, olive oil, and yeast. It will also depend on the oven temperature and time in the oven. A pure mixture experiment would vary only the ingredients while fixing temperature and time. A typical crust may consist of 71% flour, 24% milk, 2% sugar, 2% olive oil, and 1% yeast by weight.

In mixture experiment designs where the q components are unconstrained, each component can vary between 0% and 100%. Because the component proportions sum to one, the mixture experimental region is a q-1 dimensional simplex. When designing an experiment with unconstrained components, the q extreme vertices (one for each component) would be

on the list of candidate trials. In the unconstrained case, each extreme vertex represents a unary mixture, i.e., a "mixture" that consists of a single pure component. Additional trials are usually desired beyond the unary mixtures. Each pair of unary mixtures represents an edge of the simplex hypertetrahedron (e.g., a triangular pyramid in 3 dimensions when q equals 4). The next set of potential trials consists of the centroids of each edge which represent binary mixtures (i.e., mixtures of two components). The centroid of each edge is simply and exactly the arithmetic average of the two end vertices. This process can be continued by determining the centroids of the 3-dimensional faces (trinary mixtures), and so forth down to the single overall centroid of the simplex (the [q-1]-ary mixture). (Note however that the true center-of-mass centroids in general are not equal to the simple arithmetic averages of the vertices that define each face. However, in most cases the simple arithmetic average is a good approximation.) If this list of trials is very large, a subset of the q (0-dimensional) extreme vertices, the q!/(2(q-2)!) centroids of the edges, the q centroids of the q-2 dimensional faces, and the single q-1 dimensional overall centroid is chosen.

When the components are subject to both upper and lower constraints, the experimental region is no longer a simple hypertetrahedron but instead an often very complicated hyperpolyhedron. In the pizza crust example, an acceptable crust could never consist of 100% of any single ingredient, nor would the proportion of any ingredient be set to 0%. The numbers of extreme vertices, edges and so forth will vary depending of the constraints. With four or less components, the design space can be visualized in 3 or less dimensions. With five or more components, a complicated algorithm must be used to determine the set of extreme vertices.

The object of an extreme vertex design is to have the independent variables take on the lowest and highest possible values within the imposed constraints. This would allow the maximum effect (change in the response) to be observed. By limiting the independent variables via the constraints, experimental effort is directed to maximizing the information gained over the area of interest. Even without adding various centroids to the design, the constraints ensure that all the mixtures will be combinations of all components when all components have upper bounds less than 100% and lower bounds greater than 0%. Adding the various centroids helps to fill the experimental region more evenly with observations. This helps to reduce the prediction error within the experimental region, and allow fitting models which include the possibility of curvature of the response surface. Replicates of the centroids in addition provide an independent and direct estimate of the experimental error, and further provide the means to check the model for lack-of-fit.

Finding the Vertices of a Design with Lower and Upper Constraints

Determining the vertices of a mixture design with both lower and upper constraints requires the use of a relatively complex algorithm specified by Crosier [2]. This algorithm has been encoded into two Maple procedures *find_vertices* and *adjustment* shown in the Appendix. Before using these procedures, the lower and upper constraints should be checked for consistency [3], and if consistent (i.e., all the bounds are attainable), either the L or U pseudocomponents should be computed. Both the L pseudocomponents and U pseudocomponents have lower bounds of zero.

When *find_vertices* is executed it produces a list of all the vertices without duplicates. Crosier [2] presents an example for q=4 with U pseudocomponent upper bounds of 0.2, 0.8, 0.9 and 0.7 for the pseudocomponent axes Z_1, Z_2, Z_3 and Z_4, respectively. The maximum number of vertices is 12. The maximum number of edges is 18. The Maple procedure *find_vertices* produced the steps shown in Figure 1. The output exactly matches the original in Crosier [2] which identified 11 vertices (here labeled Vertex 1 to Vertex 11). Figure 2 shows the identified vertices and edges in the pseudocomponent space. Additional potential design points would include (but are not limited to) the midpoints of each edge, the centers of each face, and the center of the

hyperpolyhedron defined by all eleven vertices. (The Maple procedures *find_vertices* and *adjustment* shown in the Appendix, only determine the vertices. Additional Maple procedures, available from the author, convert from original components to pseudocomponents, find edges and faces, find the various centroids, convert from pseudocomponents to original components, and so forth.)

Outline of the Desired Experiment

An experiment was to be conducted to estimate various characteristics (e.g. viscosity) of a mineral oxide mixture as it is burned. Coal ash typically contains the following constituents: SiO_2, Al_2O_3, TiO_2, Fe_2O_3, CaO, MgO, Na_2O and K_2O, among others. The experiment was to consist of a number of trials where the proportions of these constituents were varied and the effects studied. For example, a response surface could be estimated for each dependent variable measured. These response surfaces could then be used to find an appropriate mixture that satisfies certain requirements which are advantageous for electric power plant boilers. It is desirable to determine in advance an optimal set of trials which will maximize the amount of information gained for a given number of experimental trials.

Coal Ash Constituent Constraints

Each experimental trial would consist of a sample constructed from the components whose proportions are within the ranges indicated in Table 1 below. Note that TiO_2 is fixed at 1% for all trials. The sum of the other seven components must therefore total 99%. Each trial will have the same weight, only varying in the proportion of each constituent. Because TiO_2 is fixed at 1%, after conversion to pseudocomponents it can be ignored when considering how the other

Figure 1. Output *seq([i,eval(comment[i])],i=0..j)* **from the Maple procedure** *find_vertices*.

```
[0,  [[ 0, 0, 0, 0 ], Start; adjustment = 1]],
[1,  [[ 1, 0, 0, 0 ], Does not fit]],
[2,  [[ 1/5, 0, 0, 0 ], Needs adjustment, 4/5]],
[3,  [[ 1/5, 4/5, 0, 0 ], Vertex, 1]],
[4,  [[ 1/5, 0, 4/5, 0 ], Vertex, 2]],
[5,  [[ 1/5, 0, 0, 4/5 ], Does not fit]],
[6,  [[ 1/5, 0, 0, 7/10 ], Needs adjustment, 1/10]],
[7,  [[ 1/5, 1/10, 0, 7/10 ], Vertex, 3]],
[8,  [[ 1/5, 0, 1/10, 7/10 ], Vertex, 4]],
[9,  [[ 0, 1, 0, 0 ], Does not fit]],
[10, [[ 0, 4/5, 0, 0 ], Needs adjustment, 1/5]],
[11, [[ 1/5, 4/5, 0, 0 ], Violates duplicate rule]],
[12, [[ 0, 4/5, 1/5, 0 ], Vertex, 5]],
[13, [[ 0, 4/5, 0, 1/5 ], Vertex, 6]],
[14, [[ 0, 0, 1, 0 ], Does not fit]],
[15, [[ 0, 0, 9/10, 0 ], Needs adjustment, 1/10]],
[16, [[ 1/10, 0, 9/10, 0 ], Vertex, 7]],
[17, [[ 0, 1/10, 9/10, 0 ], Vertex, 8]],
[18, [[ 0, 0, 9/10, 1/10 ], Vertex, 9]],
[19, [[ 0, 0, 0, 1 ], Does not fit]],
[20, [[ 0, 0, 0, 7/10 ], Needs adjustment, 3/10]],
[21, [[ 3/10, 0, 0, 7/10 ], Does not fit]],
[22, [[ 1/5, 0, 0, 7/10 ], Violates duplicate rule]],
[23, [[ 0, 3/10, 0, 7/10 ], Vertex, 10]],
[24, [[ 0, 0, 3/10, 7/10 ], Vertex, 11]]
```

Figure 2. Example Simplex (Based on an Example by Crosier[2]).

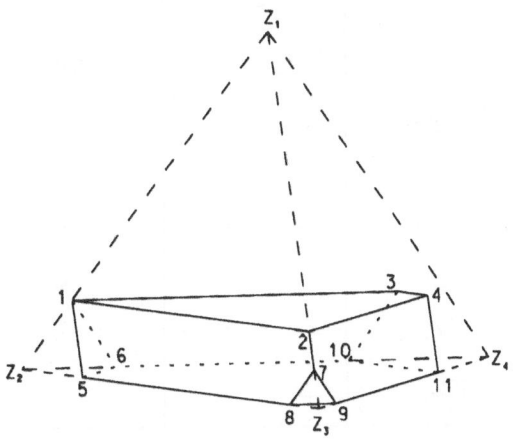

components should be varied.

A 7-Component Extreme Vertex Design

The following steps were performed:

1) Verified the consistency of the lower and upper constraints.

2) Converted the lower and upper constraints to pseudo-constraints.

3) Identified all the extreme vertices using the pseudo-constraints.

4) Determined which vertices are connected (so that edge and face centroids can be computed).

5) Determined which connected extreme vertices are close together (so that they can be replaced by a single vertex).

6) Converted the pseudocomponents back to the original component space.

To facilitate computations, the original component proportions are converted to L pseudocomponents [2]. At this point, the original component constraints were determined to be consistent (required in order to use Crosier's algorithm for finding extreme vertices). Constraint consistency means that all of the specified upper and lower bounds are attainable (although not necessarily simultaneously).

Ninety-four extreme vertices (out of a possible 140 vertices) were identified for the $q = 7$ components with the constraints shown in Table 1 (but ignoring for the moment TiO_2). The algorithm required 364

steps to locate all 94 unique vertices (and was executed with infinite precision). These results would be virtually impossible to produce by hand. The results could have been obtained by creating a C language program, but additional code would be required to handle the more tedious chores (constructing matrices, tables, and lists) that Maple handles automatically.

Another Maple V procedure was used to determine which of the extreme vertices were connected and formed edges. A total of 291 edges were found. The edge lengths were also computed. This information was used to determine 1) which vertices were close together (arbitrarily defined as lengths less than 0.1) and could be removed and replaced by an "average" vertex, and 2) which vertices were "far" apart (arbitrarily defined as lengths greater than 0.6) so that an edge centroid could be computed. The 23 longest edges were identified. There were 14 (the maximum possible) 5-dimensional faces. The vertices forming these faces were determined and centroids computed for each face. The grand average of all 94 extreme vertices was calculated and used as an estimate of the overall (6-dimensional) centroid.

The various centroids have been added to the list of experimental trials in order to be able to approximate the resulting response surfaces with quadratic or cubic models. The pseudocomponent coordinates of the extreme vertices were then converted back to original units so that they sum to 99/100 (allowing for the TiO_2 fixed at 1/100). The overall (6-dimensional) centroid will be performed 5 times, while the 5-dimensional face centroids will each be performed twice. These replications will decrease the predictive uncertainty and also provide an independent estimate of the experimental error. The total number of trials is 133 and they are summarized in Table 2 below.

Table 1. Constraints on Each Constituent

Constituent	Minimum %	Maximum %
SiO_2	35	60
Al_2O_3	15	30
Fe_2O_3	5	40
CaO	0.5	10
MgO	0.5	4
Na_2O	0.5	4
K_2O	0.5	4
TiO_2	1	1

Analysis of the Experimental Observations

The following Scheffé (full) cubic model for a given trial could be fitted to the observed data:

$$y = \sum_{i=1}^{7} \beta_i x_i + \sum\sum_{i<j} \beta_{ij} x_i x_j + \sum\sum_{i<j} \delta_{ij} x_i x_j (x_i - x_j) + \sum\sum\sum_{i<j<k} \beta_{ijk} x_i x_j x_k + \varepsilon$$

using the method of least squares [1]. The term x_i represents the proportion of component i, and y is the observed response. The random error is denoted by ε and is assumed to be distributed normally with mean equal to 0 and variance equal to σ^2. This model requires the initial estimation of 84 coefficients represented by β's and δ's. The β_i's represent the linear blending values of component i. The β_{ij}'s represent the nonadditive blending values of components i and j. The δ_{ij}'s represent the synergistic or antagonistic blending between components i and j. Finally, β_{ijk} represents ternary blending among components i, j, and k.

Alternatively, the model can be expressed as

$$Y = X\beta + \epsilon$$

where Y is an n by 1 vector of the n observed responses, X is an n by 84 design matrix (one column for each model parameter), β is an 84 by 1 vector of parameters (7 β_i's', 21 β_{ij}'s, 21δ_{ij}'s, and 35 β_{ijk}'s), and ϵ is an n by 1 vector of normally distributed random errors with mean equal to 0 and variance equal to σ^2.

The estimate of the model parameters is then given by

$$\hat{\beta} = (X^t X)^{-1} X^t Y$$

where X^t represents the transpose of X and $(X^t X)^{-1}$ represents the inverse of the matrix product $X^t X$. The predicted responses for the observed responses are given by $X\hat{\beta}$. Confidence intervals for each parameter could be determined and used to simplify the model (reduce the number of parameters) where possible.

With 133 observations and 84 parameters (initially) to be estimated, the total number of degrees of freedom for the error sum of squares will be 49 (= 133 - 84). The 49 degrees of freedom can be partitioned to allow for a lack-of-fit test for the cubic model by using the 18 degrees of freedom attributed to the replicated trials (one degree of freedom for each 5-D centroid [14 degrees of freedom] plus 4

degrees of freedom for the overall centroid). Therefore, if the cubic model is inadequate, this should be detectable. In this event, some other possibly nonlinear model may be indicated or the response variable metric should be suitably transformed.

Analysis of the 21 intercorrelations between the 7 purely linear design variables indicates that almost all intercorrelations are extremely small. Only three were larger in absolute value than 0.3. The intercorrelation matrix is shown in Table 3 below. Only the correlation between SiO_2 and Fe_2O_3 is moderately large at -0.73. A plot of SiO_2 by Fe_2O_3 shows that the addition of one or more trials with proportions of SiO_2 and Fe_2O_3 simultaneously at their maximums should reduce the intercorrelation to almost zero.

Considerations for Design Improvements

The current design should be adequate, however, improvements may be possible. In addition to adding at least one trial with simultaneously high proportions of SiO_2 and Fe_2O_3, some other modifications may improve the design. For example, the trade-off between replicating the 5-D face centroids and including more edge centroids should be studied. A much more ambitious approach would look at all extreme vertices, edge centroids, 5-D face centroids and overall centroid and choose a subset of, for example, 133 trials that produces the "optimum" design. Design improvement could be gauged by using the D-optimality criterion. D-optimality seeks to maximize the determinant of $X^t X$. This minimizes $(X^t X)^{-1}$ which is proportional to the variance of $\hat{\beta}$.

References

[1]   J. A. Cornel, *Experiments with Mixtures*, 2nd edition. New York: Wiley, 1990.

[2]   R. B. Crosier, "The geometry of constrained mixture experiments," *Technometrics*, **28**, 2, 95-102, 1986.

[3]   G. F. Piepel, "Defining consistent constraint regions in mixture experiments," *Technometrics*, **25**, 1, 1983.

Richard A. Bilonick received his Ph. D. degree in statistics from the University of Pittsburgh in 1979. For the last 15 years, he has provided statistical consultation to the research and development, engineering, marketing, and legal departments at CONSOL Inc. He can be reached at the following address:

Table 2. Number of Trials for Each Class of Vertex

Vertex Class	No. of Trials
0-D Extreme Vertices (after removing closely spaced clusters)	72
Average Vertices of Closely Spaced Clusters	5
1-D Longest Edge Centroids	23
5-D Face Centroids (including replicates)	28
6-D Overall Centroid (including replicates)	5
Total	**133**

Richard A. Bilonick
CONSOL Inc.
Consol Plaza
Pittsburgh, PA 15241-1421
USA

(412) 831-4509

**Table 3. Intercorrelations Between Independent Variables
for the Proposed Experimental Design**

	Al_2O_3	Fe_2O_3	CaO	MgO	Na_2O	K_2O
SiO_2	−0.16	**−0.73**	0.01	0.02	0.02	0.02
Al_2O_3		**−0.41**	−0.05	0.02	0.02	0.02
Fe_2O_3			**−0.34**	−0.16	−0.16	−0.16
CaO				−0.01	−0.01	−0.01
MgO					−0.00	−0.00
Na_2O						−0.00

Note: The three largest correlations are shown in **bold** type.

Appendix A -- Finding All the Extreme Vertices

The following Maple procedure determines all the extreme vertices of the hyperpolyhedron simplex without duplicates:

```
find_vertices := proc();
#
# Given a set of consistent L or U pseudocomponent constraints
#  based on an original simplex, this proc
#  finds all vertices in the q-1 dimensional simplex.
# You must supply: q - the no. of components, and
#  the 1 x q array constraints (use proc pseudo_constraints to extract
#  the L or U pseudocomponent constraints generated by proc
#  consistency).
# find_vertices produces the array vertices_nd which holds the unique
#  vertices, nv is the no. of unique vertices, j is the no. of steps
#  in the algorithm.
# Algorithm details are given in arrays work and comment and stored in the
#  table keep.
#
# This algorithm uses proc adjustment recursively
#
# First initialize some indices
  u := 0;
  n := 1;
  i := 1;
  j := 0;
  nv := 0; #no. of vertices
  work := table([]);
  work[j] := array(1..q,[seq(0,m=1..q)]); # This line and the next creates
#  first line of work table
  comment[j] := [eval(work[j]),'Start; adjustment = 1'];
  keep := table([]);

# For each of q components do the following:  for i from 1 by 1 to q
  do
     j := j + 1;
     work[j] := array(1..q,[seq(0,m=1..i-1),1,seq(0,m=i+1..q)]);
     adjust[n] := 1 - constraints[i];

# If adjust[n] is positive, recursively re-adjust until a fit occurs
     if adjust[n] > 0
     then
        comment[j] := [eval(work[j]),'Does not fit'];
        j := j + 1;
        work[j] :=
          array(1..q,[seq(0,m=1..i-1),1 - adjust[n],seq(0,m=i+1..q)]);
        keep[i,n] := copy(work[j]);
        comment[j] := [eval(work[j]),'Needs adjustment', adjust[n]];
        i0 := i;
        adjustment(i0); # This is the difficult part!

# Otherwise, if adjust[n]=0 and we have found a vertex     else
        u := u + 1;
        vertices[u] := copy(work[j]);
        nv := nv + 1;
        vertices_nd[nv] := copy(work[j]); #no duplicates
        comment[j] := [eval(work[j]),'Vertex', nv];
     fi;
  od;
  vertices_set := {seq(eval(vertices[i]),i=1..u)};
  RETURN(seq([i,eval(vertices_nd[i])],i=1..nv));
end;
```

This procedure is used by *find_vertices:*

```
adjustment := proc(i0) local k;
#
# recursive proc used by find_vertices
#
    for k from 1 by 1 to q
    do
      if work[j][k] = 0
      then
        j := j + 1;
        work[j] := copy(keep[i,n]);
        work[j][k] := adjust[n];

        if (adjust[n] - constraints[k]) > 0
        then
          comment[j] := [eval(work[j]),`Does not fit`];
          j := j+1;
          n := n + 1;  # go forward
          adjust[n] := adjust[n-1] - constraints[k];
          work[j] := copy(keep[i,n-1]);
          work[j][k] := constraints[k];
          if k < i0
          then
            comment[j] := [eval(work[j]),`Violates duplicate rule`];
          else
            keep[i,n] := copy(work[j]);
            comment[j] := [eval(work[j]),`Needs adjustment`, adjust[n]];
            k0 := k;
            adjustment(k0);
            work[j] := copy(keep[i,n-1]);
          fi;
          n := n - 1;  # go back
        else
          u := u + 1;
          vertices[u] := copy(work[j]);
          if (work[j][k] = constraints[k]) and (k < i0)
          then
            comment[j] := [eval(work[j]),`Violates duplicate rule`];
          else
            nv := nv + 1;
            vertices_nd[nv] := copy(work[j]);
            comment[j] := [eval(work[j]),`Vertex`, nv];
            work[j] := copy(keep[i,n])
          fi;
        fi;
      fi;
    od;
end;
```

MAPLE VIA CALCULUS
A Tutorial Approach
by Robert J. Lopez

*"**M**odern software tools like Maple have the potential to alter radically the way mathematics is taught, learned, and done."*

CONTENTS

With this principle firmly in mind, the author of Maple via Calculus brings to teachers and students a fresh look at the standard calculus curriculum, colored by the existence of technology like Maple — a tool that can be used in class during lectures and exams and at home while working assignments, a tool whose universal access will eventually make a software-based approach to mathematics the norm.

Whether one uses this book primarily to learn Maple or to learn calculus, the student who works through the exercises will learn a great deal about both. Drill exercises and rote manipulation are replaced here with conceptual learning activities and an exploratory interaction with mathematics not seen in traditional courses. A surprising feature is that fewer than 90 Maple commands are necessary to do all the calculus activities in the book, thus making it an ideal text for learning Maple basics. By implementing this owerful computer algebra system in their courses, teachers can clearly delineate the fundamental ideas of calculus while demonstrating the efficient use of Maple.

The techniques illustrated in this text have passed the test of student acceptance. The author has used them with great success in his own courses and demonstrated them in workshops and presentations in many locations in North America and Europe. Everyone interested in bringing the computer into the calculus classroom will find the book a valuable resource.

For Price Ordering Information: CALL: Toll-Free 1-800-777-4643.In NJ please call 201-348-4033.Your reference number is Y804.• WRITE: Birkhäuser, Marketing Dept.,675 Massachusetts Ave., Cambridge, MA 02139.• VISIT: Your local technical bookstore or urge your librarian to order for your department. Payment can be made by check, money order or credit card. Please enclose $2.50 for shipping & handling for the first book ($1.00 for each additional book) and IL, NY, NJ, MA, VT, PA, VA, TX & CA residents, please add sales tax. Canadian residents please add 7% GST. Prices are valid in North America only and are subject to change without notice. For price and ordering information outside North America, please contact Birkhäuser Verlag AG, P.O. Box 133, Klosterberg 23, CH-4010, Basel, Switzerland. Fax 61 271 7666.

Birkhäuser
Boston • Basel • Berlin

Theoretical Methods in the Physical Sciences

an introduction to problem solving using Maple V

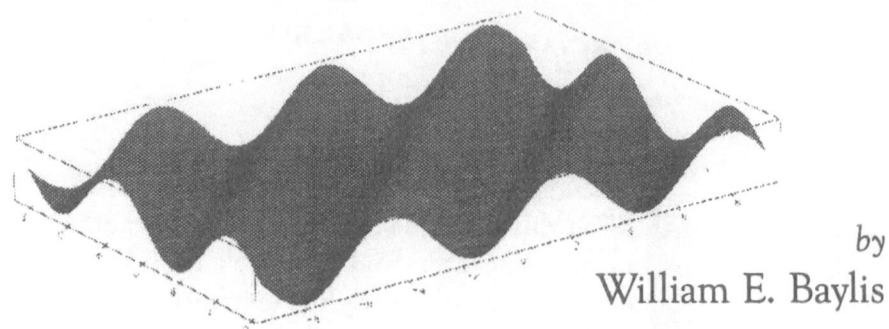

by
William E. Baylis

The way in which students in all fields of study learn mathematical skills is being fundamentally changed through the advent of inexpensive computers and the creation of powerful mathematical packages for symbolic manipulation, numerical approximation, and graphical representation. Teachers of mathematics through the entire calculus and applied mathematics sequence now have at their disposal a number of these tools, among which Maple V has become an acclaimed leader. Theoretical Methods in the Physical Sciences is a textbook that takes advantage of this technological development to teach a wider variety of more complex and realistic problems than was ever before possible.

The book is designed for a one-term course, to be taken in the first or second year after the student has completed introductory courses in physics and calculus. It uses the Maple package as an integral part of learning how to solve a range of problems taken from elementary physics, astronomy, chemistry, and geology. At the same time, it encourages the reader to think about what is being calculated, to relate results to physical experience, to make order-of-magnitude estimates of both answers and errors, and to carry out dimensional and unit checks of calculations. Above all, it uses the power of the computer, when applicable, to deepen perception and understanding. It is packaged with a 3.5-inch diskette containing Maple worksheets for each chapter and useful data files of physical constants, conversion factors, and chemical isotopes. The diskette is formatted for MS Windows, but its files can also be used on Apple and Unix platforms.

Teachers will find this book a useful resource as the main text for introductory courses in applied mathematics, mathematical physics, and theoretical science and engineering, or as an ancillary to texts in mathematics, engineering, and the physical sciences. Both students and research workers can use it for self-instruction as a quick, practical introduction to Maple.

CONTENTS: Introduction • Maple V for Physical Applications • Approximations of Real Functions • Vectors • Basic Data Analysis and Statistics • Curve-Fitting • Integration • Complex Numbers and Fractals • Vector Algebra of Physical Space • Appendices • Index

1994   304 Pages   Softcover, comb bound
3.5" high density IBM diskette
$34.50   ISBN 0-8176-3715-X

Birkhäuser
Boston • Basel • Berlin

TO ORDER, CALL 1-800-777-4643